NATURAL PRODUCTS

Essential Resources for Human Survival

NATURAL PRODUCTS

Essential Resources for Human Survival

editors

Yi-Zhun Zhu • Benny K-H Tan • Boon-Huat Bay

National University of Singapore

Chang-Hong Liu

Nanjing University, China

World Scientific

NEW JERSEY • LONDON • SINGAPORE • BEIJING • SHANGHAI • HONG KONG • TAIPEI • CHENNAI

Published by

World Scientific Publishing Co. Pte. Ltd.
5 Toh Tuck Link, Singapore 596224
USA office: 27 Warren Street, Suite 401-402, Hackensack, NJ 07601
UK office: 57 Shelton Street, Covent Garden, London WC2H 9HE

British Library Cataloguing-in-Publication Data
A catalogue record for this book is available from the British Library.

First published 2007
Reprinted 2008

NATURAL PRODUCTS
Essential Resources for Human Survival

ISBN-13 978-981-270-498-6
ISBN-10 981-270-498-1

Printed in Singapore by Mainland Press Pte Ltd

This book is lovingly dedicated to
our children,

Julia and Linda ZHU
Noel and Nigel BAY
Barnabas, Beatrice, Bernice
and Bernadine TAN
Min LIU

for their constant understanding and
encouragement in all our endeavours

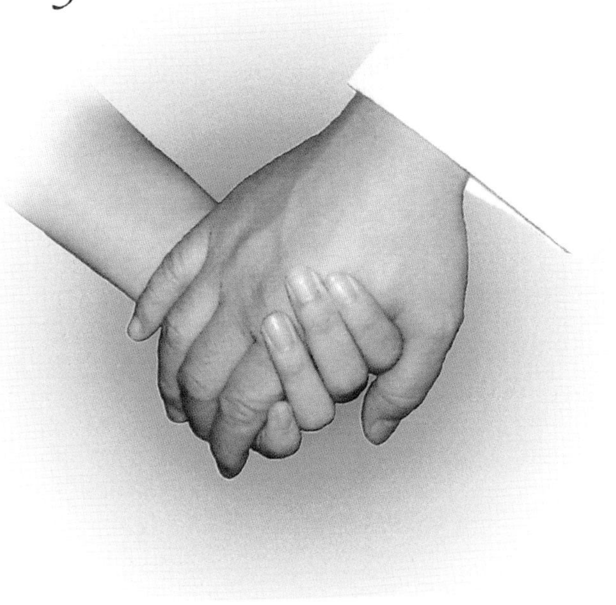

Contents

FOREWORD

At the 3rd International Conference on Natural Products co-hosted in October 2004 by the International Society for the Development of Natural Products and Nanjing University, participants from 37 countries including natural product scientists of international renown scientists came to a conclusion that the great biochemical potentials of plants and microorganisms generate various arrays of chemically complex and/or biologically active natural products, some of them being a MUST for human survival.

This book is edited to provide international readers with the major progress in natural products research. The topics include biosynthesis, chemical synthesis (or modification) and bio-transformation and the diverse range of bioactivity of natural products, and the role of biotechnology to enhance the quality of natural products and their development around the world. The on-going efforts in natural drug discovery, validation, and commercial utilization are a great plus. The editors of the book - Dr Benny K. H. Tan, Dr. Y. Z. Zhu, Dr. B. H. Bay and Dr. C. H. Liu - are leading scientists in different fields of natural product research. The publicity for this monograph certainly highlights well the importance of natural products, and their applications to human disease.

Ren Xiang TAN
Chair Professor & Associate Vice-President, Nanjing University
President, International Society for the Development of Natural Products

Note from the Editors

Recently, there has been a renewed interest in the beneficial effects of natural products for the prevention of chronic diseases. It has been estimated that 80% of the world's population rely predominantly on natural plant products which are sold as herbal/food supplements or drugs. Half of the top 50 drugs sold in European pharmacies are based on or derived from natural products. In particular, natural products have become hot spots for life science researchers since many possess unique compounds which remain to be isolated and identified. Despite this, our knowledge and understanding of how and why natural medicines work remain inadequate, thereby limiting their use in patients, especially in Western societies. The chapters in this book will provide readers with the latest overview of natural products research, reporting the chemistry and pharmacology of a range of natural products, from Chinese herbal medicine to tea extracts, microbes to marine sponges and the latest technologies to enhance the synthesis, isolation and purification of bioactive herbal compounds from these products. Additionally, chapters on new tools in information technology that can greatly facilitate the research of natural product scientists, and the technologies available for evaluating medicinal herbal products for purity add value to the book. Overall, the information to be found in this book will add to the existing body of knowledge of natural products, and further support the notion that natural products are an essential resource for human survival.

Y Z Zhu
(National University of Singapore & Fudan University)

Benny K H Tan
(National University of Singapore)

B H Bay
(National University of Singapore)

C H Lin
(Nanjing University)

The editorial assistance of Mr. Yick Tuck Yong and the artwork of Mr. Shi Qiao Huang are gratefully acknowledged. We also wish to thank the generous support of the Singapore Lee Foundation for the 3rd International Conference on Natural Products Development

Phytonutrients - The Natural Drugs of the Future

Tracy Gibbs*
Chief Scientific Officer/Pharmacognosist, NutraNomics, Inc., Salt Lake City, Utah, USA
**Corresponding email: tracyg@nutranomics.com*

For over 80 years now the allopathic industry, combined with the might and power of the pharmaceutical giants, has been trying to combat degenerative diseases. Though many advances in the diagnosing of diseases and the discovery of a myriad of new names to diseases has occurred, the battle to increase quality of life and the overcoming, curing and proper treatment of degenerative diseases has been all but lost. Allopathic medicine has not changed its treatment of cancer in over 40 years. Though new drugs keep popping up, the results are still the same and the most effective treatment for cancer is still surgery. Heart disease, diabetes, arthritis, leukemia, Alzheimer's, Parkinson's, Hodgkin's, and dozens of other disease names strike fear into our minds and as we age, many of us contemplate..."Which one will take me from this life and how will I fight it?" Today with the current standard of allopathic medicine it is only a matter of time before you are beaten by one of the many known or even unknown degenerative diseases. There are no known cures to any of theses degenerative diseases. Currently your only option is a long list of prescription drugs that may alleviate symptoms but slowly eat away your body's immunity and quality of life.

There is hope however. For more than 20 years now scientists have been researching plants and discoveries made within the last 15 years may hold some of the answers to combating many of the degenerative diseases that plague our senior years. We have known for a long time that many of these diseases can be prevented or even cured by changing our lifestyle and eating more fresh fruits and vegetables. Until recently many doctors scoffed at the thought that eating fresh fruits and vegetables would cure a disease. Even today the Food and Drug Administrations policy regarding food that is unadulterated (meaning natural foods) is that they cannot make you sick nor make you well and that no claims on labels or ads can be made to say that they are healthy for you.

However, new discoveries have proven that there are active ingredients in whole, unadulterated fruits and vegetables that can make us well and actually prevent, and in many cases, cure diseases.

Have you ever asked yourself...

- How can I reduce my risk of getting cancer?
- How can I protect myself from toxins and pollutants in the air and environment?
- How can I boost my immune systems to fight off new viral strains?
- How can I prevent premature aging and the degenerative diseases that come with it?

Your answers may lie with these new active nutrients found in simple fruits and vegetables. These new active compounds have been given the name of Phytonutrients. The word "phytonutrients" simply means nutrients that originate or are found within plants. Adding to your diet nature's most powerful phytonutrients is certainly paramount in protecting yourself and your family from the ravages of disease. The earlier we start eating fruits and vegetables, the better for our bio-cellular health. While we anxiously wait for the cure of cancer and other devastating diseases, we can arm our bodies with the most powerful protective nutrients available, Phytonutrients.

Classes of Phytonutrients

There are thousands of known phytonutrients and they have been grouped into many different classifications. Many of these you will find that you already know while others you may have never heard of before. All are of benefit and in various clinical studies, many have been proven to prevent, treat and cure degenerative diseases. I will attempt to cover the major classes of phytonutrients in order to help you understand what you are eating in those many fruits and vegetables you consume everyday. Since I began to study phytonutrients it has been hard for me to eat a fruit or vegetable without saying to myself the name of the phytonutrient found in that particular fruit or vegetable. It is very hard to eat a tomato anymore; instead it has become my daily dose of lycopene. I hope the following chapters do not have a similar effect on you but I do hope that they just might help you understand the wisdom in your mother's vocal command of "Eat your veggies!" during every meal of your childhood years.

BIOFLAVONOIDS

The term bioflavonoid refers to a large family of chemicals found throughout the plant world. Bioflavonoids are sometimes called vitamin P; however, they are not technically vitamins. So what exactly is a bioflavonoid?

Bioflavonoids are phytonutrients or plant derivatives that can have remarkable effects on biochemical pathways in human physiology. There are over 20,000 known bioflavonoids registered in chemical abstracts and over 20 million structures that fit into their chemical classification.

Bioflavonoids occur naturally in fruits and vegetables but they are subject to rapid decomposition and degradation during storage and cooking. For this reason it is important that if you choose to take a bioflavonoid dietary supplement it must be one that is made fresh, contain natural stabilizers to insure the active compounds are still active and it must be combined with all the necessary co-factors needed to activate the bioflavanoids. Bioflavonoids are considered "synergists" to vitamin C and must be combined with vitamin C for optimal benefit. For this reason I usually suggest that one stick to just eating fruits and vegetables that are high in bioflavanoids such as citrus and thus avoid the expense of supplements that may or may not work.

There are numerous bioflavonoids presently undergoing intense study in laboratories all over the world. The emerging results are exciting, to say the least. Clearly, bioflavonoids are becoming extremely impressive phytonutrient agents in cancer prevention.

While many flavonoid concentrates were used in ancient times to treat a variety of human diseases, modern medicine has failed to utilize their enormous therapeutic potential. Nutritional standards are assumed to provide us with all the vitamin C and bioflavonoids we need to be healthy. Even if these set quantities were accurate for maintaining optimal health, how many of us eat diets nutritious enough to maintain maximum health and protection? In other words, do we consume enough fruits and vegetables to afford us adequate levels of vitamin C and bioflavonoids to provide the protection we need?

"The USDA conducted a study in which they collected dietary information over the course of the year for four independent days. In that study 20% of the adult women had no fruit or juice for four days, and about 45% had no citrus fruit or citrus fruit juice in four days."

Only 9% of our population gets and eats enough fruits and vegetables on a consistent basis. Unquestionably, most of us are not getting enough vitamin C and flavonoid compounds from our diets.

In addition, it's important to remember that modern farming techniques, premature harvesting of fruits and vegetables, indefinite cold storage, freezing, canning and cooking may denature food of its vitamin C and bioflavonoid content.

Because we know that diseases are often nothing more than nutritional deficiencies, we must make adequate supplementation a priority if we want to enhance our longevity.

There are many different types of bioflavanoids. Some of the most common and thoroughly studied are listed below.

BIOFLAVINOIDS: PROANTHOCYANIDINS

For generations, certain tribes of North American Indians used bioflavonoids extracted from the bark of pine trees for a variety of disorders. Because of its marvelous healing properties, they called this pine the Annedda, or "tree of life". These Native Americans, who routinely ate deer as their primary source of protein, when confronted with a scarcity of meat asked themselves…where does the deer get its strength? They discovered that deer stripped away pine tree bark and were able to derive life-giving nutrients from its organic composition.

It was also observed that devastating diseases such as scurvy did not afflict those who ate the bark, leaves, or needles of this pine tree. In 1535, Jacques Cartier learned of the medicinal value of the bark, which remained relatively unknown until 20 or 30 years ago when scientists reviewed his notes and commenced research.

Cartier became caught in the bitter snows of Quebec while attempting to navigate the St. Lawrence River. Cartier and his crew subsisted on hard biscuits and cured meat and eventually came down with what was believed to be scurvy. Scurvy is an abhorrent disease, which causes a very slow and agonizing death.

Several of Cartier's men died before they were approached by the Quebec Indians who prepared a tea they called "Annedda" from the bark of a certain native pine tree. The men took the tea and used the pine needles as poultices. Their recovery was almost immediate. What must have seemed like a miraculous substance was technically, nothing more than vitamin C with bioflavonoids naturally inherent to the pine tree.

Cartier was resourceful enough to document the incident. Over 400 years later, a French professor, Jacques Masquelier, assigned to the University of Quebec discovered Cartier's account. Because he was already involved in bioflavonoid research he became greatly intrigued by pine tree extract. Dr. Masquelier discovered and isolated a bioactive substance known as proanthocyanidin.

After returning to France, Professor Masquelier discovered that these compounds could be extracted from the bark of the French Maritime Pine (pinus maritima) found in abundance in southern France. Subsequent intensive research by Dr. Masquelier led to the discovery of the proanthocynandin family of bioflavonoids. At the time Proanthocyanidins were thought to be the most powerful natural free radical scavenger available. This natural flavonoid has an antioxidant activity 20 times stronger than vitamin C and up to 50 times stronger than vitamin E. Proanthocyanidins are considered safe and effective in the treatment of various diseases and the maintenance of optimal health. This family of bioflavonoids is non-toxic, water-soluble and highly bioavailable.

Because proanthocyanidins scavenge free radicals so effectively, they have shown remarkable curative effects. Extensive research demonstrates that proanthocyanidins are such potent antioxidants they find and neutralize free radicals with great rapidity, allowing cells to regenerate rather than deteriorate.

Specific actions associated with proanthocyanidins include:

- Binds with Collagen and helps with skin elasticity
- Helps prevent excess wrinkling
- Protects capillaries from free radical damage which helps prevent phlebitis, varicose veins and bruising
- Acts as a powerful, natural anti-inflammatory for joint pain and injuries

- Helps control and prevent edema
- Improves the condition of the blood-brain barrier and reverses edema of the brain
- Decreases the production of histamines in Hay fever
- Treats and reduces risk of diabetic retinopathy

Proanthocyanidins have been found to benefit the following conditions

- Ulcers
- Eyesight: Increases visual acuity
- Cancers: Inhibits tumor growth
- Heart Disease
- Atherosclerosis
- Arteriosclerosis
- Multiple Sclerosis
- Colds and Flu
- Prostrate Problems
- Lupus
- Arthritis
- Memory/ Alzheimer's Disease, Senile Dementia
- Stroke
- Parkinson's Disease
- Psoriasis
- Bursitis
- Gastrointestinal Problems
- Insomnia

One of the most significant advantages of this flavonoid compound is its ability to cross and build the blood-brain barrier. Consequently, it acts as an invaluable therapeutic agent in treating depression, chronic fatigue, insomnia or loss of memory.

Other scientific tests have indicated that proanthocyanidins also possess anti-ulcer properties and may work to prevent the formation of undesirable chemicals in the stomach.

The general consensus among many experts in the health field is that proanthocyanidin supplementation is destined to become the most valuable of all the antioxidant compounds. In addition, as more scientific evidence presents itself, bioflavonoid supplementation will undoubtedly sweep the 21st century health practices.

PINE BARK PROANTHOCYANIDIN

While there is no question as to the nutritive value of the proanthocyanidin compounds, extracting them from pine bark in my own personal opinion has its disadvantages. The strong flavor components of pine bark have to be removed with chloroform, which contains some nerve damaging components. Due to their chloroform residue, some pine bark products are not the most ideal source of proanthocyanidins. In addition, extracting proanthocyanidins from pine bark is quite involved and expensive, making pine bark rather cost prohibitive. Also trees are not an easily renewable source of nutrients. Consequently, I personally prefer using OPCs found in more readily available sources such as grapes.

GRAPE SEED PROANTHOCYANIDINS

Grape seed flavonoids have undergone intensive testing and clinical studies have repeatedly supported their striking antioxidant properties. The proanthocyanidins found in these seeds can not only scavenge and remove free radicals, they can inhibit their propagation as well. The marvelous scavenging action of these compounds was confirmed by Electron Spin Resonance (ESR) spectroscopy.

In 1986 it was discovered that OPC from grape pips has an intense free radical scavenging effect (FRSE) on radical oxygen species. Recently the proanthocyanidins have been described as "the most active substances in the battle against free radicals."

What was especially exciting about these tests was the fact that grape seed flavonoids also exhibited and anti-enzyme effect which prevented the breakdown of collagen and elastin, two compounds which keep skin firm and inhibit the formation of wrinkles.

Unquestionably, studies have demonstrated that the bioflavonoids extracted from grape seed have extraordinary antioxidant properties and have proven their ability to inhibit cellular mutation.

Double blind placebo controlled studies have indicated that patients suffering form circulatory insufficiencies and diseases on the lymph system showed significant improvement in pain control and vessel elasticity after taking these flavonoids.

Other experiments concluded that this family of bioflavonoids contained in the grape seed was capable of improving night vision and initiating "a rapid and marked improvement of visual performances after glare in comparison with control group."

Grape seed proanthocyanidins are particularly valuable for anyone who suffers from water retention and edema. Studies in the Institute of Physiology at the Bulgarian Academy of Sciences confirm that the grape seed extract stabilized capillary walls, which decreased fluid leakage into tissues that cause swelling and pressure.

SAFETY

It would certainly be pointless to find a substance that while seemingly beneficial, was not safe or could not be well tolerated by its users. Studies have shown that the leucoanthocyanins and proanthocyanidins derived from grape seed and skins are practically devoid of any oral toxicity. Even extremely high dosages administered over sustained periods of time showed no toxicity or side effects.

Virtually no actual or potential risks were found with this compound. Grape seed extracted proanthocyanidins have been shown to be safe for conception, pregnant women, and the unborn fetus. They are devoid of any perinatal or postnatal toxicity. On the contrary, the compounds were found to not only benefit targeted disorders, but a variety of other conditions improved as well.

DOSES

If a serious problem or disease exists and simply eating fruits and vegetables is not sufficient to treat a disease I strongly suggest supplementation of proanthocyanidins. In various clinical studies active proanthocyanidin compounds ingested at daily doses of over 100mg were quite effective at treating a myriad of diseases. I suggest that for maintenance of good health at least 50mg be ingested daily and for treatment of disease 200mg be ingested daily.

BIOFLAVINOIDS: NARINGEN, HESPERITIN and RUTIN

These three bioflavonoids are also efficient anti-oxidants and work synergistically with vitamin C and the proanthocyanidins to scavenge free radicals. Moreover, this particular trio of flavonoids has significant anti-allergenic properties. Studies have indicated that these compounds can inhibit the release of histamine, which is the chemical cause of a whole host of miserable allergic symptoms.

Naringin and rutin in many studies have been proven to prevent the release of mastocytic histamine. We hear so much about anti-inflammatories today and they are routinely prescribed for a number of disorders. Several laboratory tests support the fact that these flavonoids can significantly decrease inflammation by preventing histamine from permeating vessel walls. Obviously, any allergic condition, edema or other inflammatory diseases would substantially benefit from this vascular action.

Many acute inflammatory conditions can be controlled by inhibitors of mediator synthesis or by antagonists...this modulation can be obtained with these three flavonoids or chemical derivatives such as S5682 (a hesperitin mix). If you bruise easily, this group of bioflavonoids is particularly desirable. Naringin and hesperitin significantly prevent capillary fragility and interstitial bleeding.

BENEFITS

- Increase capillary strength
- Inhibit viral invasion
- Natural anti-inflammatory action
- Anti-allergenic/ inhibits histamine production
- Helps control Edema

SAFETY

To date there are no known drug contraindications or side effects to the consumption of these bioflavanoids. Naringen has been found to inhibit the absorption of statin drugs. It is also advised not to take Naringen with psychotropic drugs as well as it prolongs the AVD of the drug or the time it takes to clear the blood stream.

DOSES

Most doses between 10mg and 50mg are effective. Any dose over 50mg may be unnecessary and costly for prevention. For treatment of some diseases I have found that when combined with other bioflavonoids 25mg to 50mg is sufficient.

BIOFLAVINOIDS: QUERCETIN

Quercetin is another remarkable flavonoid. Its particular antioxidant activity has been found to help reduce the risk of coronary heart disease. Quercetin helps to dilate and relax blood vessels and has a protective effect against certain types of arrhythmias. It is the major active component of Ginkgo biloba and may be responsible for the beneficial effects that Ginkgo has on brain neurons.

Quercetin has demonstrated its ability to reduce tumor incidence, which attests to its ability to neutralize oxidants within cellular material. Its anti-viral activity is particularly significant today, as we face new viral diseases capable of adjusting to various pharmacological treatments.

"Several derivatives of quercetin were found to have anti-viral activity against picornaviruses in vitro. This family includes the polio-viruses, ECHO viruses, Coxsackie viruses and rhinoviruses. The latter are the major causes of the common cold."

Quercetin is a remarkable bioflavonoid that can help to protect the body against viral or bacterial invasion if given before an infection progresses. It is invaluable as an immune system booster and a protectant against disease. For anyone who suffers from asthma, quercetin may effectively treat and help to prevent asthmatic symptoms.

Seven Chinese herbal drugs were screened for their ability to inhibit certain enzymes that cause several of the complications associated with diabetes. Quercetin was among the compounds tested and exhibited a potent action against these destructive enzymes. Anyone who suffers from diabetes should be aware of quercetin's potential benefits.

Another exciting result of laboratory tests on quercetin was its ability to help normalize hormone levels in both males and females. The effect that quercetin demonstrated on female estrogen and male testosterone levels suggests that the flavonoid is valuable in treating women with high estrogen related problems and men who suffer from prostate disorders.

While to my knowledge quercetin has not been clinically tested for its ability to treat migraine headaches, its activity as a mast cell stabilizer suggests that it may indeed be useful.

BENEFITS

- Powerful anti-oxidant
- Strengthens immune system
- Helps lower risk of coronary heart disease
- Lowers blood cholesterol
- Anti-tumor
- Has anti-viral and anti-bacterial properties
- Decreases symptoms of gout
- May help prevent migraine headaches
- Works to normalize hormone levels

SAFETY

There are no clinical tests that prove drug contraindications to quercetin and no know side effects of taking large doses. I have found that quercetin acts as a mild blood-thinning agent if taken at 100mg or more and therefore you should be concerned if you are already on a prescription blood thinner.

DOSES

For the prevention of diseases I normally recommend between 25mg and 100mg daily. For heart disease and reduction of cholesterol 150mg or more daily is sufficient.

BIOFLAVINIODS: SILYMARIN

Because the notion of prevention of disease and protection from toxins is crucial to many of us I had to include Milk Thistle and a certain bioflavonoid found within it called silymarin into this book. Because many herbs are adaptogenic, they work when the body is under stress. In other words, they affect no change unless change is required. In addition, herbs help to stimulate the immune system by sustaining higher levels of B and T cells when combined with exercise.

Milk Thistle, is an exceptional herb with a long history of use in the United States. It has undergone a number of rigorous clinical and laboratory tests and contains a special kind of bioflavonoid called silymarin.

The chemical components of silymarin are referred to as true hepato-protective, or "liver friendly" chemicals. Its biochemical activity provides a protective effect on the membranes of liver cells. Concerning patients with chronic alcoholic liver disease, tests concluded that: "the scavenger, silymarin is able to increase the antioxidant protection of the cells by ameliorating the deleterious effects of free radical reactions."

Silymarin, like the other bioflavonoids discussed, also has the ability to scavenge free radicals and has been found to be particularly valuable in treating diabetes and its side effects. Moreover, silymarin may be helpful to people suffering from high blood pressure and those that have experienced heart attacks.

Silymarin, like quercetin and the proanthocyanidins helps to control the risk of estrogen-related diseases. When combined with other ingredients, silymarin serves to enhance and complement the antioxidant supplement. I speak of silymarin when combined with supplements due to the fact that the herb is quite bitter and many people prefer to ingest it in a capsule form rather than a tea or as a garnish..

BENEFITS

- Liver protectant
- Powerful free radical scavenger
- Diabetes and its complications
- Estrogen related disorders

SAFETY

There are no cases of over toxicity of milk thistle and in many studies overdoses have only proven to cause an upset stomach. Also there are no known drug contraindications.

DOSES

At least 100mg daily of milk thistle should be consumed daily for the prevention of liver disease and for those individuals who live in over toxic environments such as large cities or developed countries. Large doses of 500mg or more have been proven to help treat hepatitis C and other viral infections.

BIOFLAVINOIDS & Natural Vitamin C

Because bioflavonoids depend on vitamin C as a co-factor I thought it wise to include some information on vitamin C. There are many forms of natural vitamin C and ascorbic acid is not one of them. As previously stressed, in order for bioflavonoids to function effectively within bio-cellular structures, vitamin C must be present. Vitamin C potentates the action of falconoid compounds. I never have liked, nor do I promote the use of ascorbic acid. Vitamin C, in its natural state as Calcium Ascorbate, is always recommended over ascorbic acid, which is derived through a man-made process. Though over 90% of the vitamin C supplements on the market contain ascorbic acid and tout it as being all natural, the truth is that ascorbic acid in large doses can actually harm the body. Instead of focusing on the harmful effects of ascorbic acid I simply wish to focus on natural vitamin C and several of its good, natural sources.

To date, the highest known natural source of vitamin C is found in the Acai berry. The Acai berry is a rare sub tropical berry originating from South America and is extremely rich in vitamin C. Acai Berry typically contains between 15% and 22% vitamin C. In South America, acai provided the people with protection from scurvy. It is very bitter to eat however and so many tribes dried the fruit and made teas from it sweetened with cane sugar or mixed into alcoholic beverages. Today it is quite easy to find the powdered form of the Acai berry and take it in a capsule form.

Other good sources of natural vitamin C are the common rose hip, acerola cherries, orange peels (make sure you eat peels that have not been sprayed with chemicals) and tropical fruits. All of these are very common and do not need to be explained further in this book.

INDOLES

Indoles are a phytochemical mix extracted from cruciferous vegetables, which include cabbage, broccoli, kale, turnips and bok choy and others. Cruciferous vegetables contain a variety of Indoles such as indole-3 carbinol, ascorbigen and others.

Indoles belong to a class of phytonutrients that have been scientifically shown to benefit the body in a number of very important ways. Recent studies are just beginning to reveal the profound value of indoles.

Indole-3 Carbinol assists in detoxifying human tissues, promotes hormone balance, boosts immunity against diseases like cancer and provides excellent cellular nourishment.

Ironically, like so many other nutrients, ancient physicians were well aware of the remarkable curative power of these indoles found in certain vegetables. 2000 years ago, Roman practitioners prescribed cabbage leaves to cure an ulcerated breast. Today, science has confirmed that certain phytochemicals contained in cabbage are considered breast cancer preventative agents. Indole-3 carbinol can actually help decrease C16 estrogen and transform estrogens into the inactive C2 estrogen, which is believed to decrease a woman's risk of getting breast cancer.

All cruciferous vegetables contain this important phytochemical. Unfortunately, only 9% of the American population eats an adequate amount of vegetables and fruits. A recent report I read stated the following..."On any given day, only one in five Americans ate a fibrous or cruciferous vegetable and only 28% ate a fruit or vegetable."

Remember that to keep C2 estrogen levels elevated consistently, phytochemicals, including indole-3 carbinol must be continually consumed. Does your menu typically feature daily portions of cruciferous vegetables?

Clearly, supplementation of bioflavonoids and indoles is warranted and can enhance longevity and improve the quality of life. Taking cruciferous vegetable extracts as part of a balanced supplement can provide as many indoles as several servings of raw cruciferous vegetables. It's important to remember that if you can't eat broccoli and cabbage every day, there are other options.

The nutritional and antioxidant properties of indoles greatly contribute to sustained health and cellular nourishment. Perhaps the best attitude we can adopt regarding disease is to intensely pursue protection and prevention rather than cure.

BENEFITS

- Increase immune function

- Anti cancer activities
- Anti oxidant and youth restoring properties

SAFETY

To date there are no known drug contraindications or side effects to the consumption of these Indoles.

DOSES

Most doses between 25mg and 50mg of purified or standardized Indole blends are effective. Any dose over 50mg may be unnecessary and costly for prevention. For treatment of some diseases I have found that when combined with other bioflavonoids 25mg to 50mg is sufficient.

XANTHONES

Xanthones consist of a group of phytochemicals that are found mostly in a certain varieties of tropical fruits. Unfortunately many of these beneficial xanthones are found in the skin or the rind of the fruit and therefore are difficult to ingest or are too bitter for ones pallet. Therefore finding xanthones in the form of a dietary supplement is ideally the best way to ingest them.

There are several varieties of xanthones that have been well researched and documented to have beneficial properties. Of these types of xanthones alpha-mangostin, gamma-mangostin, and garcinone-E have proven to be the most beneficial so far. All of these varieties are very potent free radical scavengers and the mangostins act as anti-inflammatory agents as well.

Regarding the treatment of cancer, Garcinon-E was shown in a study in Taiwan to kill all cancer cell lines and actually outperformed four of five commonly known chemotherapy drugs including cisplatin and methotrexate, in liver, lung, and stomach cancer. Also, many Japanese studies have shown that varieties of Xanthones show anitproliferative activity against human leukemia.

A group of scientists in Japan found that gamma-mangostin prevented the release of prostaglandin (an inflammation stimulator). Prostaglandin is also primarily responsible for the pain and swelling associated with most conditions of inflammation. There are several studies being conducted now to see if gamma-mangostin can prevent or treat asthma attacks in children and adults.

It has also been recently discovered that xanthones also possess some anti-viral properties. If these studies prove to be validated then the use of xanthones as an effective treatment for asthma would be highly recommended as it would act as an anti-inflammatory agent while modulating the immune system and act as an anti-microbial agent as well.

For all of these reasons I have found it wise to have my children supplement with a product containing xanthones as there is irrefutable evidence of potent anti-bacterial, anti-viral and immune boosting properties as well. For more information on Xanthones please refer to the Mangosteen Medical Reference, a book written by a good friend, Dr Frederic Templeman MD. No mother or father should be without the knowledge contained within this book.

SAFETY

To date there are no known drug contraindications or side effects to the consumption of Xanthones.

DOSES

Most doses between 25mg and 50mg are effective for disease prevention. Any dose over 50mg may be unnecessary and costly for prevention. For treatment of some diseases I have found that doses up to 500mg to 1 gram works quite rapidly as an anti-microbial and anti-cancer treatment.

CHLOROPHYLL

Chlorophyll is the molecule that catches the biggest power source in our universe...the sun. Chlorophyll acts as a photoreceptor creating the special chemical reaction in plants known as photosynthesis. Without it, plant life could not exist as we know it. Chlorophyll is found in the chloroplasts of green plants, and is what makes green plants, green. The basic structure of a chlorophyll molecule is a porphyrin ring, co-ordinated to a central atom. This is very similar in structure to the 'heme' group found in hemoglobin in blood, except that in hemoglobin the central atom is iron, whereas in chlorophyll it is magnesium.

There are actually 2 types of chlorophyll, named *a* and *b*. They differ only slightly in the composition of a side-chain of carbon atoms. Both of these two chlorophylls are very effective photoreceptors because they contain a network of alternating single and double bonds, which we do not need to get into too much detail in this book, however it is the effect of these double bonded strands of polyenes that give them the ability to produce energy from sunlight.

Chlorophyll absorbs sunlight so strongly that it can mask other less intense colors. Some of these more delicate colors (from phytochemicals such as carotene and quercetin) are revealed when the chlorophyll molecule decays in the autumn, only after this decay can we begin to see the colors of the other phytonutrients in plants. We observe this in the autumn when the leaves of the trees turn red, orange, and golden brown. One important note about chlorophyll is that it is easily damaged when the vegetation is cooked. When you cook foods that contain chlorophyll you are replacing the central magnesium core atom with a hydrogen ion. This affects the energy levels within the molecule, causing its absorbance spectrum to alter.

Chlorophyll has been well known for its anti cancer effects for over 70 years now. It is widely known that eating a diet consisting of a large amount of greens can not only prevent many types of cancer but will also assist in detoxifying the body and increase immunity and energy levels.

Most of the major problems facing Americans today in regards to obesity, cancer, diabetes and heart disease could be overcome within a generation if we simply ate more nutrient dense, dark green leafy vegetables. This simple solution alone could save taxpayers billions of dollars a year in hospice care and other related health care costs.

BENEFITS

- Increases iron and Vitamin K levels
- Powerful free radical scavenger
- Anti cancer benefits
- Enhances the absorption of Calcium

SAFETY

To date there are no known drug contraindications or side effects to the consumption of chlorophyll rich vegetables. If consumed in extremely large quantities a runny stool or symptoms such as diarrhea could develop.

DOSES

Most doses in the form of spirulina, chorella, and other types of sea greens are safe and effective at 1000mg to 5000mg a day for disease prevention. For the treatment of disease I have not found it effective at any dose except in cases of mild anemias where a simple dose of 10 grams a day can make all the difference in one's energy levels.

CAROTENOIDS

Our understanding of the value of the carotenoid family of phytonutrients has increased enormously since the 1980's. Although scientists now appreciate their value, the name "carotenoid" is still not generally a household word (it is pronounced ka-rot-ten-noid). The name is derived from "carrot" and essentially means compounds similar to the orange-pigmented nutrients found in carrots. Unfortunately, American diets are woefully deficient in these marvelous nutrients.

Exactly what is beta-carotene good for and where does it come from? This is a question I get all the time as most people seem to think that carrots are the only source. The fact of the matter is that anything with an orange or yellow pigment most likely contains carotenoids.

The objective of this section is to simply explain what carotenoids are, what they do and how they can keep you healthier longer. Also I will additionally discuss the functions of carotenoids and how they protect us from a wide array of diseases, including cancer, heart disease, cataracts, macular degeneration and many other age-related diseases.

WHAT ARE CAROTENOIDS AND CAROTENES?

Carotenoids are simply compounds having a long chain of carbon atoms with conjugated double bonds. While it is not important to know what a conjugated double bond system is, it is important to note that this particular chemical structure makes the carotenoids very highly colored, because the double bonds readily absorb most colors of the light spectrum.

Without carotenoids, it would indeed be a very dark and dreary world. Carotenoids are particularly beautiful. The yellow, orange and many red pigments in plants are usually carotenoids. These pigments are made by plants primarily to protect the stems and leaves from the powerful energy of sunlight which plants need for photosynthesis. In the autumn, when deciduous trees prepare for winter and shut off their chlorophyll production, the green fades and the carotenoids that give autumn leaves their brilliant yellows, oranges and reds are revealed before the leaves fade to brown and fall.

The beauty of nature continues as the carotenoid pigments can accumulate—or be modified— in the protective coloration and sexual attraction of some animals, especially birds, fish and crustaceans. The brilliant red plumage of the cardinal, the pink of the flamingo and the distinctive coloration of shrimp are just three of thousands of possible examples.

More importantly, without carotenoids, not only would there be less color in the world, we wouldn't be able to see anything at all. Without carotenoids, there would be no vitamin A anywhere. And without vitamin A, there would be no light-gathering pigments in the eye.

VITAMIN A AND THE CAROTENOIDS

As a child I was taught to eat my carrots because they contained vitamin A and that was good for my eyes. Most people also learned that yellow and orange vegetables as well as leafy greens contained vitamin A. The truth is that none of these vegetables contain vitamin A.

Actually, vegetables and other plants contain compounds your body can convert into vitamin A. (Vitamin A is produced by animals, not plants.) The compounds that are made into vitamin A by the body are called vitamin A "precursors" and belong to the carotenoid family. Not all carotenoids can be converted into vitamin A, but all compounds that can be converted into vitamin A are carotenoids. Beta-carotene, just one of the 600 or so members of the carotenoid family, is considered the best source of vitamin A because each molecule of beta-carotene can be made into two molecules of vitamin A in the body.

Originally, we were interested in beta carotene because we were interested in vitamin A. Then we became interested in beta-carotene because it helps prevent cancer. Beta-carotene is a stronger antioxidant than vitamin A. However, beta-carotene alone is not as effective as mixed carotenoids. In fact, beta-carotene represents less than 30 percent of the total amount of mixed carotenoids circulating in the blood. Many clinical studies have shown that beta-carotene...

- Enhances the immune system
- Promotes gap-junctional communication between cells
- Protects against cancer
- Defends against heart disease
- Protects against stroke
- Lowers total cholesterol levels
- Decreases risk of cataracts and macular degeneration
- Prevents certain photosensitivity disorders
- Reduces stress reactions
- Improves fertility
- Decreases risk of degenerative diseases
- Protects cellular integrity
- Enhances gene regulation and expression

However we now know that other carotenes such as alpha-carotene and many other carotenoids have similar properties and in many cases are more effective at preventing diseases. Several carotenoids enhance the immune system and improve our body's ability to protect us against disease and foreign invaders. Recently much knowledge has been gathered on the role of certain carotenoids in gap-junctional communication (GJC). Carotenoids inhibit tumor development via gap-junctional communication and not by antiradical action alone. Gap-functioning communication may be a new concept to readers. Some scientists believe that carotenoids achieve cancer protection by improving the chemical communication between cells, which helps cells being transformed into cancer cells revert back to normal.

Other studies on carotenoids have already elucidated the relationship between the deficiency of beta-carotene and the development of cataracts. Also, the carotenoids lutein and zeaxanthin have been shown to prevent macular degeneration, a blinding disorder of aging. Several studies have shown that carotenoids increase immune protection and may even help alleviate arthritis.

Basically the carotenoid family consists of smaller families of pigments called carotenes and xanthophylls (pronounced zan-tho-fills). Carotenes are hydrocarbons (containing only carbon and hydrogen atoms), whereas xanthophylls also contain atoms of oxygen.

The carotenes are commonly found in carrots, algae, orange fruits and green vegetables. The human diet contains about a hundred different carotenoids. Nineteen different carotenoids have been identified in human blood so far. The carotenes that are of most interest currently are beta-carotene, alpha-carotene, gamma-carotene, lycopene (pronounced lie-co-peen) and phytoene (fi-toe-een). Many of us now know the word lycopene as a Time Magazine article entitled "Is pizza good for you?" presented the fact that lycopene (contained in fruits and vegetables that are red like watermelon, pomegranates, and red peppers) was found in tomato sauce and therefore pizza was good for us. Lycopene has been proven to be a good anti-cancer agent especially for prostate cancer.

The most well studied xanthophylls are canthaxanthin (can-tha-zan-thin) found in many algae and fungi, lutein (lew-te-een or lew-teen), zeaxanthin (zee-zan-thin) found in corn, capsanthin (cap-san-thin) which gives the spice, paprika, its color, and astaxanthin (as-ta-zan-thin). Lutein, of course, is found in many dietary supplements world wide and is praised for its immune enhancing properties as well as its benefits to improving vision.

Carotenoids, are definitely phytonutrients worth our study and my purpose is not to convince you that one is better than the other. However I will say that we should ingest our carotenoids in a natural (mixed) form as we still do not know all the benefits of each individual carotenoid.

MOST PEOPLE ARE CAROTENOID DEFICIENT

Unfortunately, the vast majority of Americans are not even close to reaching the optimal dietary intake of mixed carotenoids and other antioxidant nutrients. Even those who consume adequate vitamin A can still be deficient in various carotenoids, although there is no official RDA (recommended daily allowance) for each carotenoid. The Second National Health and Nutrition Examination survey, which accurately represented the entire U.S. population, found that only 29 percent of U.S. adults ate the recommended two servings of fruit on the survey day and only 27 percent ate the recommended three servings of vegetables. Fifty-two percent did not meet either guideline. Only 9 percent of those surveyed met the official guidelines that recommended three servings of vegetables and two servings of fruit daily.

If people are not eating enough fruits and vegetables, they are not getting adequate intakes of mixed carotenoids! This trend will slowly tax America of its health care resources as we slowly bring about the disease and degeneration of our elderly populations.

THE MAJOR CAROTENOIDS

Beta-carotene

Beta-carotene is the most well-known and best-studied member of the carotenoids. Carotenoids are synthesized by plants, and as a result they have been in the human diet for as long as we have been on earth. Beta-carotene and, to a lesser extent, alpha-carotene are precursors of vitamin A. Although many other carotenoids are not converted into vitamin A in our bodies, they are antioxidants, antiradicals and singlet-oxygen quenchers and may participate in gap-junctional communication. Beta-carotene has two identical ring structures (beta-ionone rings)—one ring at each end of the molecule. Thus our bodies, with the help of dioxygenase enzymes, can split a molecule of beta-carotene into two vitamin A (retinol) molecules.

Alpha-carotene

Alpha-carotene is normally found in the same foods as beta-carotene. However, foods differ in the ratios of these two carotenes. Alpha-carotene is very similar to beta-carotene in structure and contains the same number of carbon atoms (40), the same nine conjugated double bonds backbone and the same number of ring structures (2). One of the rings is beta-ionone, identical to that found in beta-carotene. But the other ring is slightly different in that the double bond is shifted by one carbon atom. A molecule of alpha-carotene yields only one molecule of vitamin

A. However this does not undermine the fact that alpha-carotene is an essential player in the carotenoid family. Recent evidence shows that alpha-carotene may be up to ten times better than beta-carotene at preventing some types of cancer cell mutations in the liver, lungs, skin and eyes.

Lycopene

Lycopene is often the predominant carotenoid found in the standard American diet because of the prevalence of tomato products, including pizza and spaghetti, in a population that generally avoids fruits and vegetables.

Lycopene is the carotene that has the longest chain of conjugation without the involvement of ring structures. Since lycopene has no ring structures, it does not produce vitamin A. However, this structure gives lycopene the greatest ability of the carotenoids to absorb all but the longest wavelength of light and the highest efficiency in quenching singlet oxygen molecules (free radicals). Since the only color not absorbed by lycopene is the deepest of reds, lycopene imparts a deep red to those fruits and vegetables in which it is rich. For instance, it creates the color of ripe tomatoes and watermelon pulp.

Lutein

Lutein is a xanthophyll found in many leafy green vegetables, some fruits, alfalfa and egg yolks,. Lutein is of particular import because of its role in preventing macular degeneration, a leading cause of blindness in the elderly. This phytonutrients has several clinical applications for enhancing vision in elderly patients and is very simple and inexpensive to take in supplement form thus adding to its popularity.

Astaxanthin

Astaxanthin is a carotenoid found in generous quantities in plants, yeasts and marine animals. A major commercial source of astaxanthin is Antarctic krill (*Euphausia superba*). Krill are small shrimp like crustaceans that feed on algae and are, in turn, the staple diets of several sea creatures including whales.

Several studies indicate that astaxanthin has greater antioxidant activity under some conditions than beta-carotene or vitamin E and is ten times more efficient at quenching free radicals than beta-carotene. As a singlet-oxygen quencher, astaxanthin ranks just behind lycopene and gamma-carotene. Depending on the system being tested, data suggest astaxanthin may have five-to-twenty times the antioxidant activity of beta-carotene, while the data comparing astaxanthin to vitamin E suggest a range from somewhat greater antioxidant activity than that of vitamin E to nearly 1000 times more than that of vitamin E.

Unlike beta-carotene, astaxanthin is a membrane surface antioxidant. Astaxanthin can terminate free radicals at the membrane surface before they can damage the membrane or intrude into the cell. This is very important in that it may help prevent the cell wall becoming hard which is of particular importance to arterial sclerosis sufferers.

Why so many carotenoids?

In my studies I have formed the opinion that though we can question nature, we should never attempt to change her. (Similar to my wife) It may be that each carotenoid has a different efficiency in protecting against cancer depending upon the particular mechanism of carcinogenesis or other toxins involved. We do know that we need them and we do know that they are beneficial. Many of the 600 or more known and classified carotenoids may never be found in our life time to have beneficial properties. But rest assured, one day we will have answers and they may surprise even the brightest of us. In the end, we should trust nature more than we trust our evolving yet limited understanding of how nature works. These complexes in foods are varied and may have varied uses. It seems very likely to me that natural food concentrates, such as *D. salina* or palm seed oil, which have naturally occurring mixed carotenoids, may well have synergistic effects beneficial to humans as well as the plants they protect. It is important that we get adequate amounts of various carotenoids and avoid concentrating on any one carotenoid as that may inhibit the absorption or transport of the others. Nature already knows the answers and provides us with mixed carotenoids in the form of fresh fruits and vegetables.

SAFETY

To date there are no known drug contraindications or side effects to the consumption of any carotenoids.

DOSES

Doses vary due to the large variety of the carotenoid family however if taken in capsule form or liquid form a concentration of a variety of carotenoid family members would make a good antioxidant or disease prevention supplement.

SUMMARY

Though there are literally thousands of phytonutrients not covered in this brief booklet, I have attempted to touch upon the major classifications and hope that I have impressed upon your mind the importance of eating a variety of fruits and vegetables on a regular basis. Many people find changing their eating habits a hard thing to do and for this purpose I have also covered the importance of finding a good dietary supplement for these important phytonutrients.

It is my firm belief that within the next decade many new drugs will be utilized which will be chemical derivatives of many of the nutrients coved in this booklet. One day I hope that allopathic medicine will come to a realization that poor eating habits and the consumption of lifeless foods can kill us and that the consumption of wholesome, phytonutrients rich foods can be our drug; The drugs of the future.

REFERENCES

[1] Alpha mangostin induced apoptosis in human leukemia HL60 cells http://www.ncbi.nlm.nih.gov/entrez/query.fcgi?cmd=retrive&db=pubmed&list_uids=15498656&dopt=citation
[2] Anthocyanins and flavonoids www.polyphenols.com/background.html
[3] Antibacterial activity of Xanthones www.ncbi.nlm.nih.gov/entrez/query.fcgi?cmd-retrieve&db=pubmed&list_uids=8887739&dopt=abstract
[4] *Beta Carotene and other Essential Carotenoids*, Woodland Press 1996
[5] Rita Elkins M.H., *Bioflavanoids*; Woodland Press 1995
[6] Dr. Melvyn R Werbach M.D, *Botanical Influences on Illnesses*; 1994 Third Line Press Publishing
[7] Cooper, Kenneth H Antioxidant Revolution, T. Nelson Publishers, 1995
[8] *The Information Sourcebook of Herbal Medicine*; Hoffman, David, The Crossing Press 1994

[9] Masquelier, Jacques. Proanthocyanidin and free radical scavenging US patent #4,698,360 Oct 6th, 1987

[10] Templeman, Fredric MD, *Mangosteen Medical Reference*, 2005 Phytoceutical Research Inc

Research on the Supercritical CO_2 Extraction of Xinyang MaoJian Tea

Liu Zhongdong*, Yang Jing, Liu Peng, Chen Zhaotan and Meng Qinghua
Henan University of Technology, Zhengzhou Songshan Road 140#, Henan 450052, P. R. China
**Corresponding email: liuzhongdong234@163.com*

INTRODUCTION

Supercritical CO_2 Extraction or Supercritical Fluid Extraction (abbr: SFE) device is a Solid-Fluid or Fluid-Fluid extraction installation. Supercritical fluid refers to the fluid that is in the state of supercritical temperature and pressure. In this state, the fluid has the strengths of both gas and liquid. Its properties are between those of gas and liquid. Its density is several hundred times as much as the gas and close to that of liquid. Its mobility is also close to fluid. Its viscosity, however, is close to that of gas while its diffusibility is between gas and liquid and far greater than that of liquid. Supercritical fluid thus has high dissolving ability. CO_2 becomes the most commonly used media in SFE technology because it is innocuous, colorless, odorless, not corrosive, has low supercritical temperature and is convenient to obtain. In the last 30 years, extraction technology has developed very fast in its applications to the food industry [1] such as the extraction of of caffeine in coffee [2] and the extraction of DHA from fish oil [3]. This technology is now being used in the area of tea processing [4]. The extraction of aromatic tea constituents is one of the important applications. This research project selects Xinyang MaoJian as the object of study. We report that its aromatic constituents are extracted by the SFE method and analyzed by GC-MS.

MATERIAL AND METHODS

Materials

Xinyang MaoJian Tea was purchased from Xinyang, Henan Province, China. HA0231-50-2.5 SFE installation was made by Hunan

SFE Ltd., Nantong City, Jiangsu Province, P. R. China. Auto System XL Chromatograph was made by Perkin-Elmer Ltd., USA. Q-Mass 910 mass spectrometer was made by Perkin-Elmer Ltd. USA. FZ102 plants grinder was made by Qi Jiawu Scientific machine Ltd., Huanghua City, Hebei Province, P. R. China.

Methods

Supercritical CO_2 Extraction graph and technology parameters are shown in Figure 1 and Table 1, respectively. 880g dry tea powder and supercritical CO_2 are introduced into bottle A. When the air flow passed through the tea, it carried a lot of volatile compounds to form the supercritical "fluid"; the pressure at that time was 10-30 Mpa. After that, the airflow expanded to bottle B. At a pressure of 5-7 Mpa, the volatile compounds are extracted. The pure CO_2 that had no extracted compounds are released from bottle B and compressed in bottle A by a pump. After several cycles of this circulating extraction, the concentrated aromatic constituents were extracted.

GC-MS analysis of the aromatic ingredients

The system consists of a P.E Auto System XL gas chromatograph coupled with a P.E Qmass-910 mass spectrometer. The GC column is

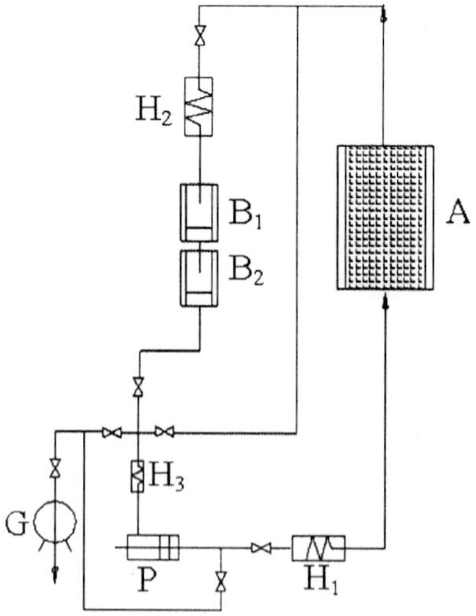

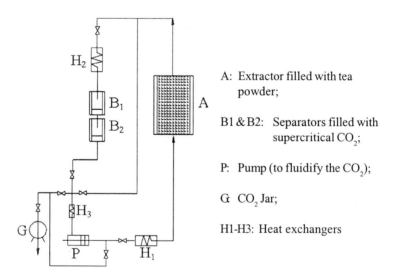

A: Extractor filled with tea powder;

B1 & B2: Separators filled with supercritical CO_2;

P: Pump (to fluidify the CO_2);

G: CO_2 Jar;

H1-H3: Heat exchangers

Figure 1. SFE installation graph

Table 1. Experimental parameters of Supercritical Fluid Extraction (SFE)

Raw Material	Extraction parameters			Separation parameters					
				Separator I			Separator II		
	Time (hr)	Temp (°C)	Press (Mpa)	Time (hr)	Temp (°C)	Press (Mpa)	Time (hr)	Temp (°C)	Press (Mpa)
Xinyang Mao Jian Tea 880g	4	30-35	20-35	4	25-35	4-8	4	25-35	4-8

SE-54 (30m×0.25mm) fused silica column.

GC-MS parameters: injector temperature: 180°C. Helium was used as the carrier gas at a flow rate of 10ml/min. Split ratio: 30:1.
Programmed temperature:

$$50^0C \xrightarrow{5^0 C/\min} 150^0C \xrightarrow{4^0 C/\min} 180^0C \xrightarrow{5^0 C/\min} 250^0C(\mathit{keep}\,10\text{min})$$

Mass spectrometer parameters: Detector temperature: 180°C. Ionizer: EI, ionization voltage: 70ev. Ionizer temperature: 240°C

RESULTS AND DISCUSSION

The results of GC-MS analysis can be seen in table 2, figure 2.

Ingredients that were identified newly as tea aroma components are Biphenylene and Dibenzofuran. Mass spectra of these compounds are shown in figures 3 & 4.

The TIC graph (figure 2) of Xinyang MaoJian tea showed more than 20 peaks. 20 compounds were identified by SE-54 column. Among them, those whose relative quantities are big are Coffeine (33.82%); Hexaecanoic acid (29.27%); 11,14,17-Eicosadienoic acid, methyl ester (19.97%); Phthalic acid butyl octyl ester (4.17%); Hexadecanoic acid, methyl ester (2.28%); Hexadecanoic acid, ethyl ester (3.51%); Diethyl phthalate (1.19%).

One of the differences between SFE and normal extraction methods are the extracted temperature. The supercritical temperature of CO_2 is 31°C and quiet close to the nature temperature. So the supercritical CO_2 extraction can be made in low temp and nature conditions. In such kind of

Table 2. The analysis of extraction compounds from Xinyang MaoJian Tea

Peak No	t_R(min)	Compound	Molecular Formula	Relative Quantity (%)
1[T]	1.653	Benzene	C_6H_6	0.38
2	1.960	1-Pentanol	$C_5H_{12}O$	0.38
3	8.393	Hexanoic acid	$C_6H_{12}O_2$	0.71
4[T]	14.093	Naphthalene	$C_{10}H_8$	0.33
5[T]	16.120	Methylnaphthalene	$C_{11}H_{11}$	0.09
6	18.780	β-Ionone	$C_{13}H_{20}O$	0.14
7	18.918	P-cresol,2,6-ditert-butyl-	$C_{15}H_{24}O$	1.04
8	19.100	Heptadecane	$C_{17}H_{36}$	0.19
9[FT]	19.813	Biphenylene	$C_{12}H_8$	0.14
10[FT]	20.600	Dibenzofuran	$C_{12}H_8O$	0.19
11	21.260	Dihydroactinidol	$C_{11}H_{16}O_2$	0.24
12[FT]	21.814	Fluorene	$C_{13}O_8$	0.24
13	22.096	Diethyl phthalate	$C_{12}H_{14}O_4$	1.19
14	23.707	Hexadecanoic acid, methyl ester	$C_{17}H_{34}O_2$	2.28
15[T]	24.480	Hexadecanoic acid, ethyl ester	$C_{18}H_{36}O_2$	3.51
16	24.914	Hexaecanoic acid	$C_{16}H_{32}O_2$	29.27
17	25.320	Phthalic acid, dissobuthy ester	$C_{16}H_{22}O_4$	1.71
18[FT]	26.733	Phthalic acid butyl octyl ester	$C_{18}H_{26}O_4$	4.17
19	27.472	Coffeine	$C_8H_{10}N_4O_2$	33.82
20	27.915	11,14,17-Eicosadienoic acid, methyl ester	$C_{21}H_{36}O_2$	19.97

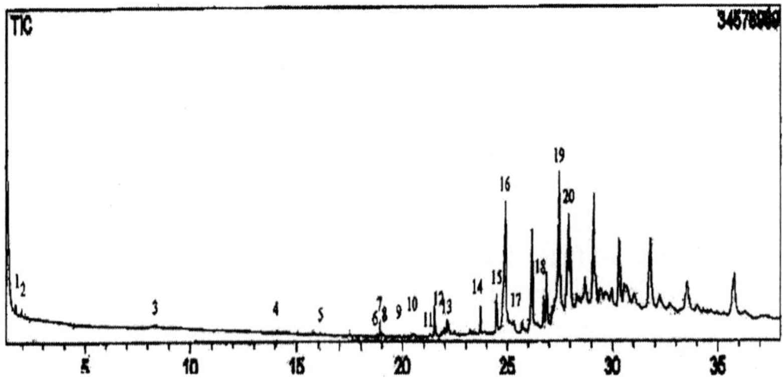

Figure 2. TIC graph of Xinyang MaoJian Tea aroma

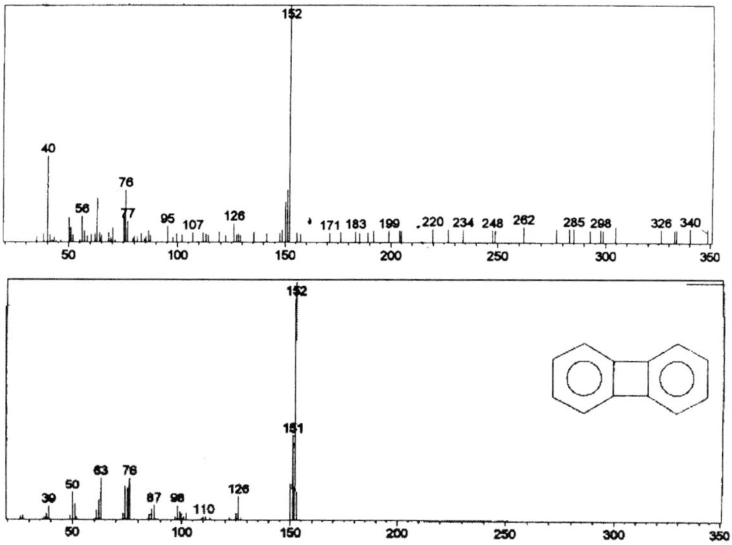

Figure 3. Biphenylene mass spectra in GC-MS (upper); Biphenylene standard mass spectra (lower)

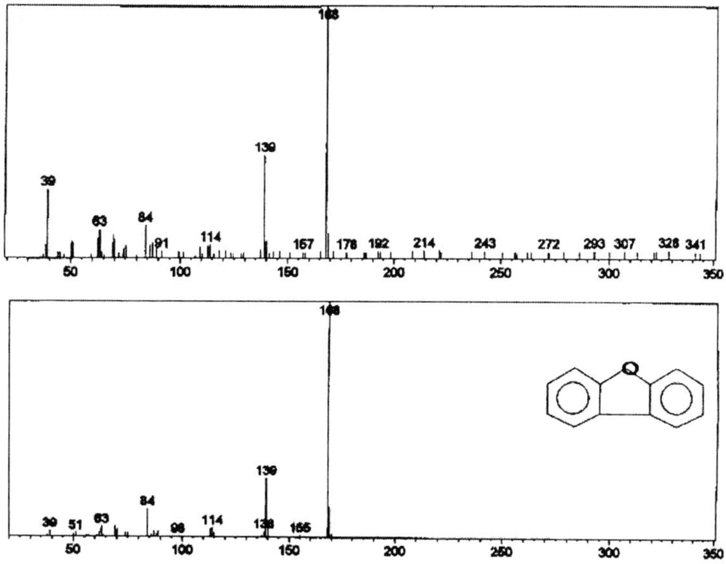

Figure 4. Dibenzofuran mass spectra in GC-MS (upper); Dibenzofuran standard mass spectra (lower)

conditions, those high boiling point compounds are easy to be extracted and separated. Nine constituents that was not identified in the normal extraction tea aroma but identified in SFE tea aroma are: Naphthalene; Methylnaphthalene; Biphenylene; Dibenzofuran, Fluorene; Dihydroactinidol; Hexadecanoic acid, ethyl ester; Phthalic acid butyl octyl ester. Coffeine has no contribution to the tea aroma, but it was identified in big relative quantity in SFE tea oil because it was easy to be extracted by supercritical CO_2. At the same time, chlorophyll and lutein were extracted, that's why the extracted oil was rather green but the normal extracted tea oil was maize.

The classified quantities of Xinyang MaoJian tea aroma constituents were shown in Table 3. There are seven esters (33.07%), two acids (29.98%), one compound has N (33.82%), two alcohols (1.42%), one ketone (0.14%), six hydrocarbons (1.37%) and another two constituents (0.19%).

Table 3. Xinyang MaoJian tea aroma constituents classified quantities

	Flower aroma constituents						Hydrocarbon	Else
	Alcohol	Ketone	Ester	Acid	Compounds have N	All		
Tea aroma Constituents classified quantities (%)	1.42	0.14	33.07	29.98	33.82	98.43	1.37	0.19

The esters are the biggest part. There are Dihydroactinidol; Diethyl phthalate; Hexadecanoic acid, methyl ester; Hexadecanoic acid, ethyl ester; Phthalic acid, dissobuthy ester; Phthalic acid butyl octyl ester;11,14,17-Eicosadienoic acid, methyl ester. These esters usually have strong aroma and most of them are flower aroma. Among them, Dihydroactinidol was degenerated from carotenoid and is rather important in the tea aroma. There are three phthalate esters, i.e. Diethyl phthalate; Phthalic acid, dissobuthy ester; Phthalic acid butyl octyl ester. These esters do not have aromas but have very strong aroma fixing ability so that the tea that contain large amount of them have strong tea aroma and long-time taste [5,6]. The high relative quantities of these compounds are consistent with the special characters of Xinyang MaoJian.

The relative quantities of acids are also very big (29.98%). Among them, the Hexaecanoic acid has very strong aroma fixing ability. The relative quantity of this compound in Xinyang MaoJian tea is rather big. It has great contribution to the aroma of Xinyang MaoJian tea.

Only one ketong was identified, it was β-Ionone. This compound has the smell of violet and is very important in the tea aroma [7].

Two alcohols are identified, they are 1-Pentanol and 2,6-ditert-butyl-P-cresol.

All of those flower aroma constituents have great contribution to the aroma of Xinyang MaoJian tea. They make the special aroma of Xinyang MaoJian tea together.

CONCLUSION

From all of above data , we conclude the special aroma constituents of Xinyang MaoJian tea are β-Ionone; Hexaecanoic acid; Dihydroactinidol; Hexadecanoic acid, methyl ester; Hexadecanoic acid, ethyl ester; Phthalic acid butyl octyl ester; 11,14,17-Eicosadienoic acid, methyl ester; 1-Pentanol and 2,6-ditert-butyl-P-cresol.

ACKNOWLEDGMENT

The authors also would like to thank Academician Yu-Fen Zhao (Department of Chemistry School of Life Science and Engineering, Tsinghua University, Beijing 100084, P. R. China.)

REFERENCES

[1] Risvi SSH, Daniels JA. Supercritical Fluid Extraction. Food Technology 1986; 40(7): 57-64

[2] James JE, Caffeine and Health. Academic Press, London, 1995

[3] Palmer MV, Ting SST. Application for Supercritical Fluid Technology in Food, Processing. Food Chemistry 1995; 52: 345-52

[4] McHugh M, et al. Supercritical Fluid Extraction. Butterworths: Butterworths Publishers, 1986

[5] Mitsuya Shimoda. et.al. Comparison of the Odor Concentrates by SDE and Adsorptive Column Method from Green Tea Infusion. J.Agric. Food.Chem. 1995; 43: 1616-20

[6] Kenji Yamaguchi et.al. Volatile Constituents of Green Tea Gyokuro. J Agri Food Chem 1981; 29: 366-70
[7] Michiko Kawakami and Tei Yamanishi. Flavor Constituents of Longjing Tea. Agri Biol Chem 1983; 47(9): 2077-83

Applications of RNAi Technology

Daina Lim, George W Yip and Boon-Huat Bay*
Department of Anatomy, Yong Loo Lin School of Medicine, National University of Singapore, 4 Medical Drive, MD10, Singapore 117597
**Corresponding author: antbaybh@nus.edu.sg*

INTRODUCTION

RNA silencing regulatory mechanism operates on two levels: transcriptional gene silencing (TGS) and post-transcriptional gene silencing (PTGS), which is generally known as RNA interference (RNAi) [1, 2]. TGS is typically associated with DNA methylation within the promoter regions [3, 4] while RNAi involves the specific degradation of a target gene. PTGS was initially known as "co-suppression" in plants. This was discovered in the study conducted by Napoli and his co-workers on increasing coloured pigment production in petunia plants using a sense chalcone synthase transgene. The introduction of a transgene surprisingly resulted in co-suppression of both the transgene and endogenous chalcone synthase gene in petunia plants [5]. The phenomenon of RNAi was first discovered in the nematode worm, *Caenorhabditis elegans*, as a result of introducing antisense or sense RNA constructs into embryos [6]. The observation that double stranded RNAs (dsRNAs) are much more effective in silencing a gene in *Caenorhabditis elegans* as compared to the traditionally used antisense strand [7] has led to the revolution of using dsRNA as a preferred method to knock down a gene. To date, RNAi has been shown in many organisms from plants to vertebrates [2].

The natural function of RNAi mechanism and its related processes is to protect the genome by silencing repetitive or transposable genetic elements [8, 9]. RNAi also acts as a defense mechanism against viral infections by knocking down the viral transcripts [10]. With the discovery of naturally occurring miRNA, RNAi has also been shown to be involved in endogenous gene expression [11] and helps to regulate growth and development [12].

MECHANISM OF RNA INTERFERENCE

The RNAi pathway can be dissected into two main phases, each involving ribonuclease activity. The initiator stage is triggered by the presence of a long dsRNA molecule, which can be exogenously or endogenously introduced. This dsRNA molecule is then processed into double stranded small interfering RNA (siRNA) molecules of 21-25 nucleotides long [13] by an RNaseIII containing enzyme, Dicer [14]. In the effector stage, the RNA induced silencing complex (RISC) guided by the siRNA binds to the target mRNA molecule [15] and cleaves it [16], thus resulting in silencing of the target gene (Figure 1).

RNaseIII enzymes can be classified into three classes [17]. Class I enzymes containing a single RNaseIII domain are found in prokaryotes. Class II and III enzymes contain two RNaseIII domains [17, 18]. Class III ribonucleases are further distinguished by the presence of a PAZ (Piwi/Agronaute/Zwille) domain [19]. The congruent model for the activity of Dicer is where the two RNaseIII domains associate intramolecularly to form a pseudo-dimer encompassing an active site between them [20].

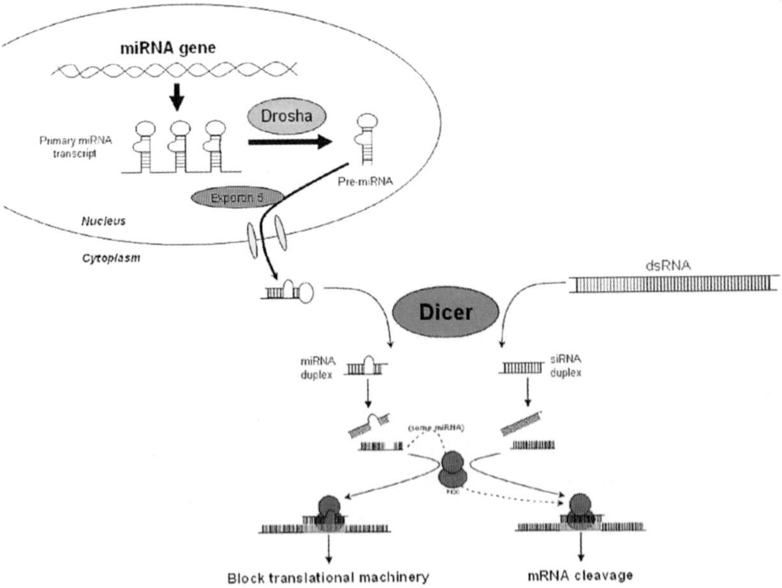

Figure 1. Biogenesis pathway of miRNA and siRNA.

Each domain then cleaves a single strand of the dsRNA molecule. Generation of the two nucleotide overhang is based on the pseudo-dimer alignment of the RNaseIII domains. The siRNA product length of 21-25 nucleotides is determined by the distance between the PAZ domain and the active site. The variation in siRNA length is due to the less efficient terminal binding of the dsRNA molecule [20].

RISC, the effector nuclease of RNAi, is a multiprotein complex with an Agronaute family member central to this complex [21]. Several members of this family have nuclease activity. Although the *Drosophila* genome encodes for at least four Agronaute family members - *AGO1*, *AGO2*, *Piwi*, and *Sting*, only *AGO2* is found to be a component of the RISC in *Drosophila* cells [22]. Ago2 is characterized to contain the PAZ and PIWI domain. Ago2 is the catalytic engine of mammalian RNA interference as it mediates RNA cleavage [23, 24]. The revelation of a RNaseH fold in the PIWI domain of Ago2 from its crystal structure further ascertain the nuclease activity of Ago2 in the RISC complex [25].

MICRORNAS

miRNAs constitute a class of noncoding small RNAs that are phylogenetically widespread in invertebrates, vertebrates and plants [25, 26]. The mature miRNA species exist in a stem-loop structure within the primary transcript. The RNaseIII enzymes Drosha and Dicer then release the mature miRNA. Drosha, a Class II RNaseIII enzyme with a pseudo-dimer catalytic core similar to Dicer, cleaves the primary miRNA transcript to release the stem-loop precursor, named pre-miRNA [27]. Pre-miRNA is then transported out of the nucleus into the cytoplasm to be processed in an siRNA-like fashion by Dicer to give miRNA duplexes. The double stranded miRNA is then incorporated by RISC. The fate of miRNAs in either cleaving directly (exact match) or repressing translation (by imperfect match) of target mRNAs is most likely determined by the degree of complementarity between miRNAs and their target mRNAs (Figure 1) [26]. In most plants, miRNA will hybridize to the target gene with perfect complementary thus resulting in mRNA cleavage of the target gene. However in most mammals, the miRNA binds to the target at the 3' UTR through imperfect complementarity at multiple sites and negatively regulates target expression at translational level. There is compelling evidence that miRNA/RISC complex blocks cap-dependant initiation of translation [28]. The targeted mRNAs are then localized to cytoplasmic mRNA processing bodies (P-bodies) for degradation [29, 30].

GENERAL APPLICATIONS OF RNAI

RNAi as a tool for gene analysis

With more complete genome sequences available, the discovery of RNAi offers the power of silencing specific genes in various organisms. This specific gene knockdown technology is able to provide us with the understanding of the functions of specific genes through phenotypic analysis. Gene knockdown can also provide information on the interacting partners of the particular silenced gene. Analysis of gene functions can also be carried out in a functional genomics approach. Kiger et al. [31] used large scale RNAi to determine the set of genes that are responsible for cell morphology and cytoskeletal organization for cell shape maintenance. Ashrafi et al. [32] and Sönnichsen et al. [33] have also utilized the ease of RNAi in conducting large scale RNAi analysis to determine the fat regulatory genes and the genes involved in early embryogenesis in *C. elegans*.

siRNA as therapeutics

There is immense excitement about using RNAi for therapy. With the advantage of a specific gene knockdown, many researchers have explored the prospect of silencing genes that play a vital role in human diseases. Initial studies of RNAi as a treatment alternative were explored in age-related macular degeneration [34]. Now, such RNAi therapy research ranges from gain-of-function hereditary diseases to cancer therapeutics and antiviral strategies. Most antiviral strategies rely on the creation of vaccines against antigenic epitopes of the virus. However vaccines against some of these diseases such as HIV have repetitively failed in clinical trials. In recent studies, several research groups have managed to suppress the replication of HIV in both continuous and primary cell lines [35-38]. Preclinical studies have shown inhibition of cell proliferation and survival in cancer cells treated with siRNA targeted against several oncogenes involved in invasion, angiogenesis and metastasis. Some of these oncogenes and their corresponding receptors include translocated oncogenes such as BCR-abl, fibroblast growth factor receptor, vascular endothelial growth factor and epidermal growth factor receptor. A study done by Sangkhathat S et al. [39] has shown that RNAi-induced silencing of beta catenin inhibits cell proliferation in paediatric hepatic tumor. Other studies have shown that RNAi combined with pre-existing cancer therapeutics can be more effective in inducing cell death in cancer cells [40, 41].

Several research groups have also done studies on siRNA as a treatment option *in vivo*. Song et al have successfully protected 80% of infected mice with fulminant hepatitis by introducing siRNA targeted against Fas through a hydrodynamic high-pressure tail vein injection [42]. Despite the promising results of this study, we have to note that this mode of siRNA administration will not be applicable to humans. Delivery of siRNA into an organism faces much challenge. The two main concerns are to reduce nuclease activity on the oligonucleotide and increase uptake of the siRNA into the cells. To increase the stability of the siRNA, modifications can be carried out on the 2' position of the pentose sugar such as 2'-deoxy-2'-fluorouridine [43]. Chiu et al. [44] has further shown that the 2 hydroxyl group is not essential for the RISC to function. This increased stability of the siRNA will then constitute to sustainability of the siRNA. In order to enhance cellular uptake of siRNA, siRNA can be packaged with lipophilic derivatives of cholesterol, lithocholic acid or lauric acid at its 5' end [45]. In addition to cholesterol labeling, siRNA can also be packaged with nanoparticles to aid the delivery into mammalian cells [46]. The additional advantage of using nanoparticles is that these siRNA can be introduced into a specific cell type [47], thus minimizing the off-target effects of siRNA on other cells.

RNAi as a tool for Nature Product Research

In plants, dsRNA-mediated RNA silencing has proven to be useful in gene knockdown-related functional studies. Plant endotoxins could also be removed if the toxin biosynthesis genes are targeted with the RNAi constructs. Recently, theobromine synthase from the coffee plant was knocked down with the hairpin construct of the transgene, which resulted in the production of decaffeinated coffee plants [48, 49]. Allen et al. [50] has created genetically modified opium poppy by the use of a chimeric hairpin construct to silence all members of a multigene COR family through

RNAi. This genetically modified opium poppy produces reticuline – a precursor of various pharmaceutically active compounds, which is usually present in small amounts in the normal poppy plant. This accumulation of reticuline is at the expense of morphine and codeine [50, 51]. Another group has increased the nutritional value of tomato fruit by suppressing an endogenous photomorphogenesis regulatory gene, *DET1*, using RNAi technology. The resultant fruit has a significantly increased carotenoid and flavonoid content [52]. The fatty acid composition in fungi was also increased by silencing one of the main enzymes required in the arachidonic acid pathway. This leads to the possibility of altering the types and amount of fatty acid in commercial strains of this fungus [53]. McDonald et al. [54] has also explored a novel way of controlling mycotoxin contamination in crops by using the option of silencing mycotoxin-specific regulatory genes using inverted repeat transgenes (IRT) containing sequences.

CONCLUSIONS

RNAi technology is still in its early stages of development but it has demonstrated great importance as a tool in basic research as well as a potential therapeutic. RNAi is already changing our ways of studying gene functions from a single gene knockdown to large scale knockdown high throughput screening. However, as a therapeutic, there are still several issues to contend with, such as delivery, sustainability and non-specific effects of the siRNA. With the promising potential of siRNA as a gene specific drug for mankind, there are still many considerations, optimizations and years of clinical trials before we will see the first siRNA therapeutic in the market. This will definitely be a wait worth waiting for.

REFERENCES

[1] Grishok A, Sinskey JL, Sharp PA. Transcriptional silencing of a transgene by RNAi in the soma of C. elegans. Genes Dev 2005; 19: 683–96

[2] Montgomery MK, Xu S, Fire A. RNA as a target of double-stranded RNA-mediated genetic interference in *Caenorhabditis elegans*. Proc Natl Acad Sci USA 1998; 95: 15502-7

[3] Mette MF, Aufsatz W, van der Winden J, Matzke M, Matzke A. Transcriptional silencing and promoter methylation triggered by double stranded RNA. EMBO J 2000; **19**: 5194–201

[4] Sijen T, Vijn I, Rebocho A, van Blokland R, Roelofs D, Mol J et al. Transcriptional and post-transcriptional gene silencing are mechanistically related. Curr Biol 2001; **11**: 436–40

[5] Napoli C, Lemieux C, Jorgensen R. Introduction of a Chimeric Chalcone Synthase Gene into Petunia Results in Reversible Co-Suppression of Homologous Genes in trans. Plant Cell 1990; 2: 279-89

[6] Guo S, Kemphues KJ. par-1, a gene required for establishing polarity in *C. elegans* embryos, encodes a putative Ser/Thr kinase that is asymmetrically distributed. Cell 1995; 81: 611-20

[7] Hannon GJ. RNA interference. Nature 2002; 418: 244-51

[8] Ketting R, Haverkamp T, van Luenen H, Plasterk R. *mut-7* of *C. elegans*, required for transposon silencing and RNA interference, is a homolog of Werner syndrome helicase and RNaseD. Cell 1999; 99: 133–41

[9] Wu-Scharf D, Jeong BR, Zhang C, Cerutti H. Transgene and transposon silencing in *Chlamydomonas reinhardtii* by a DEAH-Box RNA helicase. Science 2000; 290: 1159–62

[10] Voinnet O. RNA silencing as a plant immune system against viruses. Trends Genet 2001; 17: 449–59

[11] Voinnet O. RNA silencing: small RNAs as ubiquitous regulators of gene expression. Curr Opin Plant Biol 2002; 5: 444-51

[12] Lee RC, Ambros V. An extensive class of small RNAs in *Caenorhabditis elegans*. Science 2001; 294: 862-4

[13] Zamore PD, Tuschl T, Sharp PA, Bartel DP. RNAi: double-stranded RNA directs the ATP-dependent cleavage of mRNA at 21 to 23 nucleotide intervals. Cell 2000; 101: 25-33

[14] Bernstein E, Caudy AA, Hammond SM, Hannon GJ. Role for a bidentate ribonuclease in the initiation step of RNA interference. Nature 2001; 409: 363-6

[15] Martinez J, Patkaniowska A, Urlaub H, Luhrmann R, Tuschl T. Single-stranded antisense siRNAs guide target RNA cleavage in RNAi. Cell 2002; 110: 563-74

[16] Elbashir SM, Lendeckel W, Tuschl T. RNA interference is mediated by 21- and 22-nucleotide RNAs. Genes Dev 2001; 15: 188–200

[17] Mian IS. Comparative sequence analysis of ribonucleases HII, III, PH and D. Nucleic Acids Res 1997; 25: 3187–95

[18] Rotondo G, Frendewey D. Purification and characterization of the Pac1 ribonuclease of Schizosaccharomyces pombe. Nucleic Acids Res 1996; 23: 2377–86

[19] Cerutti L, Mian N, Bateman A. Domains in gene silencing and cell differentiation proteins: the novel PAZ domain and redefinition of the Piwi domain. Trends Biochem Sci. 2000; 25: 481-2

[20] Blaszczyk J, Gan J, Tropea JE, Court DL, Waugh DS, Ji X. Noncatalytic assembly of ribonuclease III with double-stranded RNA. Structure 2004; 12: 457-66

[21] Hammond SM, Boettcher S, Caudy AA, Kobayashi R, Hannon GJ. Argonaute2, a link between genetic and biochemical analyses of RNAi. Science 2001; 293: 1146-50

[22] Meister G, Landthaler M, Patkaniowska A, Dorsett Y, Teng G, Tuschl T. Human Argonaute2 mediates RNA cleavage targeted by miRNAs and siRNAs. Mol Cell 2004; 15:185-97

[23] Liu J, Carmell MA, Rivas FV, Marsden CG, Thomson JM, Song JJ et al. Argonaute2 is the catalytic engine of mammalian RNAi. Science 2004; 305: 1437-41

[24] Song JJ, Smith SK, Hannon GJ, Joshua-Tor L. Crystal structure of Argonaute and its implications for RISC slicer activity. Science 2004; 305: 1434-7

[25] Mourelatos Z, Dostie J, Paushkin S, Sharma A, Charroux B, Abel L et al. miRNPs: a novel class of ribonucleoproteins containing numerous microRNAs. Genes Dev. 2002; 16: 720-8

[26] Llave C, Xie Z, Kasschau KD, Carrington JC. Cleavage of Scarecrow-like mRNA targets directed by a class of Arabidopsis miRNA. Science 2002; 297: 2053-6

[27] Zeng Y, Yi R, Cullen BR. Recognition and cleavage of primary microRNA precursors by the nuclear processing enzyme Drosha. EMBO J 2005; 24: 138-48

[28] Pillai RS, Bhattacharyya SN, Artus CG, Zoller T, Cougot N, Basyuk E et al. Inhibition of translational initiation by Let-7 MicroRNA in human cells. Science 2005; 309: 1573-6

[29] Liu J, Valencia-Sanchez MA, Hannon GJ, Parker R. MicroRNA-dependent localization of targeted mRNAs to mammalian P-bodies. Nat Cell Biol 2005; 7: 719-23

[30] Sen GL, Wehrman TS, Blau HM. mRNA translation is not a prerequisite for small interfering RNA-mediated mRNA cleavage. Differentiation. 2005; 73: 287-93

[31] Kiger AA, Baum B, Jones S, Jones MR, Coulson A, Echeverri C et al. functional genomic analysis of cell morphology using RNA interference. J Biol. 2003; 2: 27

[32] Ashrafi K, Chang FY, Watts JL, Fraser AG, Kamath RS, Ahringer J et al. Genome-wide RNAi analysis of Caenorhabditis elegans fat regulatory genes. Nature 2003; 421: 268-72

[33] Sonnichsen B, Koski LB, Walsh A, Marschall P, Neumann B, Brehm M et al. Full-genome RNAi profiling of early embryogenesis in Caenorhabditis elegans. Nature 2005; 434: 462-9

[34] Check E. Firm sets sights on gene silencing to protect vision. Nature. 2004; 430: 819

[35] Jacque JM, Triques K, Stevenson M. Modulation of HIV-1 replication by RNA interference. Nature 2002; 418: 435-8

[36] Ammosova T, Berro R, Kashanchi F, Nekhai S. RNA interference directed to CDK2 inhibits HIV-1 transcription. Virology 2005; 341: 171-8

[37] Li Z, Xiong Y, Peng Y, Pan J, Chen Y, Wu X et al. Specific inhibition of HIV-1 replication by short hairpin RNAs targeting human cyclin T1 without inducing apoptosis. FEBS Lett. 2005; 579: 3100-36
[38] Ping YH, Chu CY, Cao H, Jacque JM, Stevenson M, Rana TM. Modulating HIV-1 replication by RNA interference directed against human transcription elongation factor SPT5. Retrovirology 2004; 1: 46
[39] Sangkhathat S, Kusafuka T, Miao J, Yoneda A, Nara K, Yamamoto S et al. *In vitro* RNA interference against beta-catenin inhibits the proliferation of pediatric hepatic tumors. Int J Oncol 2006; 28: 715-22
[40] Cesarone G, Garofalo C, Abrams MT, Igoucheva O, Alexeev V, Yoon K et al. RNAi-mediated silencing of insulin receptor substrate 1 (IRS-1) enhances tamoxifen-induced cell death in MCF-7 breast cancer cells. J Cell Biochem 2006; In Press
[41] Yu Y, Sun P, Sun LC, Liu GY, Chen GH, Shang LH et al. Downregulation of MDM2 expression by RNAi inhibits LoVo human colorectal adenocarcinoma cells growth and the treatment of LoVo cells with mdm2siRNA3 enhances the sensitivity to cisplatin. Biochem Biophys Res Commun 2006; 339:71-8
[42] Song E, Lee SK, Wang J, Ince N, Ouyang N, Min J et al. RNA interference targeting Fas protects mice from fulminant hepatitis. Nat. Med 2003; 9: 347-51
[43] Braasch DA, Jensen S, Liu Y, Kaur K, Arar K, White MA et al. RNA interference in mammalian cells by chemically-modified RNA. Biochemistry 2003; 42: 7967-75
[44] Chiu YL, Rana TM. siRNA function in RNAi: a chemical modification analysis. RNA 2003; 9: 1034-48
[45] Lorenz C, Hadwiger P, John M, Vornlocher HP, Unverzagt C. Steroid and lipid conjugates of siRNAs to enhance cellular uptake and gene silencing in liver cells. Bioorg Med Chem Lett 2004; 14: 4975-7
[46] Kam NW, Liu Z, Dai H. Functionalization of carbon nanotubes via cleavable disulfide bonds for efficient intracellular delivery of siRNA and potent gene silencing. J Am Chem Soc 2005; 127: 12492-3
[47] Khaled A, Guo S, Li F, Guo P. Controllable self-assembly of nanoparticles for specific delivery of multiple therapeutic molecules to cancer cells using RNA nanotechnology. Nano Lett. 2005; 5: 1797-808
[48] Ogita S, Uefuji H, Yamaguchi Y, Koizumi N, Sano H. Producing decaffeinated coffee plants. Nature 2003; 423: 823
[49] Ogita S, Uefuji H, Morimoto M, Sano H. Application of RNAi to confirm theobromine as the major intermediate for caffeine biosynthesis in coffee plants with potential for construction of decaffeinated varieties. Plant Mol Biol 2004; 54: 931-41
[50] Allen RS, Millgate AG, Chitty JA, Thisleton J, Miller JA, Fist AJ et al. RNAi-mediated replacement of morphine with the non-narcotic alkaloid reticuline in opium poppy. Nat Biotechnol 2004; 22: 1559-66

[51] Memelink J. Putting the opium in poppy to sleep. Nat Biotechnol 2004; 22: 1526-7

[52] Davuluri GR, van Tuinen A, Fraser PD, Manfredonia A, Newman R, Burgess D et al. Fruit-specific RNAi-mediated suppression of DET1 enhances carotenoid and flavonoid content in tomatoes. Nat Biotechnol 2005; 23: 890-5

[53] Takeno S, Sakuradani E, Tomi A, Inohara-Ochiai M, Kawashima H, Ashikari T et al. Improvement of the fatty acid composition of an oil-producing filamentous fungus, *Mortierella alpina* 1S-4, through RNA interference with delta12-desaturase gene expression. Appl Environ Microbiol 2005; 71: 5124-8

[54] McDonald T, Brown D, Keller NP, Hammond TM. RNA silencing of mycotoxin production in *Aspergillus* and *Fusarium* species. Mol Plant Microbe Interact 2005; 18: 539-45

Enhancing Bioactive Molecules in Medicinal Plants

M Z Abdin*

Centre for Transgenic Plant Development, Department of Biotechnology, Faculty of Science, Hamdard University, New Delhi-110 062, India.

**Corresponding email: mzabdin@rediffmail.com*

INTRODUCTION

The drugs of herbal, herbomineral and animal origin have been used by the traditional healers to maintain health and treat diseases since antiquity. According to an estimate of the world health organization (WHO), about three quarters of the world's population uses herbs and other traditional medicines for the treatment of various ailments including many refractory diseases such as arthritis, epilepsy, bronchial asthma, etc. These medicines are widely used in Africa and East Asian countries including India and China. Due to adverse side effects and development of resistant in pathogens to synthetic drugs, the use of plant-derived drugs is becoming popular also in the USA and Europe.

Herbal formulations used in Indian system of medicine (ISM) to treat diseases are derived either from a single or many herbs (Table 1). Medicinal efficacy of these formulations is however, not always consistent and mainly dependent on the endogenous levels of bioactive molecules present in the herb(s) used in these formulations (Table 2). These molecules are not uniformly distributed in various plant parts, but accumulate in specific organs such as root, stem, leaf and seed (Tables 3a & b) at a particular phenological stage of the herb such as artemisinin, arteannuin B and artemisinic acid in *in vivo* grown *Artemisia annua* L. [1] (Figure 1; Table 3a), esculin in *in vivo* as well as *in vitro* grown *Cichorium intybus* L. [2] (Table 3c) and xanthotoxin in *Ammi majus* [3] (Table 3d). Their concentrations in the plant tissues are influenced by a number of edaphic, climatic and endogenous factors including phytonutrients, growth hormones, abiotic as well as biotic stress factors and overexpression of gene(s) encoding rate-limiting enzymes of their biosynthetic pathways [4,5,6]. Medicinal quality of herbs hence, can be improved by enhancing the concentration of their bioactive molecule(s) through efficient management of these factors employing various physiochemical and biotechnological approaches as mentioned below:

Table 1. Herbal formulations used in Indian System of Medicine: constituents and uses

Herbal formulations	Composition		Uses
	Plant name	Part used	
Asrol	*Rauwolfia serpentina* (Sarpgandha)	Root	Hypotensive
Asgand	*Wathania somnifera* (Ashwagandha)	Root	Health vitalizer
Makoh	*Solanum nigrum*	Leaf, Berry	Hepato-protective and diuretic
Kasni	*Cichorium intybus* L.	Leaf, root and seed	Hepato-protective and diuretic
Adoosa	*Adhatoda vesica*	Leaf	Expectorant and cough sedative
Joshina (Syrup)	*Viola odorata* (gul Banafsha)	Flower	Cold and cough
	Onosma bracteatum (gauzaban) = *Borage officinalis*	Flower and leaf	
	Althia officinalis (Khatmi)	Seed	
	Malva rotundifolia (Khubbazi)	Seed	
	Glycirrhiza glabra (Mulethi)	Root	
	Zizyphus vulgaris (Unnab)	Fruit	
	Cordia latifolia (Sapistan)	Fruit	

Table 1 continued

Herbal formulations	Composition		Uses
	Plant name	Part used	
Jawarish Kamoni (Semi-solid)	*Carum carvi* (Zeera siyah)	Seed	Gastro-intestinal tract ailments
	Zingiber officinale (Zanjabeel)	Rhizome	
	Piper nigrum (Filfil siyah)	Fruit	
	Ruta graveolens (barg-e-sudab)	Leaf	
	Boora armani	Mineral (non-herbal)	
Triyaq-e-arb-aa	*Commiphora myrrha* (Mur)	Oleogum resin	Blood purifier, Infectious diseases
	Gentiana leutea (Juntiana)	Root	
	Habbul ghar	Fruit	
	Aristolochia longa (Zaravard taveel)	Root	

Physiochemical approaches

Through proper nutrient management

The bioactive molecules in medicinal herbs belong to three major classes of secondary metabolites namely terpenes, alkaloids and phenolics. These are synthesized through secondary metabolic pathways (Figure 2). Efficiency of these pathways depends on rate of photosynthesis for their energy and carbon requirements. Since photosynthesis is influenced by phytonutrients, especially sulphur and nitrogen as reported in earlier studies [7,8]. The concentration of bioactive molecules in medicinal herbs can thus, be enhanced through the optimization of photosynthesis with the balanced and split application of S an N [9]. In one such study, the yields

Table 2. Medicinal herbs, their bioactive molecules and medicinal uses

Name of medicinal herbs	Bioactive molecules	Clinical applications
Ammi majus (Atrilal) *Artemisia annua (Sweet Annie)* *Atropa belladonna*	Xanthotoxins Artemisinin	Vitiligo (Leukoderma) Anti-malarial
Bacopa moneirii (Brahmi) *Cassia angustifolia (Senna)* *Cichorium intybus (Kasni)* *Catharanthus roseus*	Atropine, Scopolamine Bacosides Sennosides Cichoriin and esculin Vincristine, Vinblastine Cardenolides Phyllanthin	Anodyne, antispasmodic Anticholinergic Memory vitalizer Laxative and purgative Hepatoprotective Anticarcinogenic
Digitalis purpurea *Phyllanthus amarus (Bhui Amla)* *Taxus* spp. *Rawulfia serpentina* *Withania somenifera* (Ashwagandha) *Dioscorea alata* *Aconitum bisma*	Taxol Ajmalicine, Reserpine Withanolides, Withaferins Anthocyanins Palmastine/vakotisine	Cardio-protective Hepato-protective (Hepatitis-B) Anti-carcinogenic Cardio-protective (hypotensive) Antibiotic, antirheumatic Anti-depressant Antipyretic Gastrointestinal disorders

Table 3a. Artemisinin content of different organs and structures of greenhouse- and field- grown *Artemisia annua,* determined by HPLC-EC (Source: 26).

	Artemisinin (% dw)	Artemisinin (% dw)
Organ/structure	**Greenhouse**	**Field**
Leaves	0.003-0.030	0.006-0.060
Main stems	0.0-0.0033	0.0004-0.007
Side stems	0.0	0.0004-0.0014
Roots	0.0	0.0
Flowers	0.012-0.042	0.104-0.264
Pollen	0.0	ND[z]
Seed husks	nd	0.116
Seeds[y]	0.036	0.081

[z]Not determined. [y]Containing floral debris

Table 3b. Xanthotoxin concentration in different organs of *Ammi majus* L. (Atrilal)

Plant organ	**Xanthotoxin concentration (mg g^{-1} dw)**
Root	0.058
Stem	0.045
Leaf	0.216
Seed	17.98

Table 3c. Esculin content in *Chicorium intybus* L. at different phenological stages

Phenological stage	**Esculin Content (mg g^{-1} dw)**
In Vivo	
Juvenile	1.80±0.01
Rossette	3.10±0.07
Bolting	3.75±0.05
Flowering	4.60±0.08
In Vitro	
Callus (Non Differentiating)	2.9±0.02
Callus (Differentiating)	4.40±0.01
Young Regenerant	4.80±0.07
Fully Grown Regenerant	6.75±0.05

Table 3d. Xanthotoxin yield at different phenological stages in *Ammi majus* L.

Stage	**Xanthotoxin yield (Kg ha^{-1})**
Vegetative	0.025
Pre-flowering	0.140
Flowering	0.253
Post-flowering	25.60

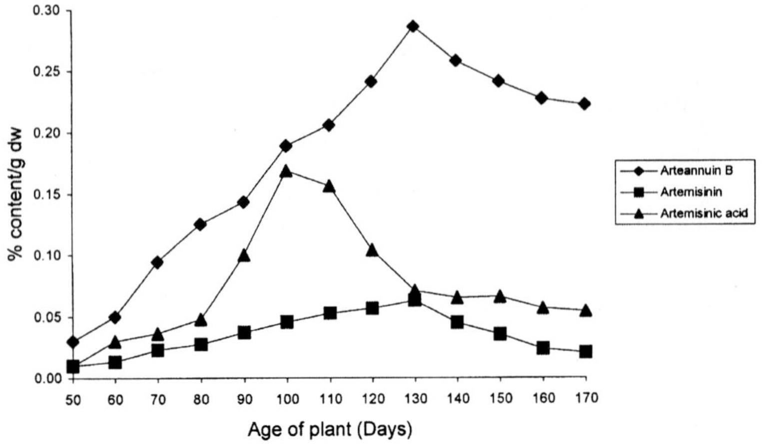

Figure 1. Age dependent variation in concentration of artemisinin and its immediate precursors.

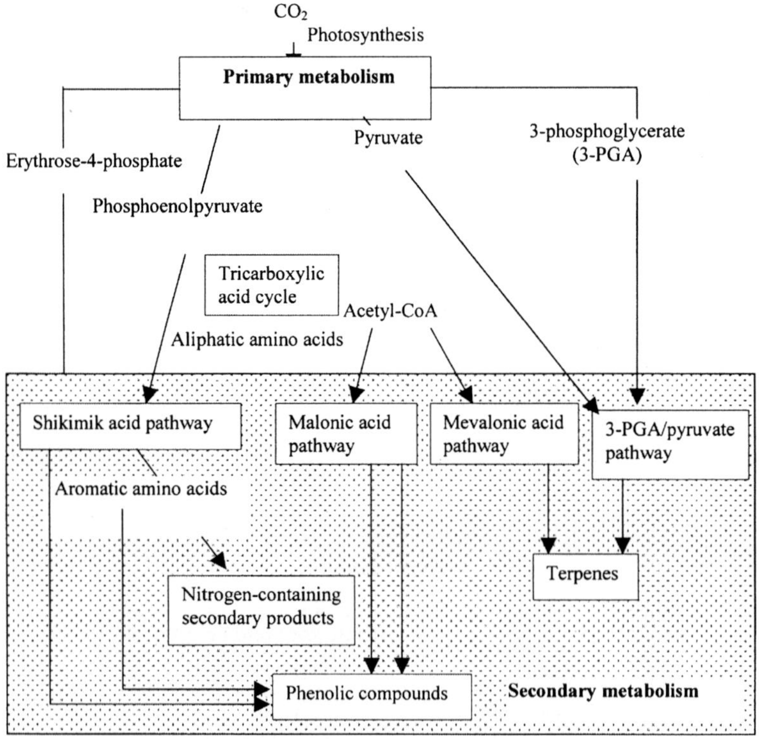

Figure 2. Biosynthesis pathways of secondary metabolites synthesis and their inter-relationship with primary metabolism.

of psoralen and xanthotoxin were increased in *Psoralea corylifolia* L. and *Ammi majus* L. (Table 4), respectively through the judicious and balanced application of these nutrients [10].

Table 4. Effect of nutrient management practices on psoralen and xanthotoxin yield in *Ammi majus* L. and *Psoralea corylifolia* L.

Treatments	**Yield (Kg ha^{-1})**	
	Psoralen	**Xanthotoxin**
Control	5.01 (0.0)	6.94 (0.0)
Manure	6.21 (24)	8.08 (16)
Fertilizers	9.66 (94)	13.4 (93)

Psoralen/Xanthotoxin Yield = Psoralen/Xanthotoxin concentration in the seed x Seed yield; Control = Without manure and fertilizer; Manure = 10 tonnes/ha; Fertilizers = 40 Kg S/ha, 40 Kg N/ha, 25 Kg P/ha, 25 Kg K/ha; Parentheses include percent variation over control.

Through hormonal treatment:

Plant hormones play a pivotal role in plant growth and development by influencing the physiological and biochemical processes of plant [11,12], and gene expression [13]. The endogenous level of auxins, gibberellins and cytokinins declines while ABA and ethylene increases with increased plant age. Since the former are required in adequate amounts for normal growth and development, these hormones are applied exogenously to bring their adequate levels in plants for their better growth and optimum economic yield.

Besides improving the growth and development, the application of growth hormones have been shown to increase secondary metabolite accumulation in plants [14,15,16]. In the studies carried out in our laboratory, combined application of growth hormones, IAA and GA_3 has led to more accumulation of artemisinin (Table 5), a novel secondary metabolite with antimalarial activity in *Artemisia annua* L.. This increase in artemisinin level was coupled with increased HMGCo-A reductase activity, an enzyme that diverts carbon flux from primary metabolic pathway to secondary metabolism for biosynthesis of secondary metabolites [17,18,19].

Table 5. Effect of growth hormones on artemisinin content in the leaves of *Artemisia annua* L.

Hormones	**Artemisinin content (%dw)**
Control	0.065 (0.0)
IAA	0.085 (30.7)
GA_3	0.087 (33.8)
IAA+GA_3	0.090 (38.4)

Through stress exposure:

Though there are few reports on increase in the levels of bioactive molecules in medicinal plants under stress, a quite convincing data has been obtained in our studies with *Artemisia annua* L. [20] and *Cassia angustifolia* Vahl. [21]. The study showed a significant increase in levels of artemisinin and sennoside accumulation (Table 6a & b) under heavy

Table 6a. Effect of abiotic stresses on artemisinin content in the leaves of *Artemisia annua* L.

Stress	**Artemisinin content (μg g^{-1} dw)**
Control	291 (0.0)
Salinity (160 mM)	576 (98)
Lead (Pb; 500 ppm)	892 (207)

The Salt (NaCl) and lead stress was given at bolting stage (100 DAS) and artemisinin was determined 10 days after the treatment; Parentheses include per cent variation

Table 6b. Effect of abiotic stresses on sennosides content in the leaves of *Cassia angustifolia* Vahl.

Stress	**Sennoside content (mg g^{-1} dw)**
Control	23 (0.0)
Salinity (160 mM)	36 (59)
Lead (Pb; 500 ppm)	33 (46)

The Salt (NaCl) and lead stress was given at 45 DAS and sennosides were determined 15 days after the treatment; Parentheses include per cent variation metal and salt stress, respectively. The increased levels of these compounds could be due to the stimulation of their biosynthetic pathways as both abiotic and biotic stress have shown to accelerate secondary metabolic pathways [22,20]. Short exposure of medicinal herbs to abiotic stresses, hence, could be useful to enhance the levels of secondary metabolites to result in better quality and medicinal efficacy of these herbs. The hormone treatments were given at vegetative stage and artemisinin was determined in the leaves collected at pre-flowering stage; Parentheses include percent variation over control.

Through Biotechnological approaches

Through *in vitro* culture

In plant tissue culture, media are always supplemented with various phytohormones either singly or in combinations. 2,4-D, IAA, IBA, NAA etc. are used as auxin source while BAP, kinetin and zeatin are used as cytokinin. At a specific combination of growth regulators, there may be an increase in accumulation and yield of bioactive molecules. One such study carried out in our laboratory showed an improvement in alkaloid yield in *Catharanthus roseus* [23].

Through genetic transformation

Genetic transformation of medicinal herbs with the gene(s) encoding the key rate limiting enzyme(s) of secondary metabolic pathways that produce bioactive molecules is of great interest in the present era. Now molecular tools are available which make the transfer and overexpression of desired trans gene(s) in medicinal herbs to enhance the level of secondary metabolites (Figure 3).

Several genetically transformed plants have been reported to accumulate increased amount of bioactive molecules. *Artemisia annua* L. transformed with farnesyl diphosphate synthase [4] and isopentyl transferase [5] showed up to 300% and 70% increase in artemisinin levels over wild type plants, respectively. *Solanum tuberosum* [24] and *Nicotiana tabacum* [25] transformed with HMGCo-A reductase gene and, *Cichorium intybus* (6) transformed with *osmotin gene* resulted in

Figure 3. Genetic transformation of *Cichorium intybus* L. with osmotin gene. A, leaf segments after cocultivation on selection medium; B and C, callus production on the base of transformed leaf segments; D, transformed regenerates on selection medium; E, full grown transformed plant in pot in natural condition.

58%, 600% and 47% increase in bioactive compounds, namely phytoalexin, sterols and esculin, respectively (Table 7).

The results of these studies thus, suggest that this approach can successfully be used to increase accumulation of bioactive compounds in medicinal herbs. Further, their accumulation can be directed in specific tissues or organs at the appropriate phenological stage by tagging the gene(s) encoding rate limiting enzyme(s) of secondary metabolic pathways with tissue/organ specific inducible promoters.

CONCLUSION

A number of experiments carried out in our laboratory and elsewhere so far suggest that concentrations of bioactive molecules in medicinal plants

Table 7. Enhanced level of bioactive molecules in genetically modified medicinal plants

Crops	Over-expressed Enzyme/ Protein	Bioactive molecules	% increase	References
Artemisia annua	Farnesyl diphosphate synthase	Artemisinin	200-300	(4)
Artemisia annua	Isopentenyl transferase	Artemisinin	30-70	(5)
Solanum tuberosum	HMG-CoA reductase	Phytoalexin	58	(24)
Nicotiana tabaccum	HMG-CoA reductase	Sterols	600	(25)
Cichorium intybus	Osmotin	Esculin	47	(6)

can be improved with the use of appropriate physiochemical and molecular approaches. These include nutrient management, short-term exposure of plants to abiotic stresses and transformation of medicinal herbs with genetically engineered gene(s) encoding rate limiting enzyme(s) of secondary metabolic pathways. The magnitude of enhancement in the levels of these metabolites however, depends on the availability of precursors and other endogenous factors including the level and kinetic properties of rate limiting enzyme(s). Nevertheless, the quality and medicinal efficacy of these herbs can be improved with the use of these approaches either alone or in combination.

ACKNOWLEDGMENTS

These studies have been funded by research grants from DBT and CCRUM, Govt. of India. The contribution of TMOP&M/CSIR, Govt. of India in terms of infra-structural research facilities is gratefully acknowledged. The author also acknowledges the contribution and help received from collaborators and research students during these studies. Special thanks are also due to Dr. M. Irfan Qureshi for his help during the preparation of this manuscript and my wife, Mrs. Shamim Fatima as well as children for their moral support and encouragement during these studies.

REFERENCES

[1] Abdin MZ, Israr M, Srivastava PS, Jain SK. In vitro production of artemisinin, a novel antimalarial compound from *Artemisia annua*. J Med Arom Plant Sci 2001; 22/4A, 23/1A: 378-84

[2] Rehman RU, Israr M, Srivastava PS, Bansal KC, Abdin MZ. *In vitro regeneration of witloof chicory (Cichorium intybus* L.) from leaf explants and accumulation of esculin. In vitro Cell Dev Biol-Plant 2003; 39: 142-46

[3] Ahmad S, Abdin MZ, Fazli IS, Jamal Arshad, Maaz M, Iqbal M. Variability in xanthotoxin concentration in organs of *Ammi majus* and yield at various phenological stages. J Med and Arom Plant Sci (In Press) 2006

[4] Chen D, Ye H, Li G. Expression of a chimeric fernasyl diphosphate synthase gene in *Artemisia annua* L. transgenic plants via *Agrobacterium tumefaciens*-mediated transformation. Plant Sci 2000; 155: 179-85

[5] Sa G, Mi M, He-Chun Y, Ben-ye L, Guo-feng L, Kang C. Effect of ipt gene expression on the physiological and chemical characteristics of *Artemisia annua* L. Plant Sci 2001; 160: 691-98

[6] Abdin MZ, Israr M, Kumar PA, Jain SK. Molecular approaches to enhance artemisinin content in *Artemisia annua* L. In: Recent Progress in Medicinal Plants, Volume 4: Biotechnology and Genetic Engineering. (Govil, J.N., Kumar, P.A., Singh, V.K., eds), 2002; Sci Tech Pub, USA, pp. 145-62

[7] Lawlor DW, Kontturi M, Young AT. Photosynthesis of flag leaves of wheat in relation to protein, ribulose bisphosphate carboxylase activity and nitrogen supply. J Exp Bot 1997; 40: 43-52

[8] Abdin MZ, Israr M, Rehman RU and Jain SK. Artemisinin, a novel antimalarial drug: Biochemical and molecular approaches for enhanced production. Planta Med 2003; 69: 289-99

[9] Ahmad S. Effect of phenology and phytonutrients on physiochemical characteristics and medicinal quality of *Ammi majus* (Atrilal) and *Psoralea corylifolia* (Babchi), 2004; Ph.D Thesis, Department of Biotechnology, Faculty of Science, Hamdard University, New Delhi, India

[10] Ahmad S, Jamal A, Fazli IS, Rehman RU, Iqbal M, Mazz M, Abdin MZ. Interactive effect of sulphur and nitrogen on economic yield and medicinal quality of Atrilal (*Ammi majus* L.). 2002; National Seminar on Recent Research Trends in Life Sciences, pp. 52-52

[11] Davies PJ. Plant hormones and their role in plants growth and development. 1987.Kluwer Academic Press, Dordrecht

[12] Cao H, Chen S. Brassinosteroid-induced rice lamina joint inclination and its relation to indole-3-acetic acid and ethylene. Plant Growth Regul 1995; 16: 189-96

[13] Ho TD and Hagen G. Hormonal regulation of gene expression. Plant Growth Regul 1993; 12: 197-05

[14] Farooqi AHA, Shukla A, Sharma S, Khan A. Effect of plant age and GA3 on artemisinin and essential oil yield in *Artemisia annua* L. J Herbs Spices Med Plants 1996; 4: 73-80

[15] Yaseen M, Tajuddin. Effect of plant growth regulators on yield, oil composition and artemisinin from *Artemisia annua* under temperate condition. J Med Arom Plant Sci 1998; 20: 1038-41

[16] Smith TC, Weathers PS, Cheetan RD. Effect of gibberellic acid on hairy root cultures of *Artemisia annua* growth and artemisinin production. *In vitro* Cell Dev Biol 1997; 33: 75-79

[17] Bach TJ, Litchenthaler HK. Application of modified lineweaver-Burk plots to studies of kinetics and regulation of raddish 3-hydroxy-3-methylglutaryl coenzyme A reductase. Biochem Biophys Acta 1984; 794: 152-61

[18] Bach TJ. Hydroxymethylglutaryl-CoA reductase, a key enzyme in phytosterol synthesis? Lipids 1986; 21: 82-86

[19] Hata S, Takagishi H, Kouchi H. Variation in the content and composition of sterols in Alfalfa seedlings treated with compactin (ML236B) and mevalonic acid. Plant Cell Physiol 1987; 28: 709-14

[20] Qureshi MI, Israr M, Abdin MZ, Iqbal M. Responses of *Artemisia annua* L. to lead and salt-induced oxidative stress. Env Exp Bot 2004; (In press)

[21] Anjum Arshi, Abdin MZ, Iqbal M. Growth and metabolism of senna as affected by salt stress. Biol Plant 2002; 45:295-98

[22] Wallart TE, Pras N, Quax WJ. Seasonal variations of artemisinin and its biosynthetic precursors in plants of *Artemisia annua* of different geographical origin: proof for the existence of chemotypes. Planta Med 2000; 61: 57-62

[23] Mujib A, Ilah A, Gandotra N, Abdin MZ. *In vitro* application to improve alkaloids yield in *Catharathus roseus*. In: Recent Progress in Medicinal Plants, Volume 4. Biotechnology and Genetic Engineering (Govil, J.N., Kumar, P.A. Singh, V.K. eds), 2002. Sci Tech Pub USA, pp. 415-40

[24] Stemer BA, Bostock RM. Involvement of 3-hydroxy-3-methylglutaryl coenzyme A reductase in the regulation of sesquiterpenoid phytoelaxin synthesis in potato. Plant Physiol 1987; 84: 404-08

[25] Schaller HB, Grausem P, Benveniste ML, Chye CT, Tan YH, Song and Chua NH. Expression of the *Hevea brasiliensis* (H.B.K.) Müll. Arg. 3-hydroxy-3-methylglutaryl-Coenzyme A reductase 1 in tobacco results in sterol overproduction. Plant Physiol 1995; 109: 761–70

[26] Ferreira JFS, Simon JE, Janick J. Relationship of artemisinin content of tissue cultured, green house and field grown plants of *Artemisia annua*. Planta Med 1995; 61: 351-55

Biotransformation of Taxanes from Cell Cultures of *Taxus* sp

Jungui Dai[a]*, Dean Guo[b], Masayoshi Ando[c] and Lin Yang[d]
[a] *Key Laboratory of Biosynthesis of Natural Products, Ministry of Public Health, Chinese Academy of Medical Sciences & Peking Union Medical College, 1 Xian Nong Tan Street, Beijing 100050, People's Republic of China*
[b] *The State Key Laboratory of Natural and Biomimic Drugs, School of Pharmaceutical Sciences, Peking University, 38 Xueyuan Road, Beijing 100083, People's Republic of China*
[c] *Department of Chemistry and Chemical Engineering, Niigata University, Ikarashi 2-8050, Niigata 950-2181, Japan*
[d] *College of Life and Environmental Sciences, The Central University for Nationalities, 27 South Zhongguancun Street, Beijing 100081, P. R. China*
**Corresponding email: jgdai@imm.ac.cn*

INTRODUCTION

Taxuyunnanine C, 2α,5α,10β,14β-tetraacetoxytaxa-4(20),11-diene (**1**), and its analogues, **2**, **3** and **4**, are the major C-14 oxygenated taxanes produced by the tissue and cell cultures of *Taxus* sp. in high yields (*ca.* 5-6% of the dry weight) [1-3]. Their high content in the cultures and their taxane-skeleton endow them with valuable potential for the semi-synthesis of paclitaxel (Taxol®), one of the most effective anticancer agents, and other structurally related bioactive agents, such as anti-MDR (multi-drug resistance) cancer agents or anti-MDR cancer reversal agents (Scheme 1) [4-6]. Unfortunately, these taxanes have fewer functional groups on the skeleton in comparison with paclitaxel and other bioactive taxoids. The regio- and stereo-selective introduction of oxygen functional groups at their C-1, C-7, C-9 and C-13 positions seems very difficult by traditional chemical methods. In this context, enzymatic conversion by using microorganisms or plant cell suspension cultures is a potential alternative, and some interesting progress has been achieved [7-14]. Furthermore, the enzymatic systems of microorganisms or plant cell cultures may be useful tools to mimic some steps of taxoid biosynthesis and can provide some useful help for the study of taxoid biosynthesis, especially for extensive oxidation of the taxane skeleton. Here, we mainly describe the specific biotransformation of these taxanes by cell cultures of a plant, *Ginkgo biloba*, and a fungus, *Absidia coerulea* IFO4011.

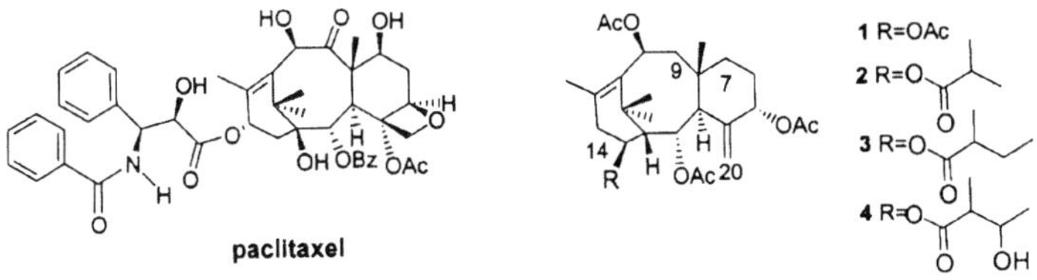

Scheme 1. Paclitaxel and taxanes from tissue and cell cultures of *Taxus*

MATERIALS AND METHODS

Melting points were measured with a Yanagimoto micro-melting point apparatus and are uncorrected. Optical rotations were obtained using a Horiba SEPA-200 polarimeter. ^{1}H NMR (500 MHz) and ^{13}C NMR (125 MHz) spectra were recorded with a Varian Unity-PS instrument using $CDCl_3$ as solvents. ^{1}H NMR and ^{13}C NMR assignments were determined by H-H COSY, DEPT, HMQC and HMBC experiments. HREIMS were carried out on a JEOL-HX 110 instrument. IR spectra were taken on a Hitachi 270-30 spectrometer in $CHCl_3$. Semi-preparative HPLCs were performed on a Hitachi L-6200 HPLC instrument with an Inertsil Prep-sil or Pre-ODS (GL Science, 25 cm × 10 mm i.d.) stainless steel column and a YRU-883 RI/UV bi-detector; the flow rate was 5 mL/min unless stated elsewhere. Silica gel (230-300 mesh) was employed for flash column chromatography; analytical TLC plates (silica gel 60 F_{254}, Merck) were visualized at UV_{254} by spraying with 10% H_2SO_4 (in EtOH) followed by heating.

Organisms, media and culture conditions

The *Ginkgo* suspension cells at an inoculum of 5 g/L of cell cultures (dry weight) were shaken in a rotary shaker at 110 rpm at 25°C in the dark. The pH values of the cultures were adjusted to 5.8 before autoclaving for 20 min at 121°C the amount of sucrose added to the medium was 30 g/L [15]. *A. coerulea* IFO4011 was purchased from Institute for

Fermentation, Osaka, Japan (IFO) and kept in solid medium containing potato (200 g/L), sucrose (20 g/L) and agar (2 %) at 4°C. The seed cultures were prepared in a 500-mL Erlenmeyer flask with 150 mL of liquid medium and incubated for 2 days. To 150 mL of medium in a 500-mL flask was added 5 mL of the seed culture followed by admixing on a rotary shaker at 110 rpm at (25±2) °C in the dark for the use of biotransformation.

Substrate

Compounds **1-4** were isolated from callus cultures (Ts-19 strain) of *T. chinensis*, and identified by chemical and spectral methods. **7-9** were obtained from hydrolysis of compound **1** [16], and compound **10** was prepared by a chemical method, according to Zhang et al. (2002) [17]. The substrates were dissolved in EtOH and diluted to 10.0 mg/mL before use as stock solution.

Biotransformation by Ginkgo cell suspension cultures
*Biotransformation of **1***

The *Ginkgo* cell suspension cultures were cultivated on a large scale using 1000 ml of flasks with 330 ml of medium. On the 15th day, 2 mL of substrate stock solution were added to each flask and the total amount of **1** administered was 1.0 g. After an additional six days of incubation, all the medium was collected, extracted with EtOAc thrice and concentrated. The residue (1.5 g) obtained was separated by silica gel chromatography (silica gel H, 5-40 mesh), eluting with acetone-petroleum ether (60-90°C) [1:5-1:1] to yield compounds **5** (500 mg) and **6** (100 mg).

Effect of substrate addition time on the biotransformation

On the 0, 3rd, 6th, 9th, 12th, 15th and 18th day during the cell culture growth period, 35 mg/L of **1** was added to 500 mL of flasks with 150 mL of medium by 3 replicates. On the 21st day, the media in each flask was collected, extracted and concentrated as described above. The residues were dissolved in the HPLC mobile phase and diluted with the same solution to give 2.0 mL and the amount of residual substrate, compounds **5** and **6**, were determined by HPLC.

Effect of substrate concentration on the biotransformation

On the 18th day, different concentrations of **1** (15 mg/L, 30 mg/L, 45 mg/L, 60 mg/L, 75 mg/L) were added in 3 replicates. On the 21st day, the media were collected, extracted and concentrated, and the amount of residual substrate, compounds **5** and **6,** were determined by HPLC.

Biotransformation kinetics

On the 18th cultural day, 60 mg/L of **1** was added in 3 replicates and after every 24 hours of incubation, the media were collected, extracted and concentrated, and the amounts of residual substrate, compounds **5** and **6**, were analyzed by HPLC.

Effects of substituents on the biotransformation yields by Ginkgo cells

Compounds **2-4** and **7-10** were used as substrates and added to 18-day-old *Ginkgo* cell cultures at a concentration of 60 mg/L, then incubated for an additional 3 days. The amounts of residual substrates and their 9α hydroxylated products were analyzed by HPLC.

Biotransformation by A. coerulea

*Biotransformation of Substrate **1***

Substrate **1** (400 mg) was dissolved in EtOH (5.4 mL), distributed among 18 Ehrlenmeyer flask cultures with pipette and incubated for 7 days, after which time the cultures were filtered and pooled; the cells were thoroughly washed with water and the filtrate was saturated with NaCl and extracted 5 times with EtOAc. All the extracts were pooled, dried with anhydrous Na_2SO_4, and concentrated under vacuum at 40°C to give 700 mg of residue. The dried cell cultures were extracted 3 times by sonication with EtOAc, the given extracts were pooled and concentrated under vacuum at 40°C to afford 136.9 mg of residue. Both the extracts were combined and chromatographed on a silica gel column eluting gradient with Hexane/EtOAc (90% Hexane and 10% EtOAc-50%Hexane and 50% -100% EtOAc) to give 5 fractions. Compound **13** (20 mg, in 5%) was obtained from fraction 3.

Effect of substituents of substrates on the biotransformation yields

Compound **3** and **4** were used as substrates and biotransformed by *A. coerulea*. The procedures were performed as described as above, and their 7β-hydroxylated products, **12** and **13,** were obtained at 10% and 15% yields, respectively.

*Effect of β-cyclodextrin on the biotransformation of **1** with A. coerulea*

First, compound **1** (500 mg in 2 mL EtOH) was added into saturated β-cyclodextrin solution (1,300 mg in 50 mL medium) and stirred until no substrate was detected in the medium; this was followed by filtration and re-suspension in 5 mL of EtOH before adding to the cell cultures, followed by incubation. The remaining procedures were the same as described above. Compounds **11**, **14-16** were obtained at 5%, 5%, 5% and trace yields, respectively.

Combined biotransformation of **1** with Ginkgo cells and A. coerulea

9α and 7β hydroxylated derivatives (**5**, **11**) were prepared as described before, and routinely acetylated. Their acetylated products (**17**, **20**) were then successively biotransformed by the fungus, *A. coerulea,* and cell suspension cultures of *Ginkgo*, respectively. Their corresponding 7β and 9α hydroxylated products (**18** and **21**) were obtained at 2% and 10% yields, respectively.

RESULTS AND DISCUSSION

Bioconversion by Ginkgo cell suspension cultures

The substrate, taxuyunnanine C (**1**), was added to 15 day-old cell cultures of *Ginkgo*, and after an additional 6 days of incubation, the two major products, compounds **5** and **6** (hydroxylated specifically at the 9α position) were obtained at ~70% and 12% yields, respectively (Scheme 2). The specific hydroxylation at the unactivated C-9 position of **1** by *Ginkgo* cells constitutes an important step in the semisynthesis from **1** to taxol and other bioactive taxoids. In an endeavor to optimize the biotransformation conditions to increase the yields of **5** and **6**, the effects of the concentration and the timing of addition of **1** as well as the kinetics of the biotransformation reaction were investigated. First, the kinetics of

Ginkgo cell growth and pH value, as well as the amount of residual **1** and the yields of **5** and **6** corresponding to different timings of addition of **1**, were investigated. The *Ginkgo* cells grew very fast under the culture conditions; the overall growth period lasted 21 days and involved three phases: (1) lag phase [0-9th day], (2) logarithmic phase [9-18th day], and (3) stationary phase [18-21st day]. The pH values remained relatively stable at a range of 4.0-6.0. The results also showed that the optimal timing for adding **1** was at the logarithmic phase (9~18th day) of the cell growth period. The substrate added at this phase, especially on the 18th day - the late logarithmic phase or the early stationary phase - was converted into the two major products efficiently. The substrate was almost completely converted, and the yields reached their highest levels, yielding approximately 70% and 20% of **5** and **6**, respectively (HPLC). These may result from different activities of the enzymes present in the respective cultural stages and the difference in sensitivity of the cultured cells to exogenous substrates.

In order to determine the optimal concentration of **1** in this experiment, the effects of different concentrations of **1** (15 mg/L, 30 mg/L, 45 mg/L, 60 mg/L and 75 mg/L, final concentration) on the bioconversion were investigated. The results showed that the optimal concentration of **1** was 60 mg/L. At this concentration, **1** was efficiently converted and residual **1** could hardly be detected; meanwhile, the two major products, **5** and **6**, reached their highest yields, ~40 mg/L and 13 mg/L, respectively. The results of this experiment also suggested that the concentration of

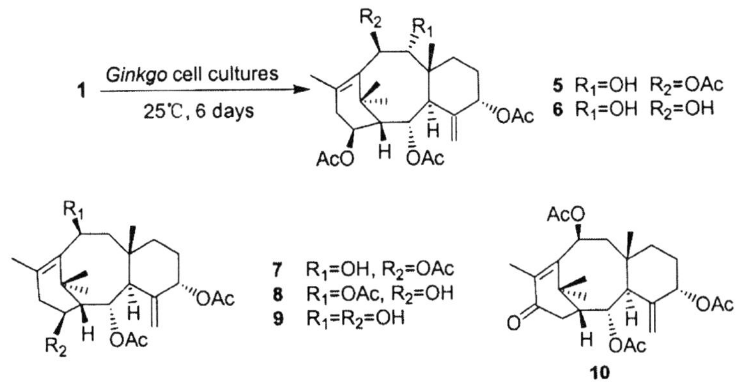

Scheme 2. 9α Hydroxylation of taxanes by *Ginkgo* cell cultures

exogenous substrate addition affects the substrate in the bioprocess to the end products.

Based on the above results, the kinetics of the substrate in the bioconversion were also investigated. 60 mg/L of **1** was added to 18-day-old cultures and incubated for 24 hours, 48 hours, 72 hours, respectively. The results of HPLC analysis (Figure 1) revealed that (1) the biotransformation ratios of **1** were about 40%, 80% and 100%, (2) the yields of **5** were about 26%, 60% and 59.5%, and (3) the yields of **6** were about 13%, 18% and 21.6%, respectively, at the above mentioned three incubation periods. Therefore, the optimal incubation time should be 48 hours as far as the production of **5** and **6** is concerned.

To investigate the specific hydroxylation ability of *Ginkgo* cells and to gain insight into the effects of different substituents on the hydroxylation, seven other taxadienes (**2-4**, **7-10**; Scheme 2) were used as substrates and converted by *Ginkgo* cells. The results showed that these taxadienes were all specifically hydroxylated at the 9α position, and different yields also resulted from substrates with different substituents (Table 1).

Plant cells *in vitro* could convert the exogenous substrates to new products. In essence, it is the enzymes produced by plant cells that play the role. If the properties of enzymes are well investigated, the biotransformation condition could be well optimized, and the yields of the

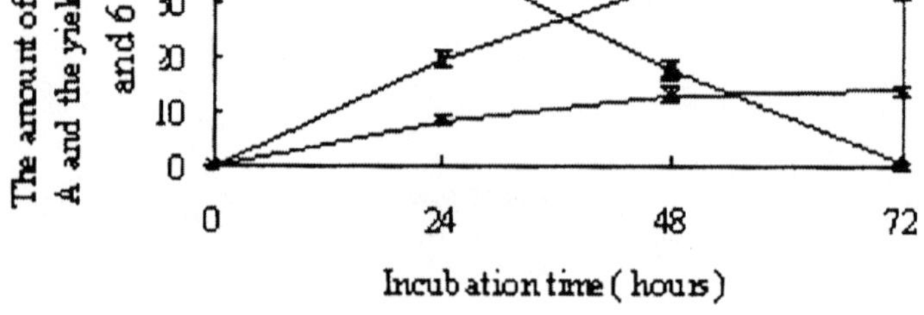

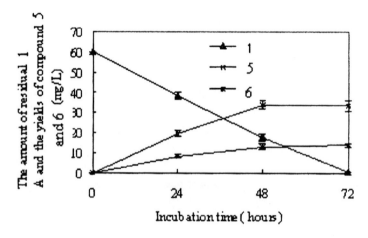

Figure 1. Time course of **1** conversion to **5** and **6** by *Ginkgo* cell suspension cultures

Table 1. The effects of different substituent groups of taxanes on the biotransformation by *Ginkgo* cell suspension cultures*

Substrates	Relative Biotransformation Ratios of Substrates	The yields of 9α-Hydroxylated Products (%)
1	100	70±6.5
2	80±5.8	60±3.8
3	50±3.5	20±1.2
4	50	20
7	100	75±3.3
8	60±6.0	40±2.5
9	50±4.6	30±1.8
10	90	50

*Each value was the mean of three replicates ± SE, the inoculum was 5 g/L cell cultures (dry weight), the concentration of the added substrates was 35 mg/L, and the substrates added on the 15th culture day, the reaction was quenched on day 21

desired products could be improved by modulating the enzymes' activities, or the enzymes could be extracted, purified and immobilized for large-scale production. In addition, the genes encoding the enzymes could be cloned and transferred into a microorganism to yield products industrially. With this view, a series of experiments (Table 2) were designed to characterize the enzymes responsible for the biocatalytic conversion of **1** to products **5** and **6**. The results of treatment **2** in which the added **1** was metabolized and compounds **5** and **6** were produced, solidly suggest that the enzymes are constitutive and extracellular, and the results from the other treatments also confirmed it.

Bioconversion by the fungus A. coerulea IFO4011

To 2-day-old cell cultures of fungus *A. coerulea* IFO4011 (obtained from Institute for Fermentation, Osaka, Japan), **1** was added, and after another week of incubation, 7β hydroxyl product **11** was obtained in 5% yield (Scheme 3). To confirm the specific hydroxylation capacity of the fungus and to gain insight into the influence of the different substrates on the biotransformation process, two other related compounds, **3** and **4**, were also used as exogenous substrates and incubated with the cell cultures. As expected, the 7β hydroxylated products (**12, 13**) were

Table 2. The basic properties of enzymes responsible for substrate biotransformation to compounds **5** and **6** [a]

Treatments	Amount of residual substrate (mg/L)	Yield of compound 5 (mg/L)	Yield of compound 6 (mg/L)	Harvested cell cultures (g/L, dry weight)
1[b]	31.6±1.40	0	0	0
2[c]	25.4 ± 2.10	4.32 ± 0.36	0.88 ± 0.11	16.4 ± 0.75
3[d]	0.78 ± 0.12	17.8 ± 2.50	7.5 ± 0.56	21.6 ± 1.12
4[e]	2.65 ± 0.32	21.5 ± 2.11	9.6 ± 1.12	18.5 ± 0.86
5[f]	1.32 ± 0.15	40.6 ± 3.22	16.8 ± 1.64	21.1 ± 0.75

[a] Each value was the mean of three repeated tests±SE, the inoculum was 5 g/L of cell cultures (dry weight), and the reaction was quenched on day 21. [b] 35 mg/L of substrate solution was added to the flask without cell cultures inoculated. [c] 35 mg/L of substrate solution was added on the 15th day to the flask from which the cell cultures were filtered out. [d] 35 mg/L of substrate solution was added on the 15th day to the flask with cell cultures. [e] 35 mg/L of substrate solution was added on the 15th day to the flask, but on the 18th day, the cell cultures were filtered out, then an additional 35 mg/L of substrate solution was added to the same flask. [f] 35 mg/L of substrate solution was added on the 15th day to the flask, and on the 18th day, an additional 35 mg/L of substrate solution was added to the same flask.

obtained under the same incubation conditions at yields of 10% and 15% (Scheme 3), respectively. It is interesting that the longer the alkyl chain of the acyloxyl groups at C-14 became, the higher the yield of 7β-hydroxylated products obtained.

In an attempt to enhance the yield of **11**, β-cyclodextrin, which has been used commonly and successfully in the biotransformation for the increase of the yield, was co-administered to the cell cultures of the fungus with **1**. The yield of the desired product **11** was ~ 5% and not increased as expected; however, very intriguingly, three other products were produced in addition to **11**. Their structures were determined as 5α,9α,10β,13α-tetraacetoxytaxa-4(20), 11-dien-14β-ol (**14**), 5α,9α,10β,13α-tetraacetoxytaxa-4(20),11-dien-1β-ol (**15**) and 5α,9α,10β,13α-tetraacetoxy- 11(15→1)abeotaxa-4(20),11-dien-15-ol (**16**) (Scheme 4). **14** and **15** were produced in about 5% yields individually, and **16** in a trace yield. The results suggested that in the presence of β-cyclodextrin, the substrate enters the organelles of the cells where

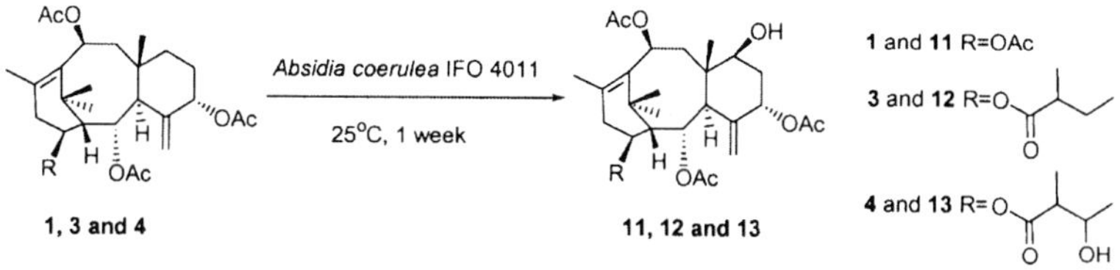

Scheme 3. 7β Hydroxylation of taxanes by *Absidia coerulea* IFO4011

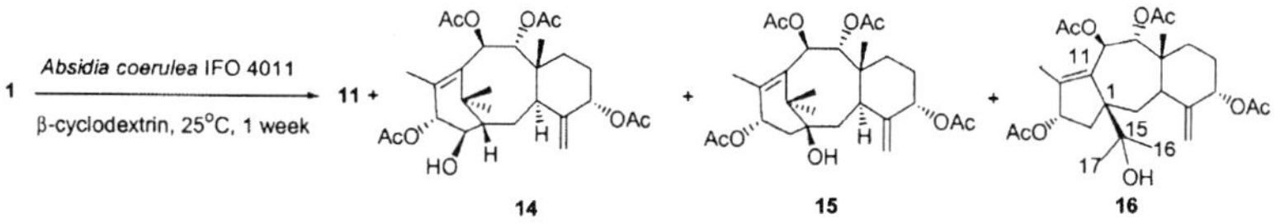

Scheme 4. Cyclodextrin effects on the biotransformation of taxanes by *Absidia coerulea* IFO4011

there are many different enzymes and these completely different reactions took place by their actions. We observed very substantial differences in the reaction modes of these biotransformations in the presence and absence of -cyclodextrin. Obviously, the difference of reaction mode means that each step of taxoid biotransformation takes place in different compartments in the cells. This biotransformation by fungus gave hypothetical biosynthetic intermediates of paclitaxel and its analogues, **11**, **14** and **15** from C-14 oxygenated taxoids, in fair yield. Since the abundance of taxanes bearing functionalized groups both at C-13 and C-14 such as **14** is very limited in yew trees, this kind of taxanes may be the intermediates between C-14 and C-13 functionalized taxanes, and even the intermediates of paclitaxel biosynthesis. The fact that the functional group at C-2 in **1** was removed after incubation with the fungal cell cultures suggested that the same reaction probably takes place in the *Taxus* plant.

Combination of the selective 9α and 7β oxidations of Taxuyunnanine C

In the previous investigation, 9α and 7β selectively hydroxylated products were obtained successfully by cell suspension cultures of *G biloba* and the fungus *A. coerulea*, respectively. However, could these two selective oxidations be combined by the above two biocatalysts? Bearing these in mind, we have been trying our further investigations and finally succeeded in combining these two reactions by subsequent biotransformation with the help of chemical acetylation. First, 9α and 7β hydroxylated products (**5**, **11**) were prepared as described before, and bioconverted directly as the substrates by the fungus *A. coerulea* and cell suspension cultures of *Ginkgo* following the method described previously, respectively, but not as expected, 7β and 9α hydroxylation did not occur. Considering the effects of the structure of the substrate (substituents, polarity, conformation, etc.) on the biotransformation, **5** and **11** were first acetylated by routinely chemical method, then their acetylated products (**17**, **20**) were successively biotransformed by the fungus *A. coerulea* and cell suspension cultures of *Ginkgo*, respectively. Intriguingly, the desired hydroxylation took place in both cases, and their corresponding 7β and 9α hydroxylated products (**18** and **21**) were obtained at 2% and 10% yields, respectively (Scheme 5). Also one by-product, compound **19** (in 4% yield), formed in the former reaction. Their structures were determined on the basis of the ^{1}H NMR, ^{1}H-^{1}H COSY, ^{13}C NMR, DEPT, HMQC, HMBC, NOE, HRMS and IR spectral data.

In conclusion, we have obtained a powerful method for preparation of 9α hydroxylated derivatives by *Ginkgo* cells, 7β hydroxylated derivatives by *A. coerulea*, and both 9α and 7β specifically oxidized derivatives by subsequent biotransformation by the two biocatalysts from readily available products from tissue and cell cultures of *Taxus*. In addition, we have reported the influences of different substituents on the biotransformation yields, and even biotransformation modes. Very interestingly, the results indicated that there were substantial differences in

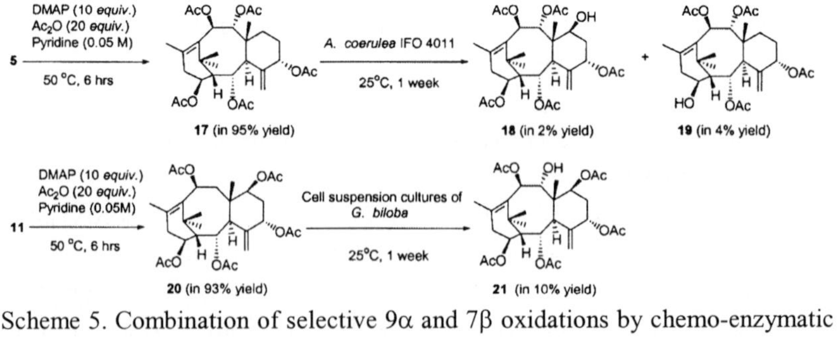

Scheme 5. Combination of selective 9α and 7β oxidations by chemo-enzymatic transformation

the presence and absence of β-cyclodextrin in the biotransformation processes of taxuyunnanine C by *A. coerulea*. All in all, these biotransformations would not only supply the very useful intermediates for the synthesis of paclitaxel and other bioactive taxoids from taxuyunanine C and its analogues, but also provide a useful tool to probe some important biosynthetic steps of taxoids in *Taxus*.

ACKNOWLEDGEMENTS

This work is supported by the National Natural Science Foundation of China (Projects No. 30100230 and No. 30572243), Natural Science Foundation of Beijing (Projects No. 706248) and the Japan Society for the Promotion of Science (Project No. 1300127).

REFERENCES

[1] Cheng KD, Chen WM, Zhu WH, Fang QC. Manufacture of taxane analog by culture of *Taxus* plants. PCT Int. Appl. WO9406, 740 (Cl. C07C35/37), 31 Mar 1994, JP Appl. 92/249, 047, 18 Sep 1992

[2] Wu YQ, Zhu WH, Lu J, Hu Q, Li XL. Selection and culture of high-yield sinenxans cell lines of Taxus spp. distributed in China. Chin. Pharm. J. 1998; 33: 15-8

[3] Menhard B, Eisenreich W, Hylands PJ, Bacher A, Zenk MH. Taxoids from cell cultures of *Taxus chinensis*. Phytochemistry 1998; 49: 113-25

[4] Suffness M. Taxol: from discovery to therapeutic use. Ann. Rep. Med. Chem. 1993; 28: 305-15

[5] Rowinsky EK, Cazenave LA, Donehower RC. Taxol: a novel investigational antimicrotuble agent. J. Natl. Cancer. Inst. 1990; 82: 1247-53

[6] Ojima I, Bounaud PY, Bernacki RJ. New weapons in the fight against cancer. Chemtech June 1998; 31-5

[7] Hu S, Tian X, Zhu WH, Fang QC. Microbial transformation of taxoids: selective deacetylation and hydroxylation of 2α,5α,10β,14β-tetra-acetoxy-4(20),11-taxadiene by the fungus *Cunninghamella echinulata*. Tetrahedron 1996; 52: 8739-46

[8] Dai J, Guo H, Lu D, Zhu W, Zhang D, Zheng J, Guo D. Biotransformation of 2α, 5α, 10β, 14β-tetra-acetoxy-4(20), 11-taxadiene by *Ginkgo* cell suspension cultures. Tetrahedron Lett. 2001; 42: 4677-9

[9] Dai J, Ye M, Guo H, Zhu W, Zhang D, Hu Q, Zheng J, Guo D. Regio- and stereo-selective biotransformation of 2α, 5α, 10β, 14β-tetra–acetoxy-4(20), 11-taxadiene by *Ginkgo* cell suspension cultures. Tetrahedron 2002; 58: 5659-68

[10] Dai J, Cui Y, Zhu W, Ye M, Guo H, Zhang D, Hu Q, Zheng J, Guo D. Biotransformation of 2α, 5α, 10β, 14β-tetra-acetoxy-4(20), 11-taxadiene by cell suspension cultures of *Catharanthus roseus*. Planta Medica 2002; 68: 1113-7

[11] Dai J, Ye M, Guo H, Zhu W, Zhang D, Hu Q, Zheng J, Guo D. Biotransformation of 4(20),11-taxadienes by *Platycodon* cell suspension cultures. J. Asian Nat. Prod. Res. 2003; 5: 5-10

[12] Dai J, Zhang S, Sakai J, Bai J, Oku Y, Ando M. Specific oxidation of C-14 oxygenated 4(20), 11-taxadienes by microbial transformation. Tetrahedron Lett. 2003; 44: 1091-4

[13] Dai J, Ye M, Guo H, Zhu W, Zhang D, Hu Q, Zheng J, Guo D. Substrate specificity for the hydroxylation of polyoxygenated 4(20),11-taxadienes by *Ginkgo* cell suspension cultures. Bioorg. Chem. 2003; 31: 345-56

[14] Dai J, Zhang M, Ye M, Zhu W, Guo J, Liang X-T. Biotransformation of 14-deacetoxy-13-oxo sinenxan A by *Ginkgo* cell cultures. Chin.Chem. Lett. 2003; 8: 804-6

[15] Dai J, Zhu W, Wu Y, Hu Q, Zhang D. Effects of precursors and fungal elicitors on GKB production in suspension cultured cells of *Ginkgo biloba*. Acta Pharma. Sin. 2000; 35: 151-6

[16] Huang G, Guo JY, Liang XT. Studies on structure modification and structure-activity relationship of new taxoids derived from sinenxan A. Acta Pharma Sin. 1998; 33: 576-86

[17] Zhang M, Yin D, Guo JY, Liang XT. Entry into major groups retaining taxol *via* sinenxan A. Chin. Chem. Lett. 2002; 13: 135-7

Biotransformation of Terpenes and Steroids by Fungi

Paul B Reese

Department of Chemistry, University of the West Indies, Mona, Kingston 7, Jamaica, West Indies

Corresponding email: paul.reese@uwimona.edu.jm

INTRODUCTION

Jamaica is a Caribbean island which boasts a number of micro-climates and this is reflected in the wide diversity of plant and microbial life. *Hyptis verticillata* (Jacq.), a member of the Lamiaceae family, has been widely used in traditional medicine [1]. Investigations into the components of the organic extract of the plant yielded a number of lignans, triterpenes and flavonoids [2-5]. The herb was also found to produce two sesquiterpenes, cadina-4,10(15)-dien-3-one (**1**) and squamulosone (**2**). Both compounds exhibit insecticidal activity against the sweet potato weevil, *Cylas formicarius elegantulus* while **1** is also a chemosterilant of the southern cattle tick, *Boophilus microplus* [6,7]. *Capraria biflora* L. (from the Scrophulariaceae) has been traditionally used in the West Indies for the treatment of numerous human ailments including fever, influenza, indigestion and diarrhoea [1,8]. Four structurally novel isomeric insecticidal sesquiterpenes were isolated from the aerial parts of the plant, and these have been designated Caprariolides A (**3**), B (**4**), C (**5**) and D (**6**) [9]. Another member of the same family, *Stemodia maritima*, yielded the antiviral and cytotoxic stemodin (**7**) along with a host of other diterpenes [10]. *Cleome spinosa*, which belongs to the Capparaceae, gave large amounts of a new cembrane (**8**) along with four congeners (**9-12**) [11]. Cembranes are known to possess diverse biological activity including potent cytotoxicity against a number of human cancer cell lines, inclusive of leukaemia, melanoma, breast and colon carcinomas [12-14].

Over the last eight years our group has exploited the ability of certain fungi to perform biocatalytic reactions. In particular we have been interested in the hydroxylation of terpene and steroid substrates. The hydroxylation reaction is performed by cytochrome P450 enzymes in the presence of oxygen and the cofactor NADPH [15]. The incubation of some of the above-mentioned terpenes with the fungi *Beauveria bassiana*, *Rhizopus arrhizus*, *Mucor plumbeus*, *Curvularia lunata* and *Aspergillus niger* has been effected. This has led to the preparation of a wide range of compounds, many which are products of hydroxylation. Some of these metabolites are novel and a number possess enhanced biological activity.

Additionally we have determined the biocatalytic potential of two locally isolated fungi using steroids as substrates. *Fusarium oxysporum* var. *cubense*, the causative agent of the Panama disease of bananas (*Musa* sp.), has been found to bring about 7α and 15α hydroxylation on steroids as well as side chain cleavage. *Exophiala jeanselmei* var. *lecanii-corni*, encountered as a contaminant of a ginger plant (*Zingiber officinale*), effected side chain degradation and 1,2- and 1,4-reduction of steroidal enones.

RESULTS AND DISCUSSION

Terpene transformations

The feeding of sesquiterpene **1** to a growing culture of the fungus *Beauveria bassiana* led to the formation of a range of compounds (**13** - **21**), the products of redox chemistry on ring A, and hydroxylation on the isopropyl group [16]. While most of these analogues showed insecticidal activity against the sweet potato weevil it was noted that **14**, **16** and **18** were more potent than the fed compound (**1**). The micro-organism *Mucor plumbeus* was also found to transform **1**. In this case twelve compounds were isolated (**14**, **20** - **30**). These included products of 10,15-epoxidation. It was interesting to note that the yields of the analogues formed could be changed quite markedly by the simple modification of the iron concentration in the medium [17]. In a similar vein *Curvularia lunata* was investigated for its potential for biocatalysis using **1** as substrate. Three cadinanes (**13**, **14** and **31**) were produced [18].

Squamulosone (**2**) was readily biotransformed by *M. plumbeus*. Terpenes **32** - **36** were formed and these were products of hydroxylation at C-2, -13 and -14 [17]. Incubation of the sesquiterpene with *C. lunata* yielded similar compounds (**32** - **35** and **37** - **39**). All products from the *Curvularia* experiment possessed insecticidal activity [19].

Stemodin (**7**) was converted by *B. bassiana* to a single compound (**40**), the product of hydroxylation at C-18 [20]. In contrast, with

Aspergillus niger, functionalisation occurred at positions 3, 7 and 16 to yield **41**, **42** and **43** respectively [21]. More recently the importance of the functional groups at C-2 and -13 of **7** in transformation experiments using *Rhizopus oryzae* was examined. The data revealed that in the absence of oxygen functions at either centre, the substrate was not transformed by the fungus. Furthermore, the best conversion (83%) was observed when the stereochemistry at C-13 was inverted [22, 23].

Nearly all of the sesqui- and diterpenes isolated from the aforementioned fermentations had not been reported previously.

Steroid transformations

Accounts of transformations by *Fusarium oxysporum* are rare and there were none for *F. oxysporum* var. *cubense*. Therefore the potential of the fungus for biocatalysis was tested using steroid substrates since the latter are generally more amenable to transformation than terpenes. 3β-Hydroxy-Δ^5-steroids, e.g. dehydroepiandrosterone (**44**) and pregnenolone (**46**), underwent allylic hydroxylation at C-7α to give **45** and **47** respectively. Testosterone (**48**) was functionalised at C-6β and -15α to yield **49** and **50**, while progesterone (**51**) was converted to **52** and **53**, products of 15α and 12β hydroxylation respectively. 17, 21-Dihydroxypregnan-20-ones, e.g. cortisone (**54**) and prednisone (**55**) underwent side chain cleavage to form 17-ketones **56** and **57** respectively [24]. Side chain cleavage [25]; and hydroxylation in the 6β [12], 12β [27] and 15α [28] positions of steroids have been previously observed in other *F. oxysporum* subspecies.

Exophiala jeanselmei var. *lecanii-corni* was observed to effect an apparent hydration of the 5,6-double bond of **44** to give **58** [29]. A similar reaction has been observed for another *E. jeanselmei* subspecies in which styrene was the substrate [30]. Epoxidation of the alkene, followed by rearrangement would give the 6-ketone. Reduction of the latter yields the 6-alcohol. Steroid **46** was seen to undergo side chain degradation to yield **59**. Δ^4-3-Ketones were sequentially reduced, first at the double bond and then at the ketone. 1, 2-Double bonds were not affected. Side chain degradation also occurred. Therefore **48** gave **60** and **61**; **54** was converted into **62** and **64**; while **55** yielded **63** and **65** [29]. The potential of these two locally isolated fungi for terpene biotransformation will be investigated in the future.

ACKNOWLEDGMENTS

The author is grateful to the researchers and co-workers listed in the references. Funding was provided by the University of the West Indies/ Inter-American Development Bank (UWI/IDB) Programme, the Canadian International Development Agency (CIDA), and the UWI Board for Graduate Studies & Research.

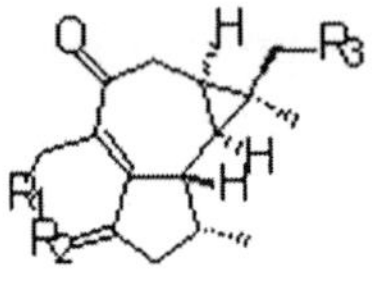

32 $R_1 = R_3 = H$, R_2 = βOH,αH
33 $R_1 = R_3 = H$, R_2 = αOH,βH
34 $R_1 = H$, $R_2 = H_2$, $R_3 = OH$
35 $R_1 = H$, R_2 = βOH,αH, $R_3 = OH$
36 $R_1 = H$, R_2 = αOH,βH, $R_3 = OH$
37 $R_1 = OH$, $R_2 = H_2$; $R_3 = H$
38 $R_1 = R_3 = OH$; $R_2 = H_2$

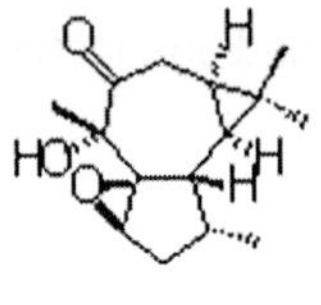

39

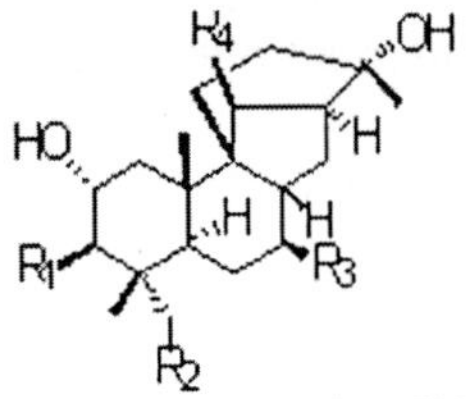

40 $R_1 = R_3 = R_4 = H$, $R_2 = OH$
41 $R_1 = OH$, $R_2 = R_3 = R_4 = H$
42 $R_1 = R_2 = R_4 = H$, $R_3 = OH$
43 $R_1 = R_2 = R_3 = H$, $R_4 = OH$

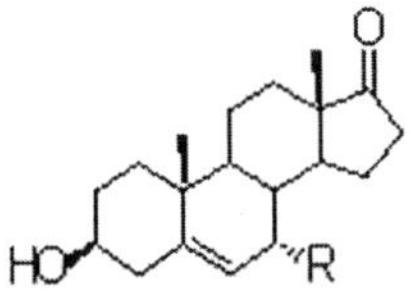

44 R=H
45 R=OH

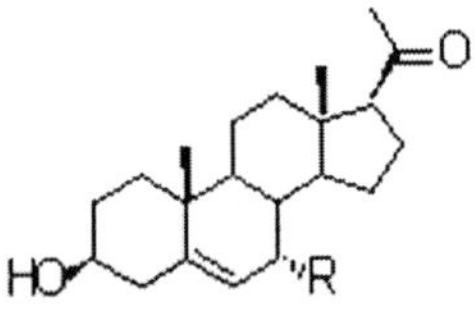

46 R=H
47 R=OH

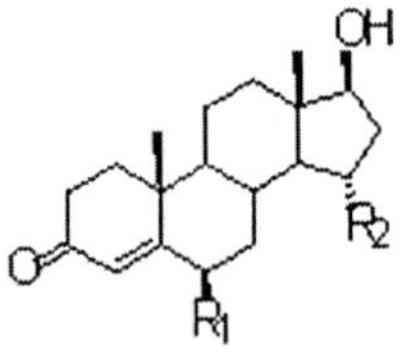

48 $R_1 = R_2 = H$
49 $R_1 = OH$, $R_2 = H$
50 $R_1 = H$, $R_2 = OH$

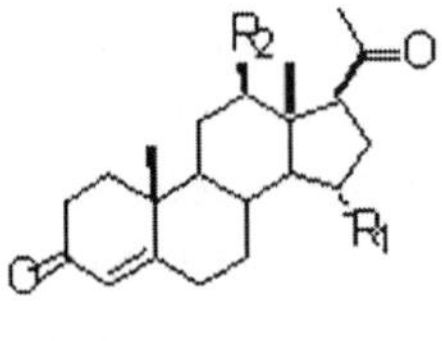

51 $R_1 = R_2 = H$
52 $R_1 = OH$, $R_2 = H$
53 $R_1 = R_2 = OH$

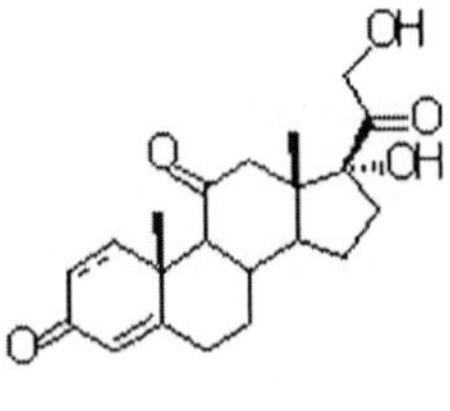

54 Δ^4
55 $\Delta^{1,4}$

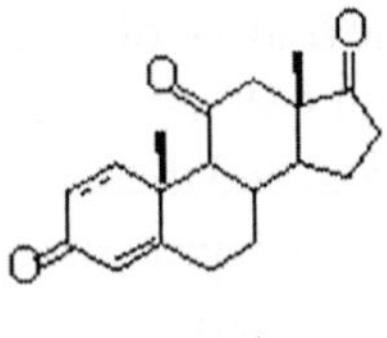

56 Δ^4
57 $\Delta^{1,4}$

REFERENCES

[1] Ayensu ES. Medicinal Plants of the West Indies. Reference Publications Inc., Algonac, Michigan, USA; 1981

[2] German VF. Isolation and characterization of cytotoxic principles from *Hyptis verticillata* Jacq. J Pharm Sci 1971; 60: 649-50

[3] Novelo M, Cruz JG, Hernandez L, Pereda-Miranda RJ. Cytotoxic constituents from *Hyptis verticillata* J Nat Prod. 1993; 56: 1728-36

[4] Kuhnt M, Rimpler H, Heinrich M. Lignans and other compounds from the Mixe Indian medicinal plant *Hyptis verticillata*. Phytochemistry 1994; 36: 485-489

[5] Porter RBR, Reese PB. Characterisation of a flavonol and several lignans from *Hyptis verticillata*. Jamaican Journal of Science & Technology 1998; 9: 17-27

[6] Porter RBR, Reese PB, Williams LAD, Williams DJ. Acaricidal and insecticidal activities of cadina-4,10(15)-dien-3-one. Phytochemistry 1995; 40: 735-8

[7] Collins DO, Buchanan GO, Reynolds WF, Reese PB. Biotransformation of squamulosone by *Curvularia lunata* ATCC 12017. Phytochemistry 2001; 57: 377-83

[8] Honychurch, PN. Caribbean wild plants and their uses. Macmillan Publishers Ltd., London; 1986

[9] Collins DO, Gallimore WA, Reynolds WF, Williams LAD, Reese PB. New skeletal sesquiterpenoids, caprariolides A - D, from *Capraria biflora* and their insecticidal activity. J Nat Prod 2000; 63: 1515-8

[10] Knoll WMJ, Huxtable RJ. Structures of stemodin and stemodinone. J Am Chem Soc 1973; 95: 2705-6

[11] Collins DO, Reynolds WF, Reese PB. New cembranes from *Cleome spinosa*. J Nat Prod 2004; 67: 179-83

[12] Duh C, Wang S, Weng Y, Chiang MY, Dai C. Cytotoxic terpenoids from the Formosan soft coral *Nephthea brassica*. J Nat Prod 1999; 62: 1518-21

[13] El Sayed KA, Hamann MT, Waddling CA, Jensen C, Lee, SK, Andersson Dunstan C, Pezzuto, JM. Structurally novel bioconversion products of the marine natural product sarcophine effectively inhibit JB6 cell transformation. J Org Chem 1998; 63: 7449-55

[14] Duh C, Hou R. Cytotoxic cembranoids from the soft corals *Sinularia gibberosa* and *Sarcophyton trocheliophorum*. J Nat Prod 1996; 59: 595-8

[15] Ortiz de Montellano, PR. Cytochrome P450: structure, mechanism and biochemistry, 2nd ed. Plenum Press, New York; 1995

[16] Buchanan GO, Williams LAD, Reese PB. Biotransformation of cadinane sesquiterpenes by *Beauveria bassiana* ATCC 7159. Phytochemistry 2000; 54: 39-45

[17] Collins DO, Ruddock PLD, Chiverton de Grasse J, Reynolds WF, Reese PB. Microbial transformation of cadina-4,10(15)-dien-3-one, aromadendr-1(10)-

en-9-one and methyl ursolate by *Mucor plumbeus* ATCC 4740. Phytochemistry 2002; 59: 479-88

[18] Collins DO, Reese PB. Biotransformation of cadina-4,10(15)-dien-3-one and 3?-hydroxycadina-4,10(15)-diene by *Curvularia lunata* ATCC 12017. Phytochemistry 2002; 59: 489-92

[19] Collins DO, Reynolds WF, Reese PB. Aromadendrane transformations by *Curvularia lunata* ATCC 12017. Phytochemistry. 2002; 60: 475-81

[20] Buchanan GO, Reese PB. Biotransformation of diterpenes and diterpene derivatives by *Beauveria bassiana* ATCC 7159. Phytochemistry 2001; 56: 141-51

[21] Chen ARM, Reese PB. Biotransformation of terpenes from *Stemodia maritima* by *Aspergillus niger* ATCC 9142. Phytochemistry 2002; 59: 57-62

[22] Martin GDA, Reynolds WF, Reese PB. Investigation of the importance of the C-2 oxygen function in the transformation of stemodin analogues by *Rhizopus oryzae* ATCC 11145. Phytochemistry. 2004; 65: 701-10

[23] Martin GDA, Reynolds WF, Reese PB. Investigation of the importance of the C-2 and C-13 oxygen functions in the transformation of stemodin analogues by *Rhizopus oryzae* ATCC 11145. Phytochemistry 2004; 65: 2211-7

[24] Wilson MR, Gallimore WA, Reese PB. Steroid transformations with *Fusarium oxysporum* f. sp. *cubense* and *Colletotrichum musae*. Steroids 1999; 64: 834-43

[25] Tom W, Abul-Hajj YJ, Koreeda MJ. Microbial oxidation of ecdysones. A convenient preparation of rubrosterone. J Chem Soc Chem Commun 1975; 24-5

[26] Petzoldt K., inventor; Schering AG, assignee. 15?-Hydroxy-4-androstene-3,17-dione. German Patent DE 3,403,862. 1985 Aug 8. Chem Abs 104: 166905q

[27] Schuepback M, Tamm Ch. Reactions with microorganisms. XII. Conversion of toad poison (bufadienolide) by microorganisms. 2. 12?-Hydroxyresibufogenin. Helv Chim Acta 1964; 47: 2217-26

[28] Defaye G, Luche MJ, Chambaz EM. Microbial 7- and 15-hydroxylations of C-19 steroids. J Steroid Biochem 1978; 9: 331-6

[29] Porter RBR, Gallimore WA, Reese PB. Steroid transformations with *Exophiala jeanselmei* var. *lecaniicorni* and *Ceratocystis paradoxa*. Steroids 1999; 64: 770-9

[30] Cox HHJ, Faber BW, van Heiningen WNM. Styrene metabolism in *Exophiala jeanselmei* and involvement of a cytochrome P-450-dependent styrene monooxygenase. Appl Environ Microbiol. 1996; 62: 1471-4

Rapid Analysis of Triterpenoid Saponins in Plant Extract Using ESI-MS^n and LC-MS^n

Bin Li, Zeper Abliz, Meijun Tang, Guangmiao Fu and Shishan Yu
Institute of Materia Medica, Chinese Academy of Medical Sciences and Peking Union Medical College, Beijing 100050, China

INTRODUCTION

Symplocos chinensis is a Chinese herb widely used as a folk medicine to treat several diseases [1]. Recent studies have further shown that triterpenoid saponins extracted from this plant exhibit strong antitumor activities [2,3]. However, it is generally difficult to isolate and identify saponins in crude extracts with conventional procedures. Electrospray ionization multi-stage tandem mass spectrometry (ESI-MS^n) plays more and more important role in the analysis of saponins over the last few years [4-6]. In this paper, the fragmentation behavior of eight symplocososides and their analogs β-escin saponin Ia and Ib were systematically investigated by positive ion ESI-MS^n. Furthermore, the bioactive extract from *Symplocos chinensis* was rapidly analyzed by ESI-MS^n and liquid chromatography / tandem mass spectrometry (LC-MS^n), and the structures of several constituents of this type were deduced.

MATERIALS AND METHODS

The pure symplocososides and crude extract were kindly provided by Professor Shishan Yu, with the structures shown in Figure 1. β-escin saponin Ia and Ib were purchased from the National Institute for the Control of Pharmaceutical and Biological Products (Beijing, China). All experiments were conducted on a QTRAP™ LC-MS/MS system from Applied Biosystems /MDS Sciex (Concord, ON, Canada). An Agilent HPLC system of 1100 Series was operated at room temperature with a YMC C_{18} column (150 × 3.0mm i.d., 5 μm). The mobile phase consisted of (A) water (0.1% formic acid) and (B) methanol at a flow rate of 300 μL/min under gradient program.

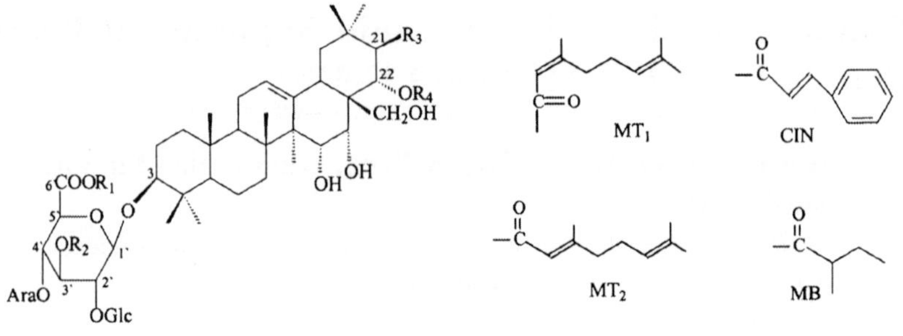

Compound	R_1	R_2	R_3	R_4	M.W.(Da)
1	butyl	H	OMT_1	MB	1266
2	butyl	H	OMT_2	MB	1266
3	butyl	H	OCIN	MB	1246
4	butyl	H	H	MT_1	1166
5	CH_3	Ac	OMT_1	MB	1266
6	H	Ac	OMT_1	MB	1252
7	H	Ac	OMT_2	MB	1252
8	H	Ac	OMT_1	benzoyl	1272

Figure 1. Structures of triterpenoid saponins from *Symplocos chinensis*
Notes: Glc, β-D-glucopyranosyl; Ara, α-L-arabinofuranosyl; Ac, acetyl.

RESULTS AND DISCUSSION

Investigation of the fragmentation behavior of pure triterpenoid saponins

In the ESI-MS spectra of all the pure compounds, a predominant $[M+Na]^+$ ion in positive mode and $[M-H]^-$ ion in negative mode were observed for molecular weight information. Subsequent studies on positive ion ESI-MS^n of eight symplocososides and their analogs β-escin saponin Ia and Ib revealed the principal fragmentation patterns of these compounds. It was found that the carboxyl group and free hydroxyl group at C-3′ position of the glucuronyl residue were the key sites for determining the fragmentation behavior. When the carboxyl group was esterified, only $C_{2\alpha}$ ion was acquired (Nomenclature of fragmentation by Costello, et al [7]), with no $B_{2\alpha}$ ion and the cationized aglycone observed. When the hydroxyl group at the C-3′ position was acetylated, the inherent process of cross-ring cleavage was hindered. However, glycosidic bond cleavage reactions occur in different product ion spectra, regardless of these crucial variations. The results provide a valuable aid for determining the structures of the same type of compounds in crude plant extract.

Analysis of the crude extract

To analyze the constituents in a crude extract from *Symplocos chinensis*, various methods of mass spectrometry were used, such as ESI-MS, ESI-MS^n, LC-MS, LC-MS^n, etc. From the positive ion full-scan mass spectrum of the saponin extract and its negative ion counterpart, it was deduced that there were at least 17 constituents in the mixture. Positive ion ESI-MS^n provides the structural information of different constituents, and it showes that components with molecular weight above 1300Da belong to another type of triterpenoid saponin different from the studied samples. Moreover, LC-MS showed that there were three pairs of isomer of triterpenoid saponins with molecular weight 1166, 1180, 1288Da, respectively. To further investigate these isomers, LC-MS^n was employed. Integrating the numerous structural information, the possible structures of 11 constituents are speculated, among which there are 10 new compounds.

CONCLUSION

In principle, ESI-MS^n is a powerful tool for the analysis of crude plant extract. However, when different isomers of the constituents are encountered, LC-MS^n can provide complementary information to differentiate these isomeric compounds. By using the two kinds of techniques together, triterpenoid saponins in crude extract from *Symplocos chinensis* can be detected rapidly and effectively, without tedious time- and material-consuming procedures in phytochemistry.

REFERENCES

[1] Li XH, Shen DD, Li N, Yu SS. Bioactive triterpenoids from *Symplocos chinensis*. J Asian Nat Prod Res 2003; 5: 49-56

[2] Tang MJ, Shen DD, Hu YC, Gao S, Yu SS. Cytotoxic triterpenoid saponins from *Symplocos chinensis*. J Nat Prod 2004; 67: 1969-74

[3] Fu GM, Wang YH, Gao S, Tang MJ, Yu SS. Five new cytotoxic triterpenoid saponins from the roots of *Symplocos chinensis*. Planta Med 2005; 71: 666-72

[4] Ackloo SZ, Smith RW, Terlouw JK, McCarry BE. Characterization of ginseng saponins using electrospray mass spectrometry and collision-induced dissociation experiments of metal-attachment ions. Analyst 2000; 125: 591-7

[5] van Setten DC, Zomer G, van de Werken G, Wiertz EJHJ, Leeflang BR, Kamerling JP. Ion trap multiple-stage tandem mass spectrometry as a pre-NMR tool in the structure elucidation of saponins. Phytochem Anal 2000; 11: 190-8

[6] Liu YZ, Liang F, Cui LJ, Xia M, Zhao LY, Yang YC et al. Multi-stage mass spectrometry offurostanol saponins combined with electrospray ionization in positive and negative ion modes. Rapid Commun Mass Spectrum 2004; 18: 235-8

[7] Domon B, Costello CE. A systematic nomenclature for carbohydrate fragmentations in FAB-MS/MS spectra of glycoconjugates. Glycoconj J 1988; 5: 397-409

HPLC-MS Analysis of Phenolic Constituents of *Phyllanthus Amarus*

Guangying Chen[b], Guoyuan Zhu[a], Guoying Li[a] and Wangfun Fong[c]
[a]Department of Biology and Chemistry, City University of Hong Kong, Hong Kong; [b]Department of Chemistry, Hainan Normal University, Haikou, P.R.China; [c]Shenzhen Applied R&D Centres of City University of Hong Kong, Shenzhen, P R China

INTRODUCTION

Phyllanthus amarus (Euphorbiaceae) is a widely distributed medicinal plant used by traditional medical practitioners for the treatment of kidney stones, viral hepatitis and other diseases. Pharmacological experiments confirm its therapeutic efficacy and safety. Intensive phytochemical examinations of this plant have been carried out. Constituents such as alkaloids, flavonoids, lignans, tannins, phenols and terpenes have been identified. Some research results showed that the therapeutic action on HBV could be attributed to the phenolic compounds [1-5]. In order to comply with regulatory requirements for the use of this plant by the pharmaceutical industry, we report here a HPLC-MS fingerprinting method for the determination of the phenols in *Phyllanthus amarus* and the identification of 9 chromatographic peaks.

MATERIALS AND METHODS

P. amarus was collected in Xi shuang ban na, Yunnan Province. The plant was dried at 50°C for 24H. CH_3OH(HPLC grade), glacial acetic acid (HPLC grade), ultrapure water from Milli-Q system with conductivity of 18 M© were used for the mobile phase preparation.

Sample preparation

About 0.3g of *P. amarus* glacial was extracted with Ultrasonic treatment in 10mL methanol for 30 min. After centrifugation, the supernatant was filtered through a Millipore filter unit. 10μL of the sample was injected into RP-HPLC.

LC-DAD and LC-MS analyses

The HPLC analysis was carried out using a Thermo Finnigan Surveyor LC system, quaternary pump, an autosampler, and a diode array detector linked to a Xcalibur data handling system. The column was Alltima C18 (5μm, 250×4.6mm I.D.). The chromatographic separation was carried out using a mobile phase with methanol as solvent A and acetic acid 1% (w/w) as solvent B at a flow-rate of 1.0 ml min^{-1}. The A:B gradient was 0-8 min, 20:80; 8-15 min, 20:30; 15-30 min, 30:50; 30-40 min, 50:70. The post-column diffluence was 0.4ml min^{-1} into mass spectrograph. The detected wavelength was 270nm. The system used for LC-MS analysis was a Thermo-Finnigan LCQ Advantage ion trap mass spectrometer equipped with Surveyor LC system. Conditions for the initial ionization in the positive and negative ionization modes included capillary voltages at +4.5kV and –6kV at 300°C. The mass data were acquired as full scan mass spectra at m/z 150-1200.

RESULTS AND DISCUSSION

A variety of solvent systems based on methanol, acetonitrile and acetic acid were tested to improve the separation of phenolic compounds in the methanol extract of *P. amarus*. A gradient system was chosen because of the matrix complexity. The mobile phase methanol:acetic acid 1% was found to be adequate. The chart of chromatogram track record in 60min revealed that there was no phenolic chromatogram peak after 40 min. So we fixed the analytical time as 40 min. There was no impurity in blank experiment except for a minor baseline excursion.

Under these chromatographic conditions, the extract had good response to negative ionization detector. The signals of positive ionization detector were weak which was consistent with the character of phenolic compounds. According to the m/z of compounds and references, we identified 9 peaks (see Figure 1, the spectra of UV and MASS) as follows: brevifolincarboxylic acid (peak 1, t_R=14.27min, m/z 291.0$[M-1]^-$), corilagin (peak 2, t_R=15.22min, m/z 633.2$[M-1]^-$), phyllanthusiin C (peak 3, t_R=17.63min, m/z925.1$[M-1]^-$), phyllanthusiin B (peak 4,

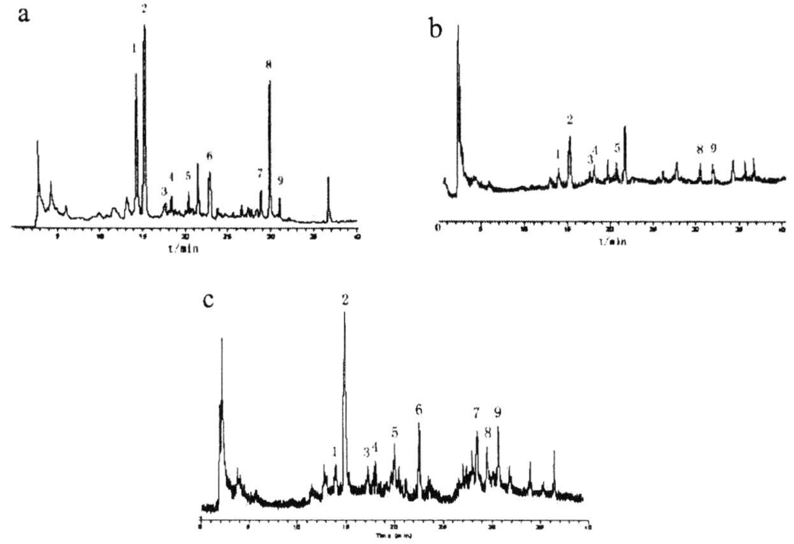

Figure 1. HPLC chromatograms of *P. amarus*
(a) LC-DAD chromatograms detected at 280nm. (b) LC-MS chromatogram of positive ions at m/z 150-1500. (c) LC-MS chromatograms of negative ions at m/z 150-1500.

t_R=18.36min, m/z 969.2$[M-1]^-$), methyl brevifolin-carboxylate (peak 5, t_R=21.44min, m/z 305.1$[M-1]^-$), phyllanthusiin A (peak 6, t_R=22.88min, m/z 951.2$[M-1]^-$), rutin (peak 7, t_R=28.82min, m/z 609.3$[M-1]^-$), ellagic acid (peak 8, t_R=29.85min, m/z 301.2$[M-1]^-$), phyllanthusiin U (peak 9, t_R=31.00min, m/z 923.1$[M-1]^-$).

The optimized HPLC analysis offers an efficient and accurate method for the determination of phenolic compounds in *P. amarus* and is useful for quality control of products derived from the plant.

REFERENCES

[1] Calixto JB, Santos ARS, Cechinel-Filho V, Yunes RA. A review of the plants of the genus *Phyllanthus*: their chemistry, pharmacology, and therapeutic potential. Med Res Rev 1998; 18: 225-58

[2] Campos AH, Schor N. *Phyllanthus niruri* inhibits calcium oxalate endocytosis by renal tubular cells: its role in urolithiasis. Nephron 1999; 81: 393–7

[3] Hussain RA, Dickey JK, Rosser MP, Matson JA, Kozlowski MR, Brittain RJ, Webb ML, Rose PM, Rernandes PA. A novel class of non-peptidic endothelin antagonists isolated from the medicinal herb *Phyllanthus niruri*. J Nat Prod 1995; 58: 1515–20

[4] Ishmaru K, Yoshimatsu K, Yamakawa T, Kamada H, Shimomura K. Phenolic constituents in tissue cultures of *Phyllanthus niruri*. Phytochemistry 1992; 31: 2015–8

[5] Wang M, Cheng H, Li Y, Meng L, Zhao G, Mai K. Herbs of the genus Phyllanthus in the treatment of chronic hepatitis B: observations with three preparations from different geographic sites. J Lab Clin Med 1995; 126: 350–2

Structure Elucidation of Norditerpene Alkaloids from Ranunculaceae Species

Peter Forgo[a], Katalin E Kövér[b] and Judit Hohmann[c]*

[a] *Department of Organic Chemistry, University of Szeged, Szeged, Hungary*

[b] *Department of Inorganic and Analytical Chemistry, University of Debrecen, Debrecen, Hungary*

[c] *Department of Pharmacognosy, University of Szeged, Szeged, Hungary*

**Corresponding email: pforgo@chem.u-szeged.hu*

INTRODUCTION

Plants of the genus *Aconitum*, *Consolida and Delphinium* produce diterpene and norditerpene alkaloids, which have attracted great interest because of their complex structure, interesting chemistry and important physiological effects[1]. Many compounds display high potency toward voltage-gated Na^+ channel[2,3] either as agonists (e.g. aconitine, mesaconitine) or antagonists (e.g. lappaconitine, 6-benzoylheteratisine). Furthermore, some norditerpene alkaloids (e.g. methyllycaconitine) demonstrated selective antagonistic activity on neuronal nicotinic acetylcholine (nACh) receptor in nanomolar concentration, and therefore are valuable lead compounds for drug development programs of Alzheimer's disease[4].

With the aim to find new biologically active compounds from Hungarian Ranunculaceae species, the methanolic extract of *Consolida orientalis* and *Aconitum vulparia* were investigated. Takaosamine, delcosine, gigactonine, 18-demethypubescenine and 14-demethyltuguaconitine were isolated from *C. orientalis*, while acovulparine, lycoctonine and delcosine were extracted from *A. vulparia*[5,6]. Although, the isolated alkaloids belong to the medium sized molecular class, a wide range of NMR experiments is required to unambiguously solve the solution structure. The NMR spectra of these compounds usually display regions with well resolved resonances providing an excellent opportunity for selective excitation. Selective experiments have become popular because the resulting spectra show correlations of the selectively excited transition exclusively. Furthermore, selective NMR experiments require shorter experimental time and yield spectra with increased spectral resolution, allowing accurate measurement of chemical shift and coupling constant values. Several protocols are available for selective excitation; among these are the well known hard pulse train techniques (DANTE[7], DANTE-Z[8,9] spin-pinging[10]), selective filtering techniques (chemical shift filters)[11], shaped pulses[12] and the combination of shaped pulses with pulsed field gradients[13-16] (DPFGSE-double pulsed field gradient spin-echo). The introduction of the DPFGSE method in several pulse sequences provide experimental techniques to study scalar (DPFGSE-COSY[15], DPFGSE-TOCSY[17]) and dipolar interactions (DPFGSE-NOESY[18], DPFGSE-ROESY[19]) of selected resonances. The techniques mentioned above assist the process of the signal assignment, however, in case of overlapping signals only a partial assignment can be reached. Heteronuclear two-dimensional experiments can aid to accomplish the complete ^{1}H-NMR assignment, moreover provide the ^{13}C-NMR (and in case of alkaloids the ^{15}N-NMR) signal assignment. There are several variants of these experiments (HSQC and HMQC) and their gradient versions are also available[16,20]. Heteronuclear correlation experiments can either be optimized to detect one-bond correlations or long-range connectivities. The former one provides chemical shift (and coupling constant) values only for carbons bearing at least one hydrogen, the latter one monitors interactions extending through two or three chemical bonds (HMBC, HSQMBC). Heteronuclear three-bond coupling constants have the important property to obey the Karplus equation, so these coupling constants can be used to predict the dihedral angle between the nuclei involved in the scalar interaction. There are several experimental procedures that have been designed to measure these important parameters, among them, the long- range HSQC experiment (HSQMBC)[21], the HSQC combined with TOCSY (HSQC-TOCSY)[16,22] and HETLOC[23] (TOCSY recorded on the ^{1}H-^{13}C satellites) are the most important ones. These experiments can provide reliable data about the ^{1}H-^{15}N scalar interactions[21] as well. Although ^{13}C-^{13}C correlation experiments are hampered by their inherent low sensitivity, the method provides primary structural information about the molecular constitution through carbon-carbon connectivity. The sensitivity of the firstly introduced carbon detected INADEQUATE experiment[24] has been greatly increased by the application of polarization transfer steps, pulsed field gradients and the introduction of proton detected ^{13}C-^{13}C correlation experiment (ADEQUATE)[25].

Present paper reports on the application of advanced NMR experiments in the structure determination of natural compounds using two nor-diterpene alkaloids (takaosamine (**1**) and gigactonine (**2**)) (Figure 1). Diterpene alkaloids represent a unique family of natural compounds featured with a structure containing a fused ring system. The family can be divided into two groups: C_{19} alkaloids are highly functionalized and show strong toxicity, while C_{20} alkaloids usually have 2-3 oxygen functions and represent less toxic biological activities. Because of their distinctive structure and significant biological activity these compounds have been studied extensively. Two prominent members of the C_{19} group are takaosamine (**1**) and gigactonine (**2**) with a complex ring system including one nitrogen atom, several hydroxyl and methoxy substituents.

EXPERIMENTAL PROCEDURE

All experiments were carried out using a Bruker Avance DRX-500 instrument (^{1}H-NMR: 500.13 MHz, ^{13}C-NMR: 125.77 MHz, ^{15}N-NMR: 50.69 MHz) equipped with a 5 mm inverse probe with *z*-pulsed field capability. NMR samples were prepared by dissolving the model compounds in $CDCl_3$. NMR measurements were carried out at room temperature. Spectra were referenced to the internal solvent signal

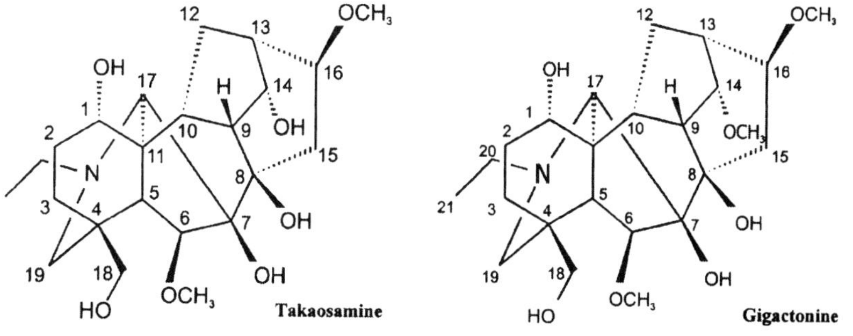

Figure 1. The structure of takaosamine (1) and gigactonine (2).

($\delta_{^1H}$=7.26 ppm, $\delta_{^{13}C}$=77.0 ppm) and to urea as external reference material ($\delta_{^{15}N}$=77.0 ppm). The lengths of the 90 degree hard pulses were 11 μs for ^{1}H, 14.3 μs for ^{13}C and 20 μs for ^{15}N. The selective excitation with the DPFGSE sequence was carried out using 80 ms Gaussian inversion pulses. The evolution time in the selective COSY experiment was varied between 30 and 50 ms, an isotropic mixing time of 60 ms was applied in the selective TOCSY experiment and 200 ms was used as anisotropic mixing in the selective NOESY experiment. Heteronuclear proton detected experiments (HSQMBC, HSQC-TOCSY and HETLOC) were carried out using sensitivity and gradient-enhanced pulse sequences. Data collecting and processing were performed according to standard protocols.

RESULTS AND DISCUSSION

Since the model compounds do not contain olefin or aromatic functional groups, their NMR spectra exhibit signals in a relatively narrow chemical shift region (Figure 1A) covering approximately 4 ppm in the high field part of the spectrum. Although signal overlapping can be noticed, a significant number of resonances are well resolved providing an excellent opportunity for selective excitation in one-dimensional correlation experiments.

The selective DPFGSE-COSY spectrum was used to identify the chemical shift of the protons, which are scalarly coupled to the selectively excited proton. Figure 2B shows an example, when H-14 (4.12 ppm) was excited and H-9 (2.96 ppm) as well as H-13 (2.37 ppm) responded with an anti-phase signal. Figure 2C and 2D show further examples with excitation of H-13 (2.37 ppm) and H-15 (2.77 ppm). Similar information can be extracted from the DPFGSE-TOCSY experiment, however scalar connectivities can be traced through the entire spin-system with an appropriately adjusted spin-lock transfer. Thus, the method allows the identification of different isolated spin systems by generating sub-spectra, which can easily be interpreted or analyzed. As examples, Figure 3B and figure 3C show two slices of DPFGSE-TOCSY spectra. The sub-spectrum of figure 3B acquired upon the selective excitation of H-14 (4.12 ppm) provides the chemical shifts of the H-9 H-10, H-12 and H-13 protons, the chemical shifts of H-20 methylen group (2.98 ppm and 2.84 ppm)

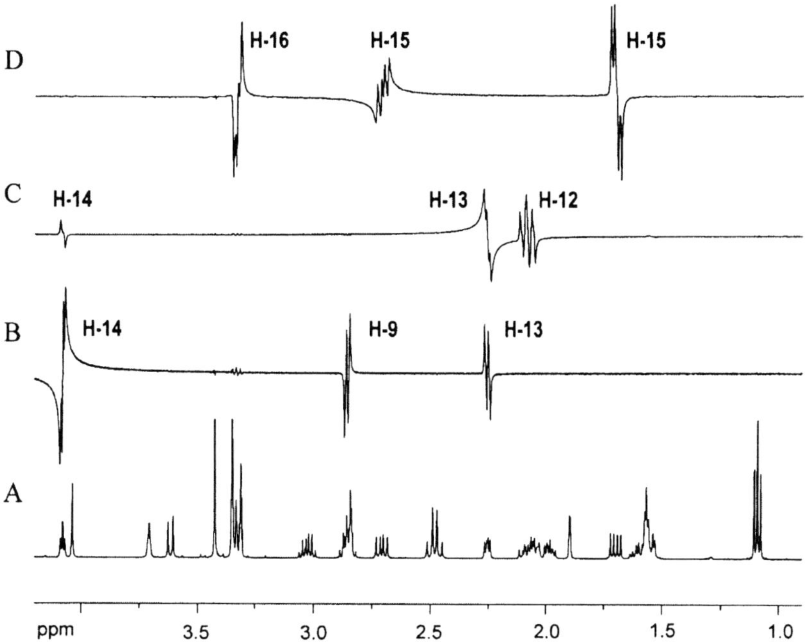

Figure 2. The DPFGSE-COSY spectra of 1.

were obtained by the selective excitation of H-21 at 1.10 ppm (Figure 3C).

The conformation in liquid phase was studied by the selective DPFGSE-NOESY experiment (Figure 4). Four spectra are shown as examples, with the excitation of H-14 (figure 4A), H-6 (figure 4B), H-1 (figure 4C) and H-18 (figure 4D).

The selective excitation of skeletal protons in DPFGSE-NOESY experiments provide important information about those protons, which reside close to the excited proton. Upon excitation of H-14 (figure 4A) H-9, H-10, H-12 and H-13 gave NOE response signal. The excitation of H-6 (figure 4B) provided NOE signals on H-18 and H-19 methylene protons and on a methoxy group attached to C-6. H-1 (figure 4C) has NOE correlations to H-3, H-10 and H-12. The excitation of one of the H-18 protons (figure 4D) provided correlations to H-6, and H-19 protons.

Since NOE interactions are based on the dipolar cross-relaxation between the corresponding nuclei, the signal enhancements strongly depend on the relevant internuclear distance. This property can be used to measure

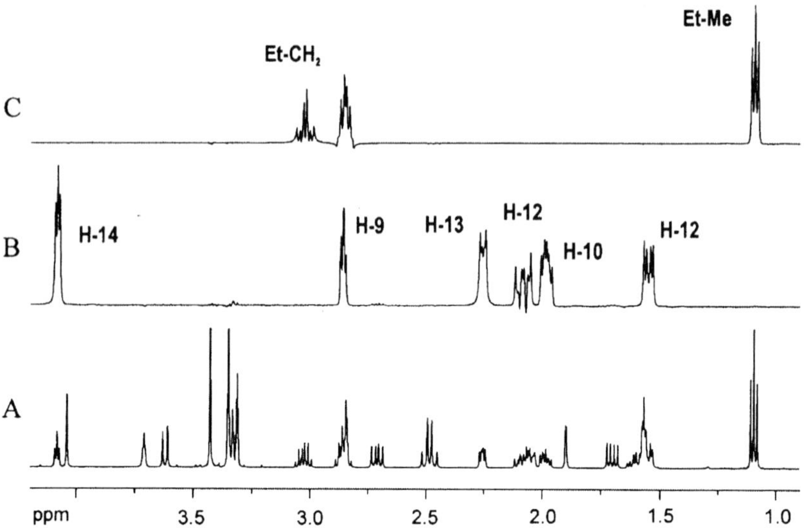

Figure 3. The DPFGSE-TOCSY spectra of 1.

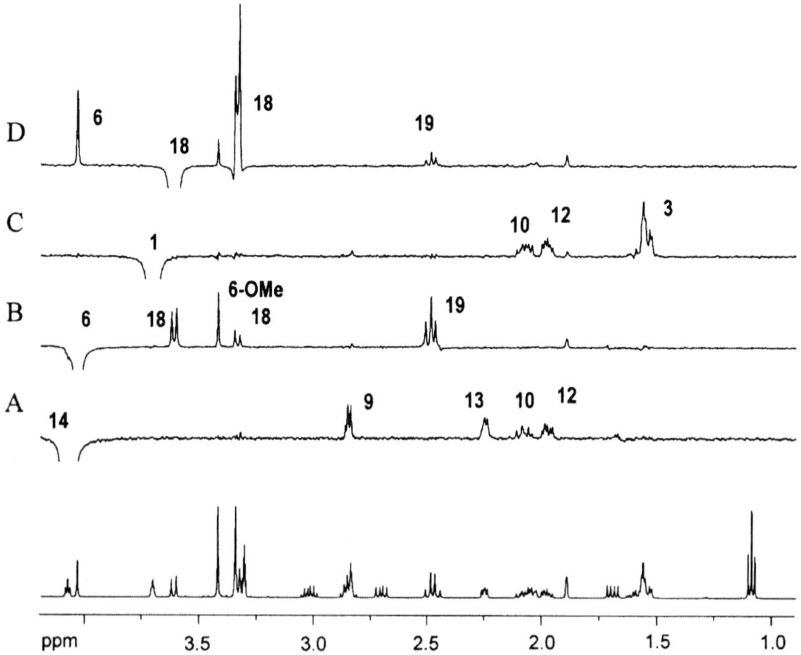

Figure 4. The DPFGSE-NOESY spectra of 1.

internuclear distances in solution. To this end, the mixing-time dependence of signal intensities is measured to obtain the cross-relaxation rates (σ) from which the unknown distances (Table 1) can be calculated using the internal rotation correlation time (τ_c=6.045*10^{-11}s) derived from ^{13}C spin-lattice relaxation time measured for skeletal carbons.

Table 1. Measured proton distances for 1.

Proton pair	Distance (Å)
H-14/H-9	2.5
H-14/H-12	2.7
H-1/H-10	2.4
H-1/H-12	2.7

Heteronuclear NMR experiments can also provide valuable information about the conformation in the liquid state. The heteronuclear three-bond coupling constants carry these informations through their dihedral angle dependence. Heteronuclear single quantum coherence experiment (HSQC) optimized for the detection of long-range interactions (HSQMBC) can be applied to extract these coupling constants accurately. The experiment yields antiphase signals and the required coupling constants can be obtained by manual or computer aided iterative fitting of the multiplet components (Figure 5.). The procedure involves the extraction of the appropriate carbon rows from the two-dimensional HSQMBC spectrum as figure 6A shows. Seven carbon rows at chemical shifts of C-3, C-4, C-10, C-11, C-16 and C-6 were selected to illustrate the appearance of the HSQC multiplets. In case of isolated (not coupled) protons the extraction of the heteronuclear coupling constants is straightforward, the frequency separation of antiphase signals gives rise to the actual value of scalar coupling. However, homonuclear couplings and signal overlapping result in a complex HSQMBC signal making the extraction of the heteronuclear coupling constants difficult.

In these cases iterative fitting of the HSQMBC multiplet is required. The simplest way to fit the HSQMBC multiplet is to use a reference proton signal (H-15 on figure 6B bottom). The reference multiplet is subtracted from itself, subsequent to a horizontal shift of one component with a trial value of the long-range coupling constant. Then the simulated antiphase multiplet (Figure 6B top), is compared to the corresponding HSQMBC slice (Figure 6B middle) and the magnitude of horizontal shift is varied to obtain the best fit of the experimental and calculated multiplets. The H-15 proton signal is used as an illustration on figure 6B, the manual analysis resulted in a 3.2 Hz long-range coupling constant between H-15 and C-16, while the iterative fitting yielded coupling constants labeled in italics on the HSQMBC spectrum of Figure 5.

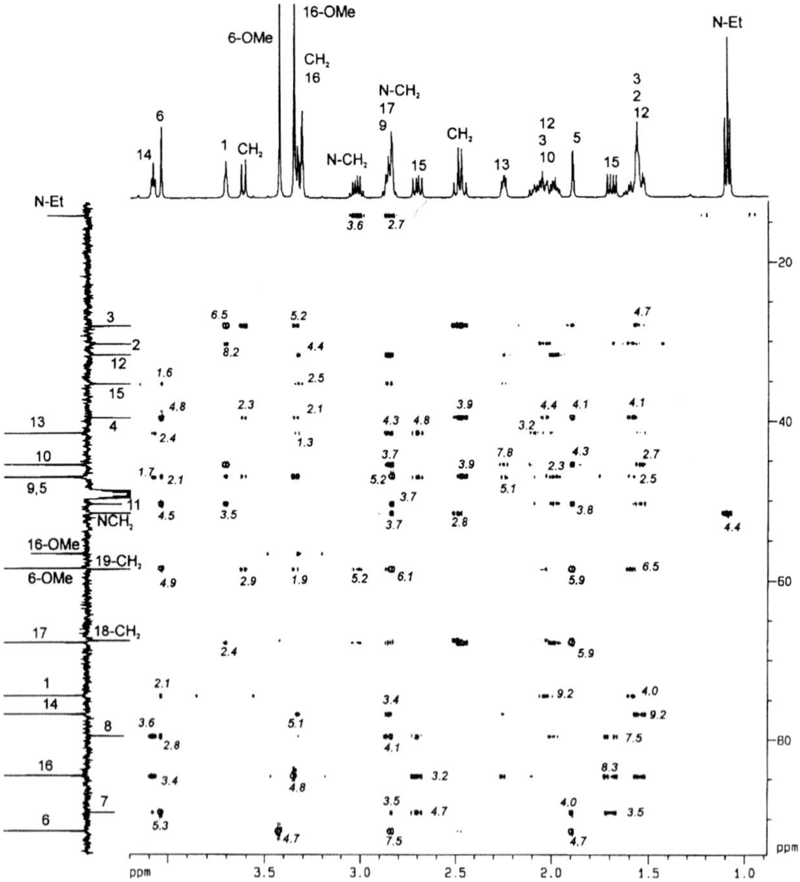

Figure 5. HSQMBC spectrum of 1.

The ^{13}C-coupled HSQC-TOCSY spectrum can also be used to extract heteronuclear long-range coupling constants. The cross-peaks exhibit in-phase proton multiplets with additional splitting due to the presence of ^{13}C in the spin-system. Figure 7A shows the HSQC-TOCSY spectrum of **1**. The fitting of the H-13 multiplet at three different ^{13}C-rows is illustrated on figure 7B (C-12 on the top, C-10 in the middle and C-9 on the bottom). An iterative fitting procedure similar to the one applied for the HSQMBC multiplets was employed here as well. However, in this case the reference multiplets were co-added subsequent to the horizontal shift of one component. Then the magnitude of the shift is varied to obtain the best fit of the simulated and experimental multiplets. This procedure yielded three coupling constants for the H-13 proton (less than 1 Hz with C-12, 7.3 Hz with C-10 and 5.2 Hz with C-9).

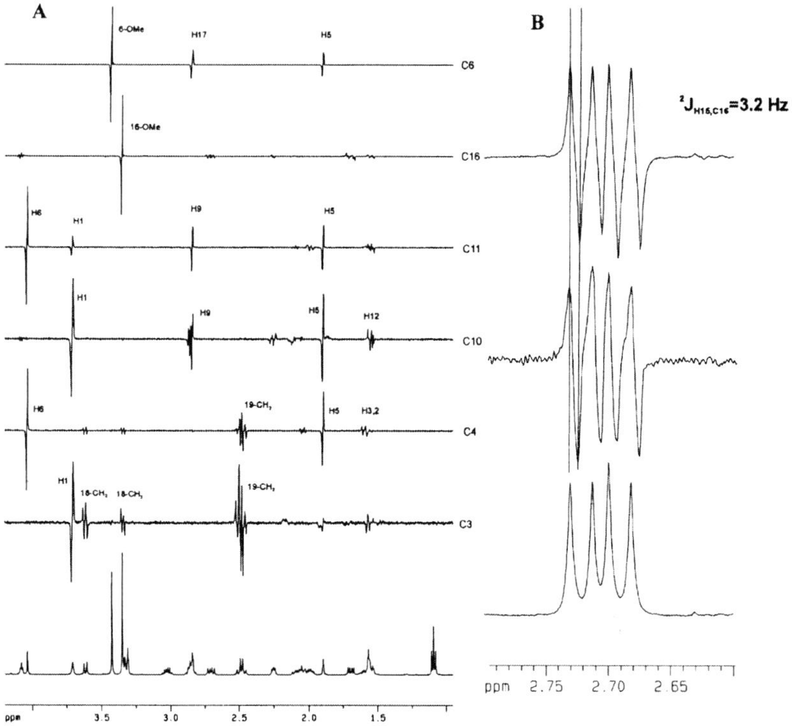

Figure 6. Extracted rows from the HSQMBC spectrum (A) and manual fitting of the H-15 multiplet (B) of 1.

The third approach used for heteronuclear coupling constant measurement employs the HETLOC sequence (Figure 8A), which is basically a homonuclear TOCSY experiment carried out on ^{13}C satellites of protons. The resulting correlation map (HETLOC spectrum) displays proton chemical shifts in both dimensions. To this end, a ^{13}C filter is inserted in the preparation part of the pulse sequence to filter out ^{12}C-connected proton magnetization. The experiment produces E.COSY multiplets. The frequency displacement of multiplet components in the F_2 dimension yields the heteronuclear long-range coupling constant, while in the F_1 dimension heteronuclear one-bond coupling constants can be measured. Figure 8 shows the HETLOC spectrum of 1 together with 12 expansions displaying the relevant heteronuclear coupling constants. The combined use of the heteronuclear experiments (HSQMBC, HSQC-TOCSY and HETLOC) provided several heteronuclear three-bond coupling constants that are essential in the evaluation of the solution state conformation (Table 2.)

^{15}N-NMR spectroscopy can be a valuable tool for the structure determination of alkaloids, because the ^{15}N chemical shifts are more disperse and more sensitive to the structural environment. Moreover, the heteronuclear three-bond ^{15}N-^{1}H coupling constants can also serve

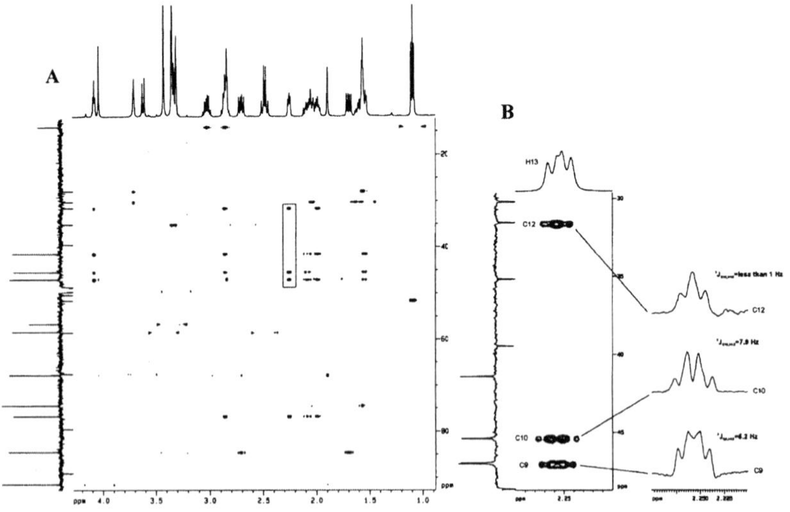

Figure 7. HSQC-TOCSY spectrum of 1 (A) and manual fitting of the H-13 multiplet (B) at different ^{13}C locations.

conformational information. However, the limited amount of the available sample, the low sensitivity and low natural abundance makes the detection of ^{15}N-NMR signals a difficult and time-consuming procedure (NOTE that to record a ^{15}N-detected NMR spectrum is practically impossible). Proton detected ^{15}N-NMR methods, however can overcome the sensitivity problem. The ^{15}N-^{1}H HSQMBC experiment can be used for the extraction of accurate ^{15}N chemical shifts and long-range heteronuclear coupling constants. The F_1 projection of the spectrum (Figure 9) yields the corresponding ^{15}N chemical shift (45.5 ppm in **1**), while the heteronuclear coupling constants were extracted with manual fitting of the F_2-multiplets. The following coupling constants, 3.5 Hz, 2.2 Hz and 3.2 Hz were extracted from the H-17, H-19 and H-21 multiplets (inset on figure 9).

The least sensitivity is associated with the homonuclear ^{13}C-^{13}C correlation experiments mainly because the low (~1%) ^{13}C natural abundance. However, the experiment is gaining high importance due to the sensitivity improvements achieved by the recent advances in NMR technology and by the design of highly sophisticated pulse sequences. Early experiments, such as INADEQUATE, used ^{13}C-^{13}C double quantum coherence followed by ^{13}C detection to acquire ^{13}C-^{13}C connectivity

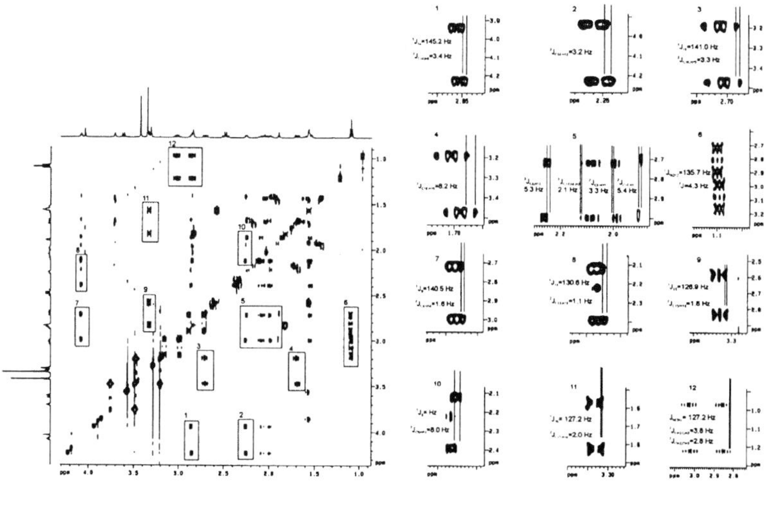

Figure 8. HETLOC spectrum of 1 (A) and insets with the heteronuclear coupling constants (B).

Table 2. Measured $^{3}J_{^{13}C,^{1}H}$ values in liquid state for 1.

^{13}C-^{1}H pair	^{3}J (Hz)
C-3/H-1	6.5
C-4/H-6	4.8
C-5/H-17	5.2
C-5/H-19	3.9
C-7/H-5	4.0
C-7/H-9	3.5
C-8/H-6	2.8
C-8/H-14	3.6
C-10/H-5	4.3
C-10/H-13	7.8
C-11/H-6	4.5
C-12/H-16	4.4
C-13/H-9	4.3
C-14/H-16	5.1
C-17/H-1	2.4
C-17/H-5	5.9
C-19/H-17	6.1

information. The method was improved by the application of polarization transfer steps and pulsed-field gradient techniques. However, proton detected methods (ADEQUATE) brought a breakthrough to the solution of sensitivity problems of ^{13}C-^{13}C correlation experiments. The polarization transfer steps of the experiment can be optimized to one-bond and to long-range interactions within the same experiment making it possible to reach correlation information through six bonds. However, ADEQUATE experiments require at least one proton attached to the carbon skeleton to be able to start the polarization transfer sequence. Even though large amount of samples and long measurement is required, the experiment provides primary information about the molecular constitution, and became a powerful tool in the NMR structure determination.

Natural products with a more complex structure like diterpene alkaloids require the application of the ADEQUATE experiment in order to derive unambiguous structure (Figure 10).

Figure 10 shows an ADEQUATE spectrum of **2**, in which ^{13}C cross peaks appear at different proton frequencies. At 3.99 ppm (H-6), for example, two correlations can be seen, one at 87.7 ppm (C-7) and 44.6 ppm (C-5), corresponding to C-6/C-7 and C-6/C-5 connectivities,

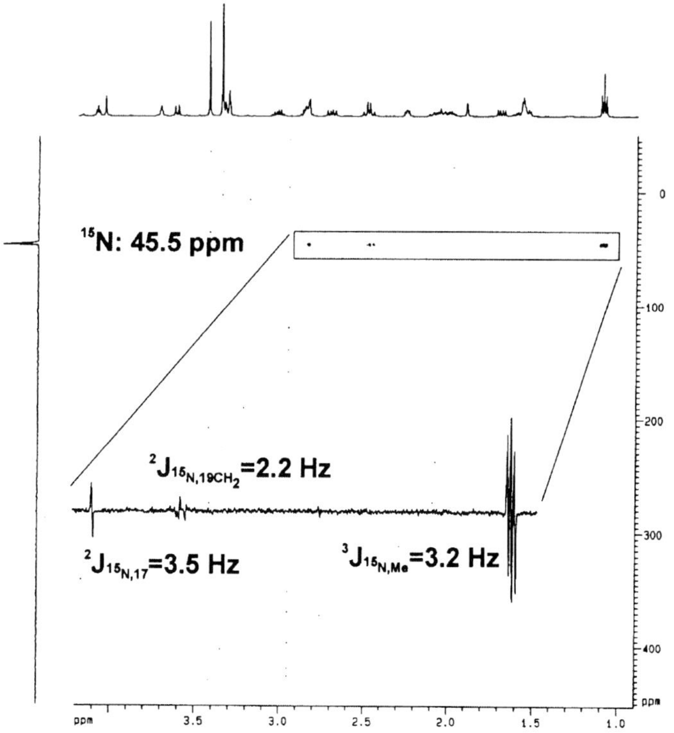

Figure 9. ^{15}N-^{1}H HSQMBC spectrum of 1.

respectively. The alkaloid skeleton can be mapped step by step reaching the assignment of the ^{13}C spectrum and verifying the proposed constitution for a given molecule.

SUMMARY

Several selective one dimensional NMR experiments were applied to detect scalar and dipolar interactions in solution of norditerpene alkaloids. The COSY and TOCSY based experiments were used to obtain the chemical shift and coupling constants of the scalarly coupled partners. The selective NOESY experiment provided conformational information and the quantitative evaluation of NOE enhancements resulted in solution state proton distances. Heteronuclear experiments, HSQMBC, HSQC-TOCSY and HETLOC were used to obtain three-bond coupling

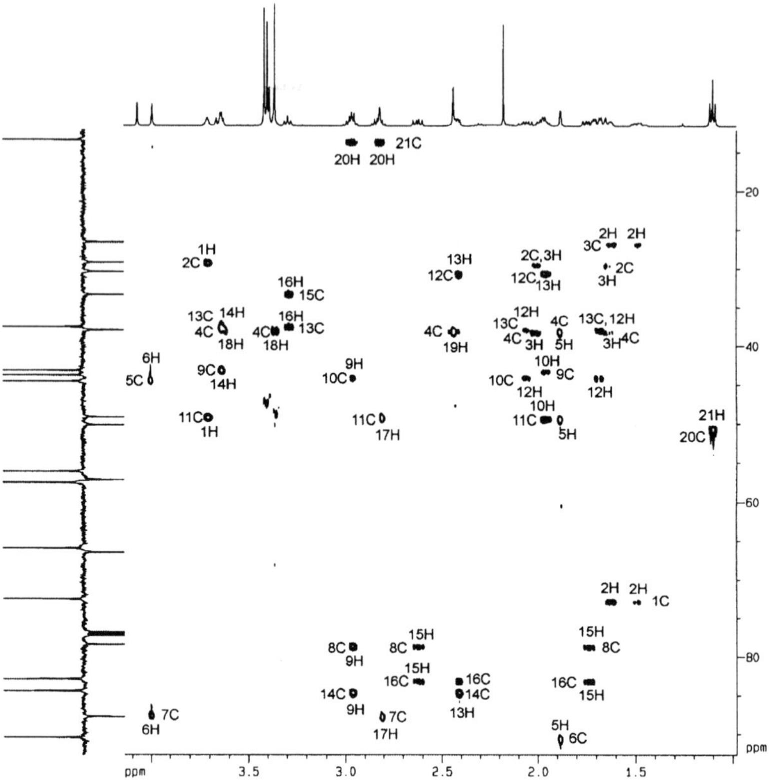

Figure 10. ^{13}C-^{13}C ADEQUATE spectrum of 2.

constants. ^{15}N chemical shifts and long range coupling constants were extracted from ^{15}N-^{1}H HSQMBC spectra. The ^{13}C-^{13}C ADEQUATE experiment was used to map the main skeletal connectivities.

ACKNOWLEDGMENTS

Financial support of the Hungarian Research Fund Agency (OTKA grant T038390) is gratefully acknowledged.

REFERENCES

[1] Ameri A. The effects of Aconitum alkaloids on the central nervous system. Prog Neurobiol 1998; 56: 211-35

[2] Friese J, Gleitz J, Gutser UT, Heubach JF, Matthiesen T, Wilffert B, Selve N. Aconitum sp. alkaloids: the modulation of voltage-dependent Na+ channels, toxicity and antinociceptive properties. Eur J Pharmacol 1997; 337: 165-74

[3] Bello-Ramírez AM, Buendia-Orozco J, Nava-Ocampo AA. A QSAR analysis to explain the analgesic properties of Aconitum alkaloids. Fund Clin Pharmacol 2003; 17: 575-80

[4] Breining SR, Recent developments in the synthesis of nicotinic acetylcholine receptor ligands. Curr Top Med Chem 2004; 4: 609-29.

[5] Hohmann J, Forgo P, Hajdú Zs, Varga E, Máthé I. Norditerpenoid alkaloids from Consolida orientalis and complete 1H and 13C NMR signal assignments of some lycoctonine-type alkaloids. J Nat Prod 2002; 65: 1069-72

[6] Csupor D, Forgo P, Máthé I, Hohmann. Acovulparine, a new norditerpene alkaloid from Aconitum vulparia. J Helv Chim Acta 2004; 87: 2125-30

[7] Morris GA, Freeman R. Selective excitation in Fourier transform nuclear magnetic resonance. J Magn Reson 1978; 29: 433-62

[8] Boudot D, Canet D, Brondeau J, Boubel JC. DANTE-Z: a new approach for accurate frequency selectivity using hard pulses. J Magn Reson 1989; 83: 428-39

[9] Roumestand C, Canet D. DANTE-Z, an alternative to low-power soft pulses. Improvement of the selection scheme and applications to multidimensional NMR studies of proteins. J Magn Reson A 1994; 106: 168-81

[10] Wu X-L, Xu P, Freeman R. A new kind of selective excitation sequence. J Magn Reson 1989; 83: 404-10

[11] Hall LD, Norwood TJ. A chemical-shift-selective filter. J Magn Reson 1988; 76: 548-54

[12] Kessler H, Mronga S, Gemmecker G. Multidimensional NMR experiments using selective pulses. Magn Reson Chem 1991; 29: 527-57

[13] Hwang TL, Shaka AJ. Water suppression that works. Excitation sculpting using arbitrary waveforms and pulsed field gradients. J Magn Reson A 1995; 112: 275-9

[14] Stonehouse S, Adell P, Keeler J, Shaka AJ. Ultrahigh-Quality NOE Spectra. J Am Chem Soc 1994; 116: 6037-8

[15] Berger S. NMR techniques employing selective radiofrequency pulses in combination with pulsed field gradients. Prog NMR Spectr 1997; 30: 137-56

[16] Parella T. Pulsed field gradients: a new tool for routine NMR. Magn Reson Chem 1998; 36: 467-95

[17] Kövér KE, Uhrin D, Hruby VJ. Gradient- and sensitivity-enhanced TOCSY experiments. J Magn Reson 1998; 130: 162-8

[18] Stott K, Keeler J, Van QN, Shaka AJ. One-dimensional NOE experiments using pulsed field gradients. J Magn Reson 1997; 125: 302-24

[19] Bauer W, Soi A, Hirsch A. Application of DPFGSE-ROE to calixarene derivatives under conditions near NOE zero-crossing. Magn Reson Chem 2000; 38: 500-3

[20] Willker W, Leibfritz D, Kerssebaum R, Bermel W. Gradient selection in inverse heteronuclear correlation spectroscopy. Magn Reson Chem 1993; 31: 287-92

[21] Williamson RT, Márquez BL, Gerwick WH, Kövér KE. One- and two-dimensional gradient-selected HSQMBC NMR experiments for the efficient analysis of long-range heteronuclear coupling constants. Magn Reson Chem 2000; 38: 265-73

[22] Kövér KE, Uhrin D, Hruby VJ. Sensitivity- and gradient-enhanced heteronuclear coupled/decoupled HSQC-TOCSY experiments for measuring long-range heteronuclear coupling constants. J Magn Reson 1997; 129: 125-9

[23] Uhrin D, Batta Gy, Hruby VJ, Barlow PN. Sensitivity- and gradient-enhanced hetero (ω1) half-filtered TOCSY experiment for measuring long-range heteronuclear coupling constants. J Magn Reson 1998; 130: 155-61

[24] Bax A, Freeman R, Kempsell SP. Natural abundance carbon-13-carbon-13 coupling observed via double-quantum coherence. J Am Chem Soc 1980; 102: 4849-51

[25] Reif B, Köck M, Kerssebaum R, Schleucher J, Griesinger J. Determination of 1J, 2J, and 3J carbon-carbon coupling constants at natural abundance. J Magn Reson B 1996; 112: 295-301.

Quantitative Detection of Isoflavones in the Extract of Red Clover by HPLC / ESI-MS

Zhang Guifeng*, Liu Tao and Su Zhiguo
National Key Lab of Biochemical Engineering, Institute of Process Engineering, Chinese Academy of Sciences, Beijing, PR China, 100080
**Corresponding author: gfzhang116@yahoo.com*

INTRODUCTION

Red clover is a rich source of polyphenolic substances that are known for potential bioactive antioxidant properties and radical scavenging capacity [1]. Pharmacological study and clinical practice have shown that the main active constituents in the red clover extracts are isoflavones that have been reported for their beneficial estrogenic effects [2]. Extracts of red clover are commercially available as isoflavone-enriched dietary supplements on the US and European markets for women suffering from menopausal complaints. The content of isoflavones is a criterion for evaluating the quality of red clover crude extract [3].

Conventional method for analyzing the content of isoflavones in commercial products is mainly based on high-performance liquid chromatography (HPLC) and thin layer chromatography (TLC) [4, 5]. Other methods include high-speed counter-current chromatography (HSCCC) and capillary electrophoresis (CE) [6, 7]. However, the isoflavones show similar UV wavelength absorbance. There are so many kinds of materials in the crude extracts that uncompleted separation will affect the results of the concentrations determined. High-performance liquid chromatography/mass spectrometry (HPLC/MS) has many advantages over traditional analytical methods, such as higher accuracy, sensitivity and speed. HPLC/MS has been successfully applied in component analysis in the research field of natural products and Chinese Traditional Medicines [8, 9].

In this study, a quantitative method was established based on HPLC/MS to determine the concentration of isoflavones in the extracts of red clover. The effect of collision energy on production of fragmental ions was investigated. Mass spectrometric parameters including capillary temperature, sheath gas and aux gas flow rate, and ionization spray voltage were optimized. Five kinds of isoflavones - daidzein, genistein, genistin, formononetin and biochanin A - were used as standards. Actual samples were analyzed to validate the method established.

MATERIALS AND METHODS

Plant materials

Five isoflavone standards - isoflavones, daidzein, genistein, genistin, formononetin and biochanin A - were purchased from Sigma-Aldrich (St. Louis, MO, USA). Crude extracts of red clover with different total isoflavone contents were obtained from China Medicines & Health Products Import & Export Corporation.

General sample preparation

5mg powder of red clover was extracted with 1ml water-methanol (2:8, v/v) solution. After ultrasonication in an ice bath for 45 min, the mixture was centrifuged at 10000 g for 10 min. Supernatant was filtered with 0.22 μm membrane and 20 μl extract was loaded for each HPLC/MS analysis. An appropriate amount of each isoflavone standard was weighed and mixed together. Stock solution was prepared by dissolving 5mg mixture in 10 ml water-methanol (2:8, v/v) solution and calibration standards were prepared by diluting the stock solution. Five different concentrations were used in the calibration curve by mass spectrometric analysis.

High-performance liquid chromatographic and mass spectrometric conditions for quantitative analysis of isoflavones

High-performance liquid chromatography was performed on Agilent 1100 system using a photodiode array and multiple wavelength detector. An Intersil ODS-3 column (4.6×250 mm, flow rate 1 ml/min) was used to separate components in the crude extracts. The mobile phase in HPLC separation was solvent A with solvent B in gradient (solvent A was 0.1% trifluoroactic acid in water and solvent B was 0.1% trifluoroactic acid in acetonitrile). The gradient was 20 to 50% B in 50 min. The wavelength of UV detection was 254 nm. The eluent from HPLC was connected with a split valve and a 100 μl/min of eluent was induced to the mass spectrometry.

The mass spectrum was measured on a LCQ Deca XP Thermo Finngian (San Jose, CA) ion trap MS equipped with an electrospray ionization source. Tandem scan (MS/MS) were performed in selective mode. Mass spectrometric conditions, capillary temperature, sheath gas and aux gas flow rate, ionization spray voltage and collision energy were optimized to achieve optimal signal.

RESULTS

Identification of isoflavones in red clover

Components of red clover extracts were separated under the chromatographic condition, as shown in Figure 1. Qualitative study has demonstrated that red clover contained many kinds of isoflavones. To quantitatively study the contents of the five kinds of isoflavones, chromatogram of samples was compared with that of standards. Five peaks of standard mixtures were genistin (peak 1, m/z433.1), daidzein (peak 2, m/z255.4), genistein (peak 3, 271.3), formononetin (peak 4, m/z269.3) and Biochanin A (peak 5, m/z285.3). These were also found in the actual sample. The retention time, protonated ions $[M+H]^+$ and characteristic fragment ions for five kinds of isoflavones corresponded with those of standards. Mass spectrum and wavelength absorbance

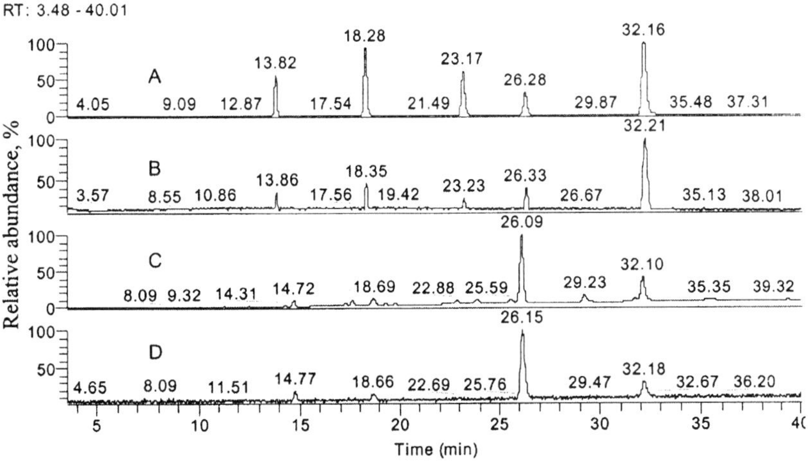

Figure 1. High performance liquid chromatogram of Red clover extract
(A) High performance liquid chromatogram of extract. (B) Current of selective ions monitoring of m/z269, 271,285. (C) Total ion current of five standards.

spectrum indicated that good peak purity was obtained for each component in the chromatogram.

Optimization of mass spectrometric conditions

Mass spectrometric conditions were modified to achieve optimal signal of each standards. The scan range was set to m/z200-500 since the molecular weight of each isoflavone is below 500 Da. The optimal signal was obtained when flow rate of sheath gas was 50 arb and no aux gas was used. The signal intensity increased with spray voltage and kept almost no change when the spray voltage was above 3.5 kV.

The thermal stabilities of some isoflavones could be affected by capillary temperature. The optimal signal was achieved when the temperature was set at 275°C. Collision energy is an important factor that can influence the intensity of fragment ions. As for daidzein, the intensity of the fragmental ion, m/z 227.2, from the ion, m/z 255, reached optimal when the collision energy was 42%, as shown in Figure 2. Genistein, formononetin and biochanin A are derivatives of daidzein (structures shown in Figure 3). They show similar collision energy, from 40% to 42%, during MS/MS detection, as indicated in Figure 4. Genistin is a glycosyl genistein and the lower collision energy (35%) may be the result of the presence of the glycosyl group. The fragment ion used to quantify daidzein is m/z227.

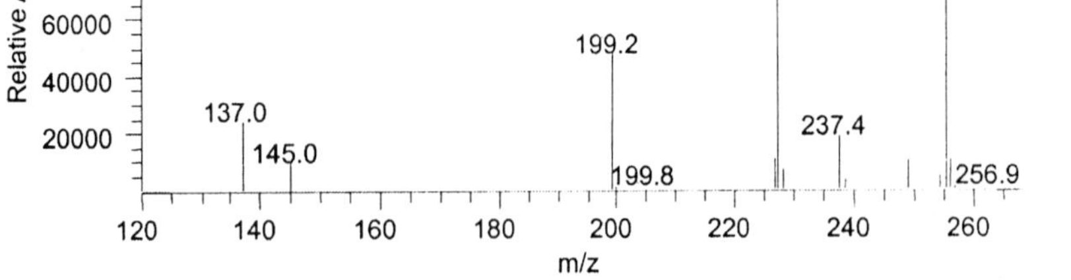

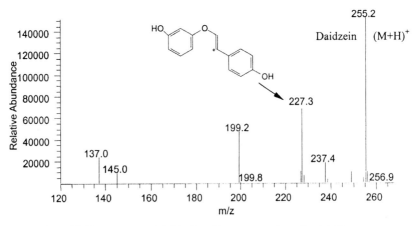

Figure 2. MS/MS spectrum of daidzein with precursor ion of m/z255.
Inset is the structure of daidzein and it's a main MS/MS fragment pathway.

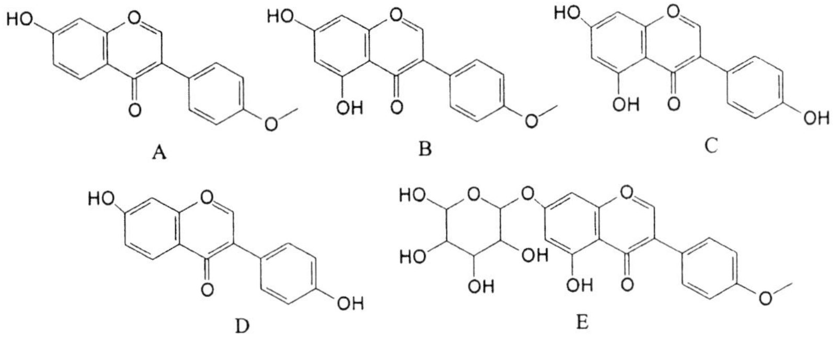

Figure 3. Chemical Structures of Formononetin, Biochanin A, Genistein and Genistin

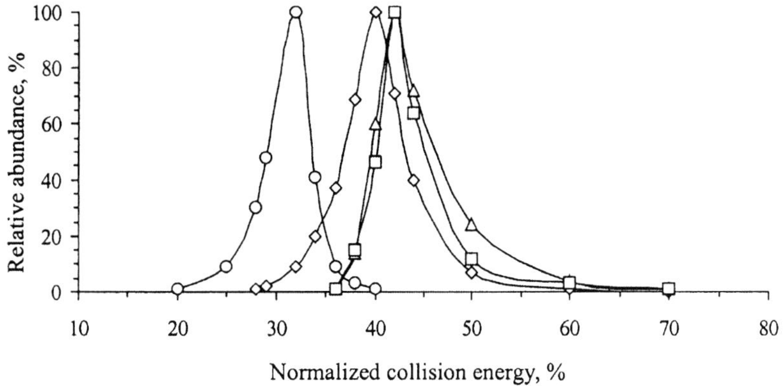

Figure 4. Relationship between collision energy and relative abundance of fragmental ions
(O) fragment ion of Genistin, m/z 271; (□) fragment ion of Genistein, m/z 256; (◇) fragment ion of Formononetin, m/z 254; (Δ) fragment ion of BiochaninA, m/z 270

Other fragment ions used were m/z256 from genistein, m/z271 from genistin, m/z254 from formononetin and m/z270 from biochanin A.

Calibration curve and detection limit

Five different concentrations of standards were analyzed three times for each analyte by HPLC/MS. As many kinds of isoflavones have the same molecular weight, eg. biochanin A (284), prunetin (284) and calysosin

(284), so typical fragment ion was used to quantitatively integrate peaks of interest. The linearity of isoflavone curves was obtained at the concentration range of 300-40,000ng/mL based upon the ratio of peak areas. The results indicate that the quantitative method reported here show good linearity and high sensitivity with correlation coefficient for each isoflavone above 0.992 (Table 1).

Methodology and analysis of actual sample

Both intra-assay and inter-assay relative standard deviations RSD %) were investigated by analyzing five replicates of the control samples and one sample in 5 analytical batches. The results are presented in Table

Table1. Regression equation and detection limit for different compounds

Component	Fragment ion, m/z	Regression equation	Correlation coefficient (r^2)	Linear range (ng)	Detection limit (ng)
Daidzein	227	Y=-111590+9873x	0.997	0.45-339.6	0.11
Genistein	256	Y=-125792+8974x	0.993	0.72-381.3	0.12
Genistin	271	Y=-145872+10089x	0.994	0.31-321.8	0.11
Formononetin	254	Y=-107825+12124x	0.998	0.44-398.4	0.12
BiochaninA	270	Y=-142369+11982x	0.994	0.52-362.6	0.11

Table 2. Results of method evaluation

Component	Repeatability RSD (%,n=5)	Stability RSD (%,n=5)
Daidzein	1.9	2.0
Genistein	2.1	1.8
Genistin	2.0	1.9
Formononetin	2.2	2.5
BiochaninA	1.9	1.7

2. In order to validate this method, known amounts of daidzein, genistein, daidzin and genistin were added into the samples, and their recoveries were determined. The overall recoveries of these compounds were above 97%. Two commercial samples, A and B, were analyzed and the results showed that sample A and B separately contained 0.51μg/ml and 0.38μg/ml of daidzein, 0.96μg/ml and 1.05μg/ml of genistein, 0.18μg/ml and 0.15μg/ml of genistin, 1.42μg/ml and 1.29μg/ml of biochanin A, 6.06μg/ml and 7.12μg/ml of formononetin.

CONCLUSION

High-performance liquid chromatography/electrospray ionization-mass spectrometry has been applied to develop a method for quantitative analysis of isoflavones in the crude extract of red clover. The molecular ions and specific fragment ions were used to identify specific isoflavone and quantitatively determine its contents. The mass spectrometric conditions were optimized and optimal signal was achieved. Analysis of actual sample indicated the method developed is valid for practical application.

REFERENCES

[1] He XG, Lin LZ, Lian LZ. Analysis of flavonoids from red clover by liquid chromatography-electrospray mass spectrometry. J Chromatogr A 1996; 755: 127-32

[2] Setchell KDR and Cassidy A. Dietary isoflavones: biological effects and relevance to human health. J Nutr 1999; 129: 758-67

[3] Mullner C, Sontag G. Determination of some phytoestrogens in soybeans and their processed products with HPLC and coulometric electrode array detection. Fresenius J Anal Chem 1999; 364: 261-5

[4] Penalvo JL, Nurmi T, Adlercreutz H. A simplified HPLC method for total isoflavones in soy products. Food Chem 2004; 87: 297-05

[5] Carbone V, Montoro P, Tommasib ND, Pizza C. Analysis of flavonoids from *Cyclanthera pedata* fruits by liquid chromatography/electrospray mass spectrometry. J Pharm Biomed Ana 2004; 34: 295–304

[6] Yang FQ, Ma Y, Ito Y. Separation and purification of isoflavones from a crude soybean extract by high-speed counter-current chromatography. J Chromatogr A 2001; 928: 163–70

[7] Shihabi ZK, Kute T, Garcia LL, Hinsdale M. Analysis of isoflavones by capillary electrophoresis. J Chromatogr A 1994; 680: 181-5

[8] Rijke E, Zappey H, Ariese F, Gooijer C, Brinkman UAT. Liquid chromatography with atmospheric pressure chemical ionization and electrospray ionization mass spectrometry of flavonoids with triple-quadrupole and ion-trap instruments. J Chromatogr A 2003; 984: 45–58
[9] Twaddle NC, Churchwell MI, Doerge DR. High-throughput quantification of soy isoflavones in human and rodent blood using liquid chromatography with electrospray mass spectrometry and tandem mass spectrometry detection. J Chromatogr B 2002; 777: 139–45

Measurement of Bioactive Constituents in Traditional Chinese Medicines by CE with Electrochemical Detection

Gang Chen[a,*] and Xiaohong Chen[b,*]
[a]*Department of Chemistry, Fudan University, Shanghai 200433, China*
[b]*Shanghai Medipharm Biotech Co., Ltd., Shanghai 201203, China*
**Corresponding emails: gangchen@fudan.edu.cn or chen_rainbow@yahoo.com*

INTRODUCTION

People are becoming more and more interested in traditional Chinese medicines (TCMs) because of their low toxicity and good therapeutic effects. Chinese people have utilized herbs and plants to treat various diseases for more than 8000 years [1, 2]. TCM has been modified to some extend in Korea and Japan and has attracted considerable attention even in European and North American countries [3]. Chinese Pharmacopoeia has recorded more than 1146 examples of TCMs that are either crude drugs or composite pharmaceutical preparations [4]. In order to study their pharmacological properties and to control their quality, the identification and quantification of the active constituents in herbal medicines have become increasingly urgent [5, 6]. In addition, TCMs are usually in the form of multicomponent prescriptions, and each component may exhibit a complicated profile of constituents, the analysis of the active constituents in TCMs is often very difficult. Therefore, efficient and selective methods, including the extraction techniques are highly required for identification and quantitative analysis of the active compounds or drug standardization [7].

Liquid chromatography (LC) and gas chromatography (GC) are the main techniques applied in this field due to their powerful separation efficiency combined with sensitive detection [8–12]. Since capillary electrophoresis (CE) in its modern form was first described by Joegenson and Luckas in 1981 [13, 14], its application for the separation and determination of various samples has become increasingly widespread because of its low cost, minimal sample volume requirement, short analysis time, and high separation efficiency. CE has been investigated for many years to analyze various constituents in TCMs mainly in capillary zone electrophoresis (CZE) [15–17] and micellar electrokinetic capillary chromatography (MECC) [18–20] modes. It seems extremely suitable for the measurement of the constituents in TCMs based on the fact that the separation capillary is easy to be regenerated by flushing with suitable solution after it was contaminated by the coexistent interferences in TCMs. The interferences may deteriorate or even damage LC column completely.

As with other analysis systems for TCMs, sensitive and selective detection techniques are also required for CE. Ultraviolet (UV) absorption, laser-induced fluorescence (LIF), mass spectroscopy (MS), and electrochemical detection (ECD) have been used in combination with CE for the detection of the TCM constituents. LIF and MS are two important detection methods for CE with the detection limit ranges of 1–100 pM and 1–100 nM, respectively. However, both methods need sophisticated and expensive instrumentation. LIF typically requires pre- or postcapillary derivatization of the sample with a fluorophore and are limited to fluorescent analytes and analyte derivatives. UV detector is the most widely used method for the detection of the constituents in TCMs after separation by CE [21–23]. Because the absorbance path length of the capillary (the inner diameter, 25 to 100 μm i.d.) is very short, the low sensitivity of the UV detector used results in low detection limit (typically 10 μM). Usually, the content of the constituents in the TCMs is very low. High sensitive detection methods are highly demanded. Recently, electrochemical detection (ECD) has been coupled with CE for the sensitive detection of TCMs [24–28]. ECD offers great promise for such CE systems, with features that include remarkable sensitivity, inherent miniaturization of both the detector and control instrumentation, independence of sample turbidity or optical path length, low cost, minimal power demands, and high compatibility with the CE separation system [29–31]. CE-ECD offers effective alternatives to analytical methodologies currently in use for characterizing many constituents in TCMs. ECD can also provide higher selectivity as only electroactive substances can be detected so that the electropherograms are simplified, which is important for the analysis of medicinal plants because the constituents in them are usually complex. ECD can be classified into three general modes, conductimetry, potentiometry, and amperometry. However, amperometry is the only reported ECD for the detection of the constituents in TCMs after CE separation [32–34]. It is accomplished by applying a constant potential to the detection electrode and measuring the resulting current that is proportional to the concentration of analytes oxidized or reduced at the electrode surface. Many different electrode materials have been employed for CE-ECD. However, only carbon and copper disc electrodes have been employed for the detection of the constituents in TCMs in oxidation mode [25–28]. The disc electrode was made of a piece of 300 μm diameter graphite rod [26] or copper wire [27] coated with epoxy from polishing technique. Flavonoids, anthraquinones, some alkaloids, phenolic acids, and glycosides of phenols, phenols and polyphenols in TCMs can be detected on carbon electrode in oxidation mode. In the case of amino acid, sugars and constituents that contain multiple hydroxyl groups, copper electrode can be employed for their detection in diluted NaOH aqueous solution. In 2000, U.S. Food and Drug Administration (FDA) published a draft of Guidance for Industry Botanical Drug Products [35]. Before a plant drug becomes legally marketed, its spectroscopic or chromatographic fingerprints and chemical assay of characteristic markers are required. CE-ECD should find more applications in this area.

This paper focuses on recent advances in the application of CE-ECD for analyzing bioactive constituents in TCMs. CE-ECD has been employed for the analyses of both crude herbal drugs and their multicomponents pharmaceutical preparations. The following sections will cover the separation techniques employed for the analysis of TCMs, sample pretreatments of TCMs, determination of bioactive constituents in Chinese medicinal materials and their preparations by CE-ECD, and differentiation of traditional Chinese medicines based on CE-ECD.

SEPARATION TECHNIQUES AND INSTRUMENTS

CE encompasses a family of related separation techniques that use narrow-bore fused-silica capillaries to separate a complex array of large and small molecules. Depending on the types of capillary and electrolytes used, the technology of CE can be segmented into several separation techniques such as capillary zone electrophoresis (CZE), micellar electrokinetic capillary chromatography (MECC or MEKC), capillary gel electrophoresis (CGE), capillary electrochromatography (CEC), etc.

Among them, CZE and MECC are the two commonly used methods for the separation of the various constituents in the extracts of TCMs [36, 37].

CZE, also known as free-solution CE (FSCE), is the simplest form of CE. The separation mechanism is based on differences in the charge-to-mass ratio of the analytes. The separation relies principally on the pH-controlled dissociation of acidic groups on the analyte or the protonation of basic functions on the analyte. CZE is the most widely employed method for the separation of the TCM constituents presented in ions partially in the buffer for CE. If several analyses in TCMs are neutral, MECC has to be employed for their separation. MECC is a mode of electrokinetic chromatography in which surfactants are added to the buffer solution at concentrations that form micelles. It was first developed by Terabe in 1984 [38]. The commonly used surfactant for the separation of TCMs is sodium dodecylsulphate (SDS) [39]. Separation of nonionic analytes is based on their different distribution coefficients between the aqueous phase and the micellar phase [40]. MECC has great utility in separating mixtures that contain both ionic and neutral species.

The CE-ECD system employed for the analyses of TCMs consists of a ±30 kV high-voltage dc power supply, a piece of fused silica capillary coated with a layer of polyimide, a amperometric detector, data acquisition system, a three-dimensional manipulator, and detection reservoir containing a three-electrode system consisting of a 300-μm diameter carbon or copper disc detection electrode, a platinum auxiliary electrode, and a reference electrode such as saturated calomel electrode (SCE) [41]. The high-voltage dc power supply provided an injection or a separation voltage (typically 9–20 kV) between both the ends of the capillary. The inlet of the capillary was held at a positive potential and the outlet of capillary was maintained at the ground potential. The separations were usually carried out in a 40–80 cm length of 25 μm i.d. and 360 μm o.d. fused silica capillary.

SAMPLE PREPARATIONS

The bioactive constituents in TCMs have to be extracted into solutions before CE analysis [42, 43]. The commonly employed extraction approaches were hot solvent extraction (HSE) [43, 44] and ultrasonic extraction (USE) [45]. The free constituents in the crude drugs can be easily extracted by using suitable solvents or solutions. Before the solid–liquid extraction, the crude drugs were dried and pulverized into powder with the particle size ranged from 50 to 200 meshes. An accurately weighed amount of the powder was then dispersed in the selected extraction solvents or solutions. In order to enhance the extraction efficient, the mixture was performed under heat or sonication. Usually, the extraction was performed repeatedly for more than 3 times so that most target constituents can be extracted. Prior to CE analysis, the extracts have to be diluted with background electrolyte (BGE) and were introduced into the separation capillary subsequently.

There are a great deal of bioactive constituents presented in TCMs that can be classified into phenols and polyphenols, flavonoids, anthraquinones, coumarins, alkaloids, organic acids, sugars and glycosides, etc. Flavonoids, phenolic acids, phenols, and polyphenols can be extracted with methanol, ethanol, aqueous alcohol solutions, and alkaline aqueous solutions. Some organic acids and sugars can be directly extracted with water while acidified water can be employed for the extraction of some alkaloids. Anthraquinones were usually extracted with chloroform.

DETERMINATION OF BIOACTIVE CONSTITUENTS IN CHINESE MEDICINAL MATERIALS

Some recent developments in the major applications of CE-ECD for analyzing bioactive constituents in Chinese medicinal materials and their pharmaceutical preparations are reviewed in the following sections. In addition, the application of CE-ECD in the differentiation and the quality control of TCMs have also been highlighted.

Phenols

It has been proved that small phenols in TCMs have a broad range of physiological activities, such as anti-oxidation, anti-bacteria, anti-inflammation, anti-allergy, alleviating pain, and enhancing the immune system [5, 45–47]. They have found wide applications in medicine, incense, and chemistry. For example, paeonol (**1**) has been used clinically as a therapeutic medicine for myalgia, rheumatic pain, neuralgia, coetaneous pluribus, etc [45]. The Pharmacopoeia of China requires the content of **1** in Moutan Cortex to be no less than 1.2% [45]. We have established a method based on CZE-ECD for the identification and determination of **1**, benzoyloxypaeoniflorin (**2**), and oxypaeoniflorin (**3**) in Moutan Cortex [45]. trans-Resveratrol (**4**), a phenol, is the main active constituent in Rhizoma Polygoni Cuspidati [47]. Gao [47] and Chu [48] have established CE-based methods for its analysis with the limit of detection (LOD) of 59.6 ng/ml ($S/N = 3$).

Oligomeric stilbenes are important bioactive constituents in *Caragana* species [46]. CZE-ECD has been employed for the separation and detection of hypaphorine ((**5**), an alkaloid), and four oligomeric stilbenes (including pallidol (**6**), kobophenol A (**7**), miyabenol C (**8**), and (+)-α-viniferin (**9**)) in Radix seu Cortex Caraganae Sinicae, the dried root of *Caragana sinica* (Buc'hoz) Rehd. Figure 1 illustrates the electropherograms of (A) a standard mixture solution of **5**, **6**, **7**, **8**, and **9** and (B) the diluted extracts from the crude drug. The four oligomeric

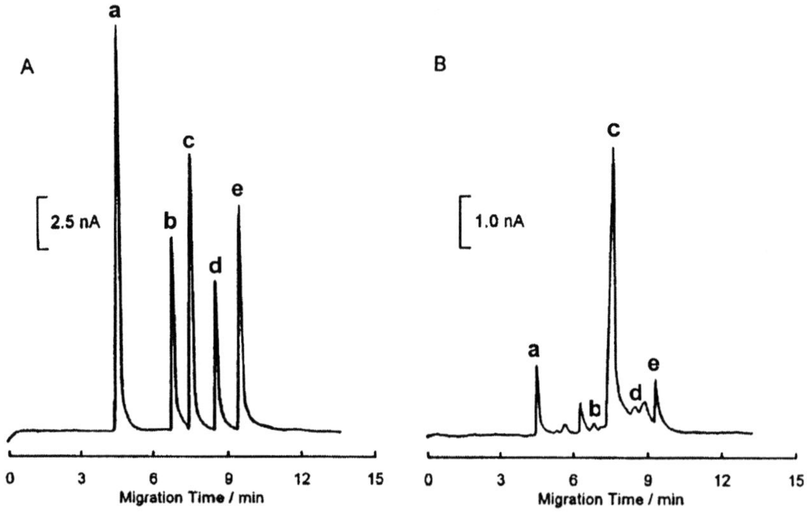

Figure 1 Electropherograms of (A) a standard mixture solution containing 5.26×10^{-2} g/L of **5** (a), 3.55×10^{-2} g/L of 6 (b), 6.80×10^{-2} g/L of 7 (c), 4.81×10^{-2} g/L of **8** (d) and 7.80 10^{-2} g/L of **9** (e) and (B) the diluted extracts from Radix Caraganae Sinicae. Fused-silica capillary, 25 μm i.d. × 40 cm; detection electrode, 300 μm diameter carbon disc electrode; detection potential, +0.90 V (*vs*. SCE); BGE, 100 mM BB (pH 10.0); separation and injection voltage, 12 kV; injection time, 8 s. Reprinted with permission from ref. [46].

stilbenes are all polyphenols. Pharmacological studies showed that these oligomeric stilbenes have the analogous effects of estrogen that can promote the adsorption of calcium. So, they may be a class of potential medicines for the treatment of osteoporosis for women after menopause. In addition, Cao et al. have determined 4-hydroxybenzyl alcohol (**10**), vanillyl alcohol (**11**), 4-hydroxybenzylaldehyde (**12**), and vanillin (**13**) in Rhizoma Gastrodiae and its preparations [5]. All the four phenols can be detected on carbon electrode at moderate potential with high sensitivity.

Flavonoids

Flavonoids are widely presented in herbal drugs and have a broad range of physiological activities such as anti-inflammatory, anti-tumor, anti-oxidant, anti-bacteria, etc. Flavonoids usually contain several phenolic hydroxyl groups and are easy to be oxidized on carbon electrodes at moderate detection potential ranged from +0.9 to +1.0 V (*vs*. SCE). The BGE for the CE separation of flavonoids was usually alkaline borate buffer because the nearby hydroxyl groups in some flavonoids could react with borate to form anionic complex so that the separation might be improved [15].

Cao et al. have determined the content of farrerol (**14**), quercetin (**15**), and other four phenolic acids in Folium Rhododendri Daurici by CZE-ECD [25]. Under the optimum conditions, the analytes were baseline separated within 16 min in a borate buffer (pH 8.7). Notably, excellent linearity was obtained over two orders of magnitude with detection limits ($S/N = 3$) ranged from 0.9 to 3 μM for all analytes. Chinese traditional medicine, Scutellariae Radix, is the dried root of *Scutellaria baicalensis* Georgi. Baicalin (**16**) and baicalein (**17**) are its two active components in Radix Scutellariae. A method based on CZE-ECD has been developed for the determination of **16** and **17** in Scutellariae Radix and its pharmaceutical preparations [26]. The contents of **16** and **17** in the investigated samples of Radix Scutellariae were determined to be 44.4–85.5 and 10.7–20.6 mg/g, respectively. We have also developed a method based on CZE-ECD for the detection of rutin (**18**) and **15** in Flos Sophorae [44]. The assay results are well in agreement with a previous report.

In addition, CZE-ECD has been developed for the determination of puerarin (**19**) and daidzein (**20**) in Pueraria Radix and the vines of *Pueraria lobata* (Wild.) Ohwi. [28]. Figure 2 shows the

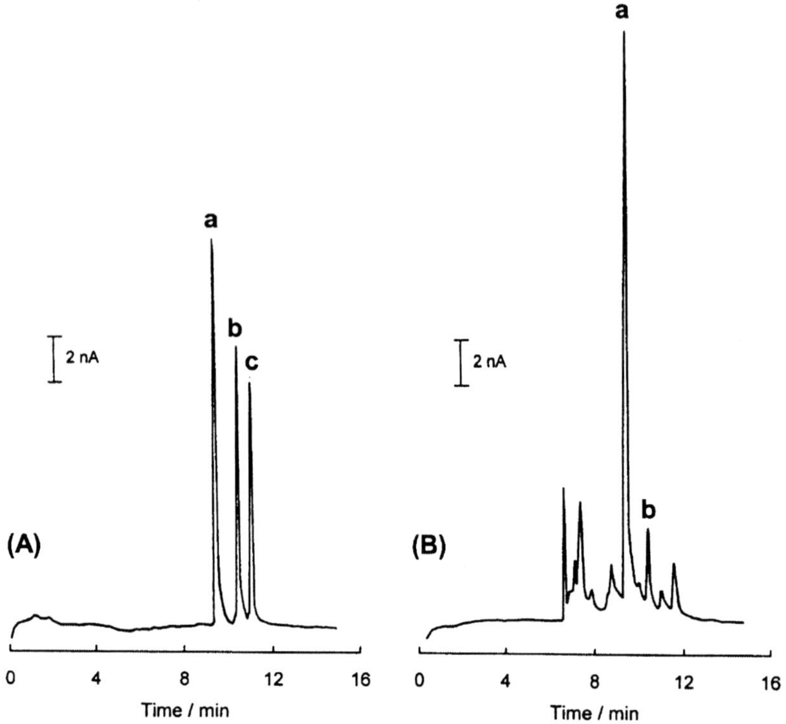

Figure 2 Electropherograms of (A) a standard mixture solution containing 0.2 mM of **19** (a), 0.1 mM **20** (b), and 0.2 mM **18** (c) and (B) the diluted extracts from Puerariae Radix under the optimum conditions. BGE, 50 mM BB (pH 9.0); separation and injection voltage, 9 kV; injection time, 6 s. Other conditions are the same as in Figure 1. Reprinted with permission from ref. [28].

electropherograms of (A) the standard mixture solution of 0.2 mM of **19**, 0.1 mM of **20** and 0.2 mM of **18** and (B) the diluted extracts from Puerariae Radix. It has been successfully applied for the determination of puerarin and daidzein in the crude drugs.

Peng et al. have determined six flavonoids (including **14**, scopoletin (**21**), umbeliferone (**22**), hyperoside (**23**), kaempferol (**24**), and **15**) in Folium Rhododendri Daurici by CZE-ECD [49]. Under the optimum conditions, the analytes could be separated on base-line in a 70 mM borate buffer (pH 9.2) within 20 min. CZE-ECD has been employed by Chu et al. for the separation and detection of seven flavonoids (**17**, naringenin (**25**), scopoletin (**26**), kaempferol (**27**), apigenin (**28**), scutellarin (**29**), luteolin (**30**)) and two phenolic acids (caffeic acid (**31**) and protocatechuic acid (**32**)) in Herba Erigerontis [50]. The nine bioactive constituents in the crude drugs could be well separated and detected within 24 min, indicating the powerful separation performance and sensitive detection capability of CE-ECD.

Anthraquinones, coumarins and alkaloids

CZE-ECD has been employed to analyze **4** and three anthraquinones (including emodin (**33**), chrysophanol (**34**), and rhein (**35**)) in Rhizoma Polygoni Cuspidati and its pharmaceutical preparations [48]. Operated in a wall-jet configuration, a carbon disc electrode was used as the detection electrode, which exhibits good response at +0.95 V (*vs.* SCE) for the four analytes. Under the optimum conditions, these analytes were well separated within 27 min in a borate buffer (pH 8.7).

Coumarins are another important class of bioactive constituents in some TCMs. You et al. have determined aesculin (**36**) and aesculetin (**37**), two coumarins, in Cortex Fraxini by CZE-ECD with LODs of 60 nM for aesculin and 300 nM for aesculetin ($S/N = 3$), respectively [24]. CE-ECD has also been employed for the determination of umbelliferone (**38**) in Flos Chrysanthemum [51], **26** in Herba Erigerontis [50], psoralen (**39**) and isopsoralen (**40**) in Radix Sileris and Fructus Psoraleae [52], respectively. Because the coumarins are difficult to be oxidized, the detection potentials were as high as +0.95 to +1.0 V (*vs.* SCE) to acquire higher sensitivity.

Alkaloids represent a crucial class of physiologically active constitutions in the crude drugs. We have quantified the content of hesperidin (**41**) in Pericarpium Citri Reticulatae [53], theophylline (**42**) in Folium Camelliae [54], and hypaphorine (**43**) in Radix Caraganae Sinicae [46] by CZE-ECD using carbon detection electrode.

Organic Acids

CE-ECD has been used for the determination of the phenolic acids in TCMs because they can be oxidized on the carbon electrode at moderate detection potentials [25, 55–59]. Chlorogenic acid (**44**) widely presents in many herbal drugs that owns the therapeutic functions of cleaning away toxic heat, detoxicating, dispelling wind-heat, promoting blood circulation, etc. The contents of **44** and **31** in a series of crude drugs have been successfully determined by CE-ECD [51, 56–60]. By using CE-ECD, Cao et al. have determined syringic acid (**45**), vanillic acid (**46**), 4-hydroxybenzoic acid (**47**), and **32** in Folium Rhododendri Daurici by CZE-ECD [25]. The four phenolic acids could be well separated from the coexistent flavonoids within 16 min. In addition, CZE-ECD has been applied in the determination of ferulic acid, rosmarinic acid, caffeic acid in Folium Perillae and Fructus Perillae, i.e. the dried leaves and the seeds of Perilla frutescens L., with satisfactory assay result [55].

Sugars

Carbohydrates are neutral compounds within the physiological pH range. Strongly alkaline separation medium (50–100 mM NaOH) was used to keep all the analytes in anionic form. The separation of these analytes by CE is based on their degrees of dissociation. Carbohydrates and their derivatives are not normally electroactive at carbon electrodes, the most commonly used detection electrode in ECD. Usually, a copper disc electrode was employed for the electrochemical detection of sugars and compounds containing nearby hydroxyl groups at constant applied potentials (+0.60 – +0.65 V (*vs*. SCE)) in 50 or 75 mM NaOH aqueous solution [27].

Cao et al. have determined gastrodin (**48**), sucrose (**49**), glucose (**50**), fructose (**51**) in Rhizoma Gastrodiae by CZE coupled with copper detection electrode [5]. The four analytes could be well separated within 17 min. Mannitol (**52**) is an active constituent in Fructus Ligustri Lucidi. Recently, a method based on CZE-ECD has been developed for the detection of the content of **52**, **49**, **50**, and **51** in Fructus Ligustri Lucidi and the leaves of *Ligustrum lucidum* Ait. [27]. It has been demonstrated that there were increases in the contents of **52** and sugars when the fruits of *Ligustrum lucidum* Ait. became ripe. CZE-ECD has also been employed for the determination of sucrose, glucose, fructose in Radix Angelicae Dahuricae, Radix Codonopsis Pilosulae, and Radix Astragali [61]. The content of paeoniflorin (**53**), **49**, **50**, and **51** in Radix Paeoniae Alba have been determined with satisfactory results by CZE-ECD [62]. Figure 3 shows the electropherograms for (A) a mixture containing 0.5 mM of **53**, **49**, **50**, and **51** and (B) the diluted extracts from Radix Paeoniae Alba. Recently, a method based on CZE-ECD has also been used to determine **53** and **49–51**, in Moutan Cortex [63]. The assayed content

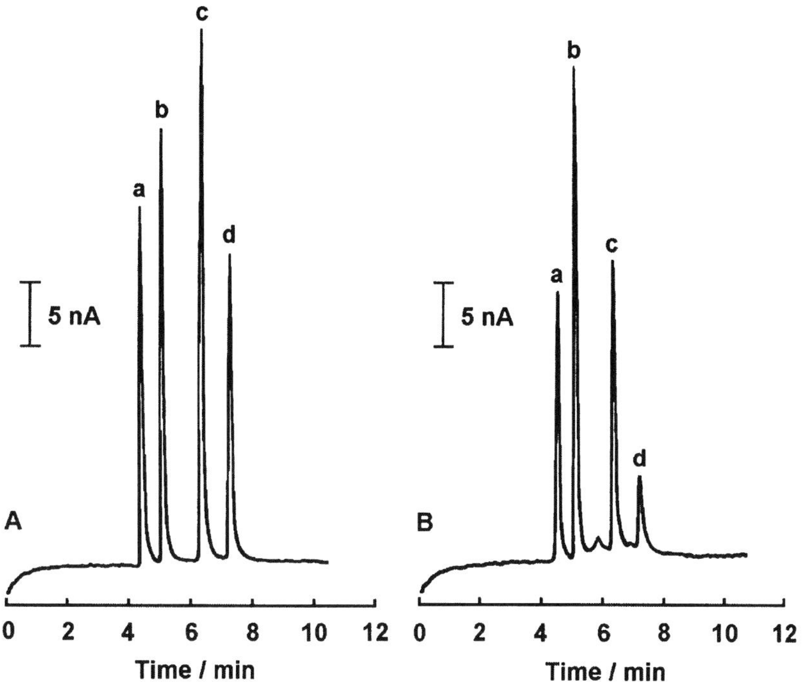

Figure 3 Electropherograms for (A) a mixture containing 0.5 mM of **53** (a), **49** (b), **50** (c), and **51** (d) and (B) the diluted extracts from Radix Paeoniae Alba. Fused-silica capillary, 25 μm i.d. × 40 cm length; detection electrode, 300 μm diameter copper disc electrode; detection potential, +0.60 V (*vs*. SCE); BGE, 75 mM NaOH; separation and injection voltage, 12 kV; injection time, 6 s. Reprinted with permission from ref. [62].

of **53** in Moutan Cortex is well in agreement with a previous report [63].

DIFFERENTIATION OF TCMS

Some bioactive compounds in TCMs were usually employed as markers for the quality evaluation. The peaks of the marker compounds in the electropherograms are the main character in the fingerprints of the crude drugs and can be employed for the purposes of identification and differentiation.

Scutellariae Radix is the root of *Scutellaria baicalensis* Georgi [43]. **16** and **17** are the main active constituents isolated from Scutellariae Radix and can be used as marker for its identification in its preparations

[43]. However, Scutellariae Radix is usually confused with Radix Astragali due to their shape, color, and also the Chinese names are similar. Radix Astragali is the root of Astragalus membranaceus (Fisch.) Bge. Var. mongholicus (Bge.) Hsiao or Astragalus membranaceus (Fisch.) Bge. and their therapeutic functions are completely different. No peak of baicalin and baicalein can be found in the electropherograms of Radix Astragali. Therefore, Radix Scutellariae can be differentiated from Radix Astragali by comparing their electropherograms. There should be apparent peaks of **16** and **17** in the electropherograms of Radix Scutellariae and its pharmaceutical preparations [43].

Recently, a method based on CZE-ECD has been developed for differentiation of Swertia Mussotii Franch from Artemisiae Capillaris Herba by Cao et al. [58]. Swertia Mussotii Franch contains a great deal of swertiamarin ((**54**), an iriodoid glycoside) and mangiferin (**55**) that are not presented in Artemisiae Capillaris Herba, whereas Artemisiae Capillaris Herba contains abundant **44**. Therefore, the two crude herbs can be differentiated by determining **54**, **55** and **44** in them. This method was successfully used to analyze and identify the crude drugs.

DETERMINATION OF BIOACTIVE CONSTITUENT IN CHINESE MEDICINAL PREPARATIONS

Nowadays, some preparations of traditional Chinese medicines have been prepared in some pharmaceutical factories in the concentrated forms such as oral liquids, capsules, and tablets for the convenience of the patients. Because many crude drugs are mixed together and decocted, the composition of the compound preparations is much more complex than that of any crude drug inside. CE has shown powerful separation performance for the analyses of the preparations of TCMs.

16 and **17** in Shuang-Huang-Lian oral liquid has been identified and determined by CZE-ECD [26]. Shuang-Huang-Lian oral liquid is a concentrated composite herbal preparation that contains Radix Scutellariae, Flos Lonicerae Japonicas, and Fructus Forsythiae. The peaks of baicalein and baicalin can be found in the electropherogram of the preparation, indicating the oral liquid contains two main constituents in Radix Scutellariae. Recently, the contents of **16**, **17** and **44** in a similar oral liquid made from Radix Scutellariae and Flos Lonicerae Japonicas has also been determined by CZE-ECD with satisfactory results [64]. Ye et al. have determined flavonoids in a series of TCM preparations by CZE-ECD [33, 65–68]. The preparation formulations are granule, tablet, and capsule. Rhizoma Gastrodiae is an important Chinese herbal drug that has been used to prepare Rhizoma Gastrodiae capsule as an ingredient. CE coupled with carbon and copper electrodes have been employed by Cao et al. for the determination of phenolic and multiple hydroxyl compounds, respectively [5]. Pan et al. has established a method based on CZE-ECD for the determination of protocatechuic aldehyde (**56**) and **32** in Radix Salviae Miltiorrhizae tablets [60]. The mean recoveries of **56** and **32** were 97.4% and 103.3%, respectively. More recently, three anthraquinones (aloe-emodin (**57**), **33**, and **35**) and two organic acids (**44**, **31**) in Huang-Dan-Yin-Chen granule have been successfully determined by CE-ECD [69]. Huang-Dan-Yin-Chen granule is a compound preparation of Radix et Rhizoma Rhei and Herba Artemisiae Scopariae and shows therapeutic performance toward liver diseases.

CONCLUSIONS AND PERSPECTIVES

Analytical methods that are simple, rapid and environmentally friendly will be highly demanded and are likely to play an important role in the development and modernization of TCMs. CE-ECD is an alternative approach to meet this requirement because of its high separation efficiency and detection sensitivity. Another important advantage of CE as an analytical tool for the analysis of plant drugs is that capillary is much easier to be regenerated to acquire high reproducibility. The ongoing maturation of CE and further developments in ECD offer a promising technique for the separation, quantitation, and identification of the bioactive constituents in TCMs, such as those described in this paper. It can be concluded that CE-ECD is simple, efficient and sensitive, providing not only a way for evaluating the quality of TCMs in marketplace, but also an alternative technique for the constituent analysis and the fingerprint investigation of TCMs.

ACKNOWLEDGMENTS

This work was financially supported by NSFC (20405002),

Shanghai Science Committee (051107089 and 2004ZR140150212), and State Education Ministry of China.

REFERENCES

[1] Nyiredy S. Separation strategies of plant constituents–current status. J. Chromatogr. B 2004; 812, 35–51

[2] Drasara P, Moravcovaa J. Recent advances in analysis of Chinese medical plants and traditional medicines. J. Chromatogr. B 2004; 812: 3–21

[3] Bent S, Ko R. Commonly used herbal medicines in the United States: A review. Am. J. Med. 2004; 116, 478–85

[4] Committee of National Pharmacopoeia. Pharmacopoeia of People's Republic of China, Chemical Industry Press: Beijing, 2005; Vol. 1, pp. 1-2

[5] Cao YH, Zhang X, Fang YZ. Ye JN. Determination of the active ingredients in Gastrodia rhizoma by capillary electrophoresis with electrochemical detection. Analyst 2001; 126, 1524–28

[6] Lianga YZ, Xie PS, Chan K. Quality control of herbal medicines. J. Chromatogr. B 2004; 812: 53–70

[7] Huang XD, Kong L, Li X, Chen XG, Guo M, Zou HF. Strategy for analysis and screening of bioactive compounds in traditional Chinese medicines. J. Chromatogr. B 2004; 812: 71–84

[8] Li FM, Sun SY, Wang J, Wang DW. Chromatography of medicinal plants and Chinese traditional medicines. Biomed. Chromatogr. 1998; 12: 78-85

[9] Lin G, Li P, Li SL, Chan SW. Chromatographic analysis of Fritillaria isosteroidal alkaloids, the active ingredients of Beimu, the antitussive traditional Chinese medicinal herb. J. Chromatogr. A 2001; 935: 321-38

[10] Fu PP, Yang YC, Xia QS, Chou MW, Cui YY, Lin G. Pyrrolizidine alkaloids-Tumorigenic components in Chinese herbal medicines and dietary supplements. J. Food Drug Anal. 2002; 10: 198-211

[11] Wang X, Kapoor V, Smythe GA. Extraction and chromatography-mass spectrometric analysis of the active principles from selected Chinese herbs and other medicinal plants. Am. J. Chin. Med. 2003; 31: 927-77

[12] Hadacek F. Secondary metabolites as plant traits: Current assessment and future perspectives. Crit. Rev. Plant Sci. 2002; 21: 273-322

[13] Jorgenson JW, Lukacs KD. Zone electrophoresis in open-tubular glass capillaries. Anal. Chem. 1981; 53: 1298-302

[14] Jorgenson JW, Lukacs KD. High-resolution separations based on electrophoresis and electroosmosis. J. Chromatogr. 1981; 218: 209-16.

[15] Morin P, Vollard F, Dreux M. Borate complexation of flavonoid-O-glycosides in capillary electrophoresis : II. Separation of flavonoid-3-O-glycosides differing in their sugar moiety. J. Chromatogr. 1993; 628: 161-9

[16] McGhie TK, Markham KR. Separation of flavonols by capillary electrophoresis-the effect of structure on electrophoretic mobility.

Phytochem. Anal. 1994; 5: 121-6
[17] Liu YM, Sheu, SJ. Determination of ephedrine alkaloids by capillary electrophoresis. J. Chromatogr. 1992; 600: 370-72
[18] Bjergegaard C, Michaelsem S, Mortensen K, Sørensen H. Determination of flavonoids by micellar electrokinetic capillary chromatography. J. Chromatogr. 1993; 652: 477-85
[19] Cen HR, Sheu SJ. Determination of glycyrrhizin and glycyrrhetinic acid in traditional Chinese medicinal preparations by capillary electrophoresis. J. Chromatogr. 1993; 653: 184-88
[20] Moodley VE, Mulholland, DA, Raynor M. Micellar electrokinetic capillary chromatography of limonoid glucosides from citrus seeds. J. Chromatogr. A 1995; 718: 187-93
[21] Ku YR, Chag LY, Ho, LK Lin J H. Analysis of synthetic anti-diabetic drugs in adulterated traditional Chinese medicines by high-performance capillary electrophoresis. J. Pharm. Biomed. Anal. 2003; 33: 329-34
[22] Liu HT, Wang KT, Chen X.G, Hu Z. Determination of rhein, baicalin and berberine in traditional Chinese medicinal preparations by capillary electrophoresis with two-marker technique. Biomed. Chromatogr. 2004; 18: 288-92
[23] Pan ZW, Chen XG, Hu ZD. Continuous capillary electrophoresis with flow injection and its application for determination of Ephedrine and pseudo-ephedrine in Chinese medicinal preparations. Biomed. Chromatogr. 2004; 18: 581-88
[24] You TY, Yang XR, Wang EK. End-column amperometric detection of aesculin and aesculetin by capillary electrophoresis. Anal. Chim. Acta 1999; 401: 29-34
[25] Cao YH, Lou, CG, Fang YZ, Ye JN. Determination of active ingredients of Rhododendron dauricum L. by capillary electrophoresis with electrochemical detection. J. Chromatogr. A 2002; 943: 153-57
[26] Chen G, Zhang HW, Ye JN. Determination of baicalein, baicalin and quercetin in Scutellariae Radix and its preparations by capillary electrophoresis with electrochemical detection. Talanta 2000; 53: 471-79
[27] Chen G, Zhang LY, Wu XL, Ye JN. Determination of mannitol and three sugars in Ligustrum lucidum Ait. by capillary electrophoresis with electrochemical detection. Anal. Chim. Acta 2005; 530: 15-21
[28] Chen G, Zhang JX, Ye, JN. Determination of puerarin, daidzein and rutin in Pueraria lobata (Wild.) Ohwi by capillary electrophoresis with electrochemical detection. J. Chromatogr. A 2001, 923: 255-62
[29] Wang A, Fang Y. Z. Applications of capillary electrophoresis with electrochemical detection in the pharmaceutical and biomedical analyses. Electrophoresis 2000; 21: 1281-90
[30] Lacher NA, Garrison KE, Martin RS, Lunte SM. Microchip capillary electrophoresis/electrochemistry. Electrophoresis 2001; 22: 2526–36
[31] Martin RS, Ratzlaff KL, Huynh BH, Lunte SM. In-channel electrochemical

detection for microchip capillary electrophoresis using an electrically isolated potentiostat. Anal. Chem. 2002; 74: 1136–43
[32] Chen G, Luo HF, Ye JN, Hu CQ. Identification and determination of oligomeric stilbenes in the roots of Caragana species by capillary electrophoresis. Plant. Med. 2001; 67: 665-8
[33] Peng YY, Liu FH, Ye JN. Identification and determination of oligomeric stilbenes in the roots of Caragana species by capillary electrophoresis. Electroanalysis, 2005; 17: 356-62
[34] Chu QC, Fu L, Wu T, Ye JN. Simultaneous determination of phytoestrogens in different medicinal parts of Sophora japonica L. by capillary electrophoresis with electrochemical detection. Biomed. Chromatogr. 2005; 19: 149-54
[35] U.S. Food and Drug Administration. Guidance for Industry Botanical Drug Products, Rockville, 2004, pp.1–48
[36] Ostergaard J, Heegaard NHH. Capillary electrophoresis frontal analysis: Principles and applications for the study of drug-plasma protein binding. Electrophoresis 2003; 24: 2903-13
[37] Manetto G, Crivellente F, Tagliaro F. Capillary electrophoresis: a new analytical tool for forensic toxicologists. Therap. Drug Monit. 2000; 22: 84-8
[38] Terabe S, Otsuka K, Ichikawa K, Tsuchiya A, Ando T. Electrokinetic separations with micellar solutions and open-tubular capillaries. Anal. Chem. 1984; 56: 111-3
[39] Cao YH, Lou CG, Zhang X, Chu QC, Fang YZ, Ye JN. Determination of puerarin and daidzein in Puerariae radix and its medicinal preparations by micellar electrokinetic capillary chromatography with electrochemical detection. Anal. Chim. Acta. 2002; 452: 123-28
[40] Terabe S, Otsuka K, Ando T. Electrokinetic chromatography with micellar solution and open-tubular capillary. Anal. Chem. 1985; 57: 834-41
[41] Ye JN, Baldwin RP. Amperometric detection in capillary electrophoresis with normal size electrodes. Anal. Chem. 1993; 65, 3525-27
[42] Holland LA, Leigh AM. Amperometric and voltammetric detection for capillary-electrophoresis. Electrophoresis 2002; 23: 3649–58
[43] Chen G, Ying XY, Ye JN. Differentiation of Scutellariae Radix from Astragali Radix by capillary electrophoresis with electrochemical detection. Analyst 2000; 125: 815-8
[44] Chen G, Zhang HW, Ye JN. Determination of rutin and quercetin in plants by capillary electrophoresis with electrochemical detection. Anal. Chim. Acta 2000; 423: 69-76
[45] Chen G, Zhang LY, Yang PY. Determination of three bioactive constituents in Moutan Cortex by capillary electrophoresis with electrochemical detection. Anal. Sci. 2005; 21: 1161-5
[46] Chen G, Luo HF, Ye JN, Hu CQ. Determination of hypaphorine and oligomeric stilbenes in the root of Caragana sinica by capillary electrophoresis with electrochemical detection. Talanta 2001; 54: 1067-76

[47] Gao LY, Chu QC, Ye JN. Determination of active ingredients of Polygonum cuspidatum Sied. et Zucc. by capillary electrophoresis with electrochemical detection. Food. Chem. 2002; 78: 255-60
[48] Chu QC, Peng YY, Ye JN. Determination of active ingredients of Polygonum cuspidatum Sied. et Zucc. by capillary electrophoresis with electrochemical detection. Electroanalysis 2004; 16: 1434-38
[49] Peng YY, Liu FH, Ye JN. Determination of bioactive flavonoids in Rhododendron dauricum L. by capillary electrophoresis with electrochemical detection. Chromatographia 2004; 60: 597-602
[50] Chu QC, Wu T, Fu LA, Ye, JN. Simultaneous determination of active ingredients in Erigeron breviscapus (Vant.) Hand-Mazz. by capillary electrophoresis with electrochemical detection. J. Pharm. Biomed. Anal. 2005; 37: 535-41
[51] Chu QC, Fu L, Guan YQ, Ye JN. Determination and differentiation of Flos chrysanthemum based on characteristic electrochemical profiles by capillary electrophoresis with electrochemical detection. J. Agri. Food. Chem. 2004; 52: 7828-33
[52] Wu F, Wang AF, Zhou Y, Zhou TS, Jiang HL, Fang YZ. Determination of psoralen and isopsoralen in traditional Chinese medicines by capillary zone electrophoresis with amperometric detection. Chromatographia 2005; 61: 157-60
[53] Chen G, Zhang LY, Zhao JQ, Ye JN. Determination of hesperidin and synephrine in Pericarpium Citri Reticulatae by capillary electrophoresis with electrochemical detection. Anal. Bioanal. Chem. 2002; 373: 169-73
[54] Chen G, Chu QC, Zhang LY, Ye JN. Separation of six purine bases by capillary electrophoresis with electrochemical detection. Anal. Chim. Acta. 2002; 457, 225-33
[55] Peng YY, Ye JN, Kong JL. Determination and differentiation of two seemingly identical medicinal herbs by capillary electrophoresis with electrochemical detection. J. Agri. Food.Chem. 2005; 53: 8141-47
[56] Chu QC, Cao YH, Peng YY, Fu L, Ye JN. Determination and differentiation of two seemingly identical medicinal herbs by capillary electrophoresis with electrochemical detection. Chromatographia, 2004; 60: 125-30
[57] Chu QC, Wu T, Fu L, Ye JN. Determination and comparison of phytoestrogens in both crude and parched Flos sophorae immaturus by capillary electrophoresis with electrochemical detection. Microchim. Acta. 2004; 148: 311-15
[58] Cao YH, Wang Y, Ye JN. Differentiation of Swertia Mussotii Franch from Artemisiae Capillaris Herba by capillary electrophoresis with electrochemical detection. J. Pharm. Biomed. Anal. 2005; 39: 60-65
[59] Pan YL, Zhang L, Chen GN. Separation and determination of protocatechuic aldehyde and protocatechuic acid in Salivia miltorrhrza by capillary electrophoresis with amperometric detection. Analyst 2001; 126: 1519-23
[60] Peng YY, Yuan JJ, Liu FH, Ye JN. Determination of active components in

rosemary by capillary electrophoresis with electrochemical detection. J. Pharm. Biomed. Anal. 2005; 39: 431-7

[61] Hu Q, Zhou TS, Hu G, Fang YZ. Determination of sugars in Chinese traditional drugs by CE with amperometric detection. J. Pharm. Biomed. Anal. 2002; 30: 1047-53

[62] Chen G, Zhang LY, Yang PY. Determination of paeoniflorin and three sugars in Radix Paeoniae Alba by capillary electrophoresis Anal. Sci. 2005; 21: 247-51

[63] Chen G, Zhang LY, Zhu YZ. Determination of glycosides and sugars in Moutan Cortex by capillary electrophoresis with electrochemical detection J. Pharm. Biomed. Anal. 2006; 41: 129-34

[64] Peng YY, Ding XH, Chu QC, Ye JN. Determination of baicalein, baicalin, and chlorogenic acid in Yinhuang Oral Liquid by capillary electrophoresis with electrochemical detection. Anal. Lett. 2003; 36: 2793-803

[65] Chu QC, Qu WQ, Peng YY, Cao QH, Ye, J. N. Determination of flavonoids in Hippophae rhamnoides L. and its phytopharmaceuticals by capillary electrophoresis with electrochemical detection. Chromatographia 2003; 58: 67-71

[66] Cao YH, Zhang X, Fang YH, Ye JN. Analysis of flavonoids in Ginkgo biloba L. and its phytopharmaceuticals by capillary electrophoresis with electrochemical detection. Mikrochim. Acta. 2001; 137: 57-62

[67] Cao YH, Chu QC, Fang YZ, Ye JN. Analysis of flavonoids in Ginkgo biloba L. and its phytopharmaceuticals by capillary electrophoresis with electrochemical detection. Anal. Bioanal. Chem. 2002; 374: 294-99

[68] Cao JH, Wang Y, Ji C, Ye J. N. Determination of liquiritigenin and isoliquiritigenin in Glycyrrhiza uralensis and its medicinal preparations by capillary electrophoresis with electrochemical detection. J. Chromatogr. A 2004; 1042: 203-09

[69] Wang AF, Zhou Y, Wu F, He PG, Fang YZ. Determination of active ingredients in Huangdan Yinchen Keli by CZE with amperometric detection. J. Pharm. Biomed. Anal. 2004; 35: 959-64

Chemical Constituents of *Aegiceras corniculatum*

Zhang Daojing, Zhang Si*, Wu Jun and Yang Jin
Guangdong Key Laboratory of Marine Materia Medica, South China Sea Institute of Oceanology, the Chinese Academy of Sciences, 164 West Xingang Road, Guangzhou 510301, China
**Corresponding email: sizhmd@scsio.ac.cn*

INTRODUCTION

Aegiceras corniculatum, a mangrove plant of the family Myrsinaceae, is a small tree widely distributed in Asia and Australia. In mainland China, *A. corniculatum* is one of two species of the genus and it's widely distributed along the coastline and in tropical and subtropical areas. The bark and seeds of *A. corniculatum* can be used to poison fish. Previous chemical investigations on this plant demonstrated the presence of saponins, sterols, flavones, terpenoids and hydroquinones [1- 6]. This paper deals with the isolation and structural elucidation of twelve compounds from the stem bark of *A. corniculatum*. A new compound (**1**) was isolated and elucidated as 16α, 28-dihydroxyl-3-oxo-12-oleanene on the basis of spectroscopic techniques. The assignment of ^{1}H and ^{13}C NMR data for protoprimulagenin (**2**) was confirmed by detailed NMR analysis, while those of ^{1}H and ^{13}C NMR data for other known compounds were based on literature [5, 7-12].

RESULTS AND DISCUSSION

The EtOH extract of powdered dried bark of *A. corniculatum* was evaporated to dryness, then suspended in water and partitioned with EtOAc. The EtOAc-soluble extract was subjected to column chromatography over silica gel and Sephadex LH-20 to afford compounds **1-12**, of which one was determined as a new natural product. The ^{1}H and ^{13}C NMR resonance assignments of new compound and some known compounds were made by acquiring 1D ^{1}H, 1D ^{13}C, DEPT, ^{1}H-^{1}H COSY, HSQC, and HMBC datasets as required.

The EI-MS molecular ion *m/z* 456 [M^{+}], combined with ^{13}C NMR data and a DEPT experiment established the molecular formula of $C_{30}H_{48}O_3$ for **1**. NMR spectral data were closely related to those of the earlier reported primulagenin A [8, 9]. However, the ^{13}C NMR spectrum of **1** showed a ketone at δ 217.7 and lacked one of the oxygenated methine carbons present in the spectrum of primulagenin A [8, 9]. The ketone at δ 217.7 had the distinct HMBC correlations with Me-23 (δ 1.08), Me-24 (δ 1.06) and H-2 (δ 2.43, 2.51), which suggested **1** was an oxidized derivative of primulagenin A at position 3. The presence of a double bond was evident in the spectra. It involved a quaternary carbon atom, since the only signal was observed in the ^1H NMR at δ 5.34, appearing as a triplet (*J*=3.5 Hz), and was thus assigned between C-12 and C-13, like in primulagenin A [8, 9]. All the other assignments were based on primulagenin A and corroborated by 2D NMR experiments.

Experimental Section

1. General Experimental Procedures

NMR spectra were recorded on the Bruker AVANCE 500 Spectrometer (^1H: 500 MHz and ^{13}C: 125 MHz) with TMS as internal standard. The EI-MS was carried out on a QP5050A Spectrometer. Melting points were determined on an X_4 apparatus and are not uncorrected. Optical rotations were measured with an AA-10R digital polarimeter. IR spectra were measured on a Nicolet 5DX-FTIR spectrophotometer. Silica gel (200-300 mesh) from Qingdao Haiyang Chem. Ind. Co. Ltd. and Sephadex LH-20 from Pharmacia were used for column chromatography (CC). The solvent systems for chromatography: I, n-hexane-EtOAc (99:1; 49:1; 20:1; 10:1; 5:1; 2:1); II, n-hexane-EtOAc-MeOH (200:100:1; 200:100:2; 200:100:5; 200:100:10; 200:100:20; 200:100:50); III, n-hexane-$CDCl_3$-MeOH (1:1:2); IV, 100% MeOH. TLC was visualized by spraying 5% H_2SO_4 or 5% phosphomolybdic acid in 95% ethanol.

2. Plant Material

The stem bark of *A. corniculatum* was collected in October 2002 from Sanya of Hainan Province, China. The plant sample was identified by Prof. Si Zhang, at South China Sea Institute of Oceanology, the Chinese Academy of Sciences (CAS). A voucher sample (GLMMM004) is kept in the Herbarium of South China Sea Institute of Oceanology, CAS.

3. Extraction and Isolation

The powdered dried bark (3.0 kg) of *A. corniculatum* was soaked with 95% EtOH (4 L) at room temperature (72 h), and the process was repeated three times. After evaporating the solvent, the residue (240 g) was suspended in water and defatted with petroleum ether (bp. 60-90 °C). The aqueous layer was further extracted with EtOAc (3 L x 3 times) and *n*-BuOH (3 L x 3 times), respectively. The EtOAc fraction (24 g) was subjected to CC (Φ 4 x L 120 cm) of silica gel (720 g) eluted with n-hexane/ EtOAc systems from system I-II to give 150 fractions (each 400 mL). Fractions 64 and 67 were subjected to Sephadex LH-20 column (Φ 2 x L 45 cm) with solvent IV to yield **1** (10 mg) and **5** (10 mg), respectively. Fraction 71 was repeatedly subjected to Sephadex LH-20 column (Φ 2 x L 45 cm) with solvent IV to yield **2** (15 mg) and **6** (30 mg) from the 8th and 10th subfraction (each 10mL). Fractions 57 to 60 were combined and separated by Sephadex LH-20 column (Φ 2 x L 100 cm), eluted with solvent III to afford **3** (8 mg) from the 6th subfraction (each 10 mL). Fractions 74 to 76 were combined and separated by Sephadex LH-20 column (Φ 2 x L 100 cm), eluted with solvent IV to afford **7** (8 mg) from the 10th subfraction (each 10 mL). Fractions 70, 101 and 135 were subjected to Sephadex LH-20 column (Φ2 xL 45 cm) with solvent IV to give compounds **4** (6 mg), **8** (10 mg) and **9** (20 mg), respectively. Fractions 46, 108 and 119 were subjected to Sephadex LH-20 column (Φ2 x L 45 cm) with solvent IV to give compounds **10** (15 mg), **11** (20 mg) and **12** (40 mg), respectively.

4. Identification

16α, 28-dihydroxyl-3-oxo-12-oleanene (**1**). Colorless needles ($CHCl_3$). IR (KBr) cm^{-1}: 3395(OH), 2934, 2856, 1704 (C=O), 1642 (C=C), 1591, 1461, 1383, 1281, 1080, 1022, 799. ^{1}H NMR (500 MHz, $CDCl_3$): δ 0.91-0.92 (6H, s, Me-29, Me-30), 0.98 (3H, s, Me-26), 1.06 (6H, s, Me-24, Me-25), 1.08 (3H, s, Me-23), 1.35 (3H, s, Me-

27), 1.21-1.92 (m, 18H), 2.07 (1H, dd, *J*=14.9, 2.1 Hz, H-19), 2.43 (1H, ddd, *J*=15.5, 7.5, 4.0 Hz, H-2), 2.51 (1H, ddd, *J*=17.5, 7.5, 2.5 Hz, H-2), 3.35 (2H, s, H-28) 4.05 (1H, t, *J*=4.5 Hz, H-16) 5 . 3 4 (1H, t, *J*=3.5 Hz, H-12). ^{13}C NMR (125 MHz, $CDCl_3$): see Table 1. EIMS m/z (rel. int., 70 eV): 456 $[M]^+$, 439 $[M\text{-}OH]^+$, 248, 235, 203.

Protoprimulagenin (**2**). Colorless needles ($CHCl_3$). mp. 269-271 !, [±]25 D25 D+16.0° ($CHCl_3$, c 0.75). IR (KBr) cm^{-1}: 3410 (OH), 2925, 2859, 1448, 1384, 1363, 1301, 1259, 1182, 1117 (C-O-C), 1098 (C-O-C), 1040 (C-O-C), 985, 940, 880. ^{1}H NMR (500 MHz, $CDCl_3$): δ 0.68 (1H, d, *J*=11.0 Hz H-5), 0.77 (3H, s, Me-24), 0.87 (3H, s, Me-25), 0.91 (3H, s, Me-30), 0.94 (1H, m, H-1), 0.98 (6H, s, Me-29 and Me-23), 1.12 (1H, m, H-15ax), 1.16 (3H, s, Me-27), 1.21 (2H, m, H-9 and H-7), 1.21 (3H, s, Me-26), 1.23-1.77 (m, 13H), 1.89 (2H, td, *J*=13.5, 5.0 Hz, 2H-21), 1.97 (1H, m, H-12), 2.17 (1H, brdd, *J*=14.5, 5.5 Hz, H-15eq), 2.26 (1H, dd, *J*=14.9, 2.1 Hz, H-19), 3.17 (1H, d, *J*=7.5 Hz, H-28), 3.21 (1H, dd, *J*=11.4, 5.0 Hz, H-3), 3.50 (1H, d, *J*=7.5 Hz, H-28), 3.98 (1H, d, *J*=5.3 Hz, H-16). ^{13}C NMR (125 MHz, $CDCl_3$): see Table 1. EIMS m/z (rel. int., 70 eV): 458 $[M]^+$, 441 $[M\text{-}OH]^+$, 426 $[M\text{-}CH_2OH]^+$, 248, 410, 409, 385, 249, 236, 220, 219, 207, 189. According to the HMBC correlation of C-7 with H-5 (δ 0.68), C-12 with H-11 (δ 1.59), and C-22 with H-16 (δ 3.98) and H-28ax (δ 3.17), the assignments of C-7, C-12 and C-22 were corrected. The assignments of C-24 and C-25 were not correct in the literature [9, 13-15]. The assignments of Protoprimulagenin (**2**) were based on the HMBC, HSQC and ^{1}H-^{1}H COSY experiment and literatures [7-8, 16].

Embelinone (**3**). Colorless needles ($CHCl_3$). mp. 257-259 °C, $[\alpha]^{25}_{D}$-4° ($CHCl_3$, c 0.6). IR (KBr) cm^{-1}: 2949, 2859, 1704, (C=O), 1436, 1384, 1233, 1121 (C-O-C), 1080 (C-O-C), 1045 (C-O-C), 990. ^{1}H NMR (500 MHz, $CDCl_3$): δ 0.86 (3H, s, Me-30), 0.94 (3H, s, Me-29), 1.00 (3H, s, Me-25), 1.04 (6H, s, Me-23 and Me-27), 1.08 (3H, s, Me-24), 1.15-1.19 (2H, m, H-12 and H-22), 1.21-1.26 (2H, m, H-9 and H-21), 1.28 (3H, s, Me-26), 1.35-1.60 (m, 10H), 1.78 (1H, ddd, *J*=17.9, 13.5, 4.9 Hz, H-11ax), 1.88 (1H, d, *J*=16.0 Hz, H-15), 1.96 (1H, dd, *J*=11.3, 2.5 Hz, H-18), 1.98-2.00 (2H, m, H-7, H-1), 2.16 (1H, ddd, *J*=13.3, 5.1, 2.5 Hz, H-22eq), 2.42 (1H, ddd, *J*=15.7, 7.2, 4.2 Hz, H-2eq), 2.53 (1H, ddd, *J*=15.7, 10.3, 7.4 Hz, H-2ax), 2.72 (1H, d, *J*=16.0 Hz, H-15), 3.47 (1H, d, *J*=8.3 Hz, H-28), 3.89 (1H, d,

J=8.3 Hz, H-28). ^{13}C NMR (125 MHz, $CDCl_3$): see Table 1. EIMS m/z (rel.int., 70 eV): 454 $[M]^+$, 424 $[M-CH_2O]^+$, 383, 269, 248, 235, 219, 203. The assignment of C-22 was confirmed in the literature [9], while other data were identical with those of embelinone [9].

Aegicerin (**4**). Colorless needles ($CHCl_3$), mp. 252-254 °C, $[\alpha]^{25}_{D}$ -24.5° ($CHCl_3$, c 0.86). IR (KBr) cm^{-1}: 3382(OH), 2951, 2864, 1701, 1456, 1386, 1216, 1122 (C-O-C), 1077 (C-O-C), 1033 (C-O-C), 990, 757. ^{1}H NMR (500 MHz, $CDCl_3$): δ 0.67 (1H, d, *J*=11.0 Hz, H-5), 0.77 (3H, s, Me-25), 0.78 (3H, s, Me-30), 0.86 (3H, s, Me-24), 0.89 (1H, m, H-1), 0.91(3H, s, Me-29), 0.94 (3H, s, Me-23), 1.03 (3H, s, Me-27), 1.12 (1H, m, H-12), 1.16 (2H, m, H-9 and H-22), 1.23 (3H, s, Me-26), 1.23-1.62 (m, 11H), 1.69 (1H, ddd, *J*=18.1, 11.0, 3.6 Hz, H-11ax), 1.75 (1H, m, H-1ax), 1.86 (1H, d, *J*=16.1 Hz, H-15), 1.96 (1H, ddd, *J*=17.3, 12.5, 3.5 Hz, H-7ax), 1.97 (1H, dd, *J*=11.5, 3.0 Hz, H-18), 2.16 (1H, ddd, *J*=13.5, 5.0, 2.4 Hz, H-22eq), 2.71 (1H, d, *J*=16.1 Hz, H-15), 3.20 (1H, dd, *J*= 11.0, 4.5 Hz, H-3), 3.46 (1H, d, *J*=8.3 Hz, H-28), 3.88 (1H, d, *J*=8.3 Hz, H-28). ^{13}C NMR (125 MHz, $CDCl_3$): see Table 1. EIMS m/z (rel. int., 70 eV): 456 $[M]^+$, 439 $[M-OH]^+$, 426 $[M-CH_2O]^+$, 248, 235, 217, 202. These data were identical with those of aegicerin [9].

ACKNOWLEDGEMENTS

This work was financially supported by a grant (code: KZCX3-SW-216) from Important Project of CAS, and a grant (code: 2003-1) from the Guangdong National Fundation. Mass spectra were recorded in the Institute of Chemistry, CAS.

REFERENCES

[1] Rao KV, Bose PK. J Indian Chem. Soc. 1959; 36: 358-60.
[2] Rao KV, Bose PK. Tetrahedron. 1962; 18: 461-4.
[3] Rao KV. Tetrahedron. 1964, 20: 973-7.
[4] Hensens OD, Lewis KG. Aust J Chem. 1966;19: 169-74.
[5] Gomez E, Cruz-Giron ODL. J. Nat. Prod. 1989; 52: 649-51
[6] Xu MJ, Deng ZW, Li M, Li J. J. Nat. Prod. 2004; 67: 762-6
[7] Jia Z, Koike K, Ohmoto, T. Phytochemistry.1994; 37: 1389-96
[8] Ohtani K, Mavi S, Hostettmann K. Phytochemistry. 1993; 33: 83-6

[9] Machocho AK, Kiprono PC, Grinberg S, Bittner S. Phytochemistry. 2003; 62: 573-7
[10] Sadtler Research Laboratories. Sadtler Stardard NMR Spectra [Z]. Pennsylvania, USA: Sadtler Research Laboratories, 1980, p. 3158
[11] Li J L, Li JS, Wang AQ. Chin. Tradit. Herb. Drugs. 1998; 29: 721-5
[12] Markham KR, Ternal B, Stanly R. Tetrahedron. 1978; 34: 1389-94
[13] Hisashi K, Noriko S, Akiko H, Haruo O. Phytochemistry. 1990; 29: 2351-5
[14] Kohda H, Takeda O, Tanaka S. Chem. Pharm. Bull. 1989; 37: 3304-5
[15] Mahato SB, Kundu AP. Phytochemistry. 1994; 37: 1517-75
[16] HegdeV R, Siiver J, Das PR. J. Nat. Prod. 1995; 58: 1492-7

Table 1. ^{13}C NMR spectral data of **1-4** (125MHz, $CDCL_3$)*

C No	**1**	**2**	**3**	**4**	**2**[9,13-15]	**2**[7-8,. 16]
1	39.7 *t*	39.0 *t*	39.7 *t*	39.1 *t*	38.7 *t*	39.9 *t*
2	34.1 *t*	27.4 *t*	34.1 *t*	27.5 *t*	27.3 *t*	28.8 *t*
3	217.7 *s*	79.0 *d*	217.6 *s*	79.0 *d*	79.0 *d*	78.1 *d*
4	47.6 *s*	38.9 *s*	47.4 *s*	39.0 *s*	38.9 *s*	39.6 *s*
5	55.2 *d*	55.2 *d*	54.9 *d*	54.6 *d*	55.1 *d*	55.8 *d*
6	19.4 *t*	17.8 *t*	18.9 *t*	17.7 *t*	17.7 *t*	18.3 *t*
7	32.2 *t*	34.1 *t*	31.7 *t*	31.7 *t*	31.0 *t*	31.9 *t*
8	40.5 *s*	42.1 *s*	42.6 *s*	42.8 *s*	42.0 *s*	42.6 *s*
9	48.8 *d*	50.2 *d*	49.3 *d*	49.8 *d*	50.1 *d*	50.7 *d*
10	36.9 *s*	37.0 *s*	36.7 *s*	37.0 *s*	36.9 *s*	37.3 *s*
11	24.1 *t*	18.7 *t*	19.0 *t*	18.6 *t*	18.6 *t*	19.3 *t*
12	122.6 *d*	32.4 *t*	33.0 *t*	33.8 *t*	34.0 *t*	34.6 *t*
13	145.3 *s*	86.5 *s*	86.2 *s*	86.3 *s*	86.4 *s*	86.5 *s*
14	42.7 *s*	44.0 *s*	49.7 *s*	49.8 *s*	43.9 *s*	44.7 *s*
15	35.3 *t*	36.9 *t*	45.4 *t*	45.5 *t*	36.8 *t*	37.0 *t*
16	74.8 *d*	78.2 *d*	213.3 *s*	213.4 *s*	77.4 *d*	77.2 *d*
17	41.8 *s*	44.2 *s*	56.1 *s*	55.3 *s*	44.1 *s*	44.7 *s*
18	48.3 *d*	50.8 *d*	54.5 *d*	54.6 *d*	50.6 *d*	51.6 *d*
19	46.5 *t*	38.8 *t*	40.0 *t*	40.1 *t*	38.8 *t*	39.1 *t*
20	31.6 *s*	31.6 *s*	31.7 *s*	31.7 *s*	31.5 *s*	31.9 *s*
21	37.0 *t*	36.7 *t*	35.2 *t*	35.3 *t*	36.6 *t*	36.9 *t*
22	31.3 *t*	31.0 *t*	24.8 *t*	24.8 *t*	32.3 *t*	33.0 *t*
23	28.2 *q*	28.0 *q*	26.5 *q*	28.1 *q*	27.9 *q*	28.7 *q*
24	15.7 *q*	15.4 *q*	21.1 *q*	16.0 *q*	16.1 *q*	16.6 *q*
25	15.4 *q*	16.2 *q*	15.7 *q*	15.3 *q*	15.3 *q*	16.4 *q*
26	18.0 *q*	18.2 *q*	18.2 *q*	18.6 *q*	18.1 *q*	18.7 *q*
27	27.3 *q*	19.4 *q*	21.5 *q*	21.7 *q*	19.4 *q*	19.6 *q*
28	70.8 *t*	77.5 *t*	75.2 *t*	75.2 *t*	78.1 *t*	78.0 *t*
29	33.8 *q*	33.5 *q*	33.3 *q*	33.3 *q*	33.4 *q*	33.8 *q*
30	25.2 *q*	24.5 *q*	23.6 *q*	23.6 *q*	24.3 *q*	24.8 *q*

*Multiplicities were determined by DEPT experiment.

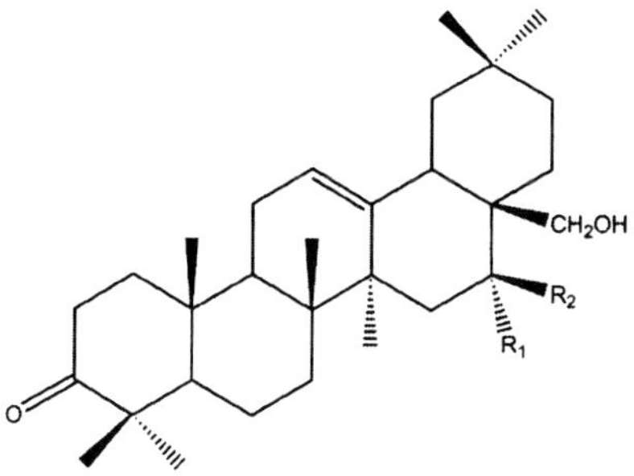

1 R_1=OH; R_2=H;

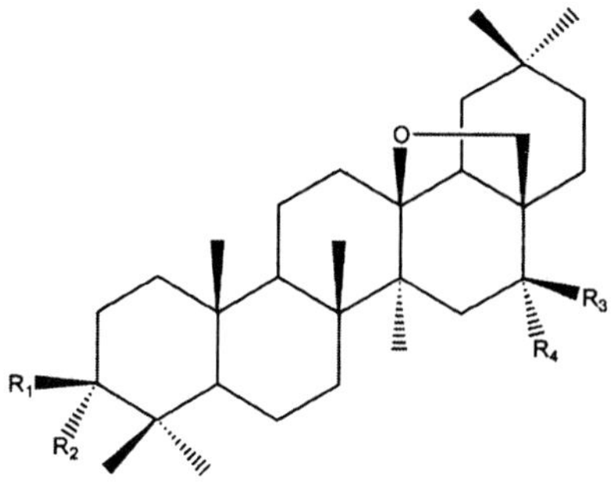

2 R_1= R_4=OH; R_2= R_3=H
3 R_1+R_2=R_3+R_4=O
4 R_1=OH; R_2=H; R_3 +R_4=O

Figure 1. Structures of Compounds **1-4**

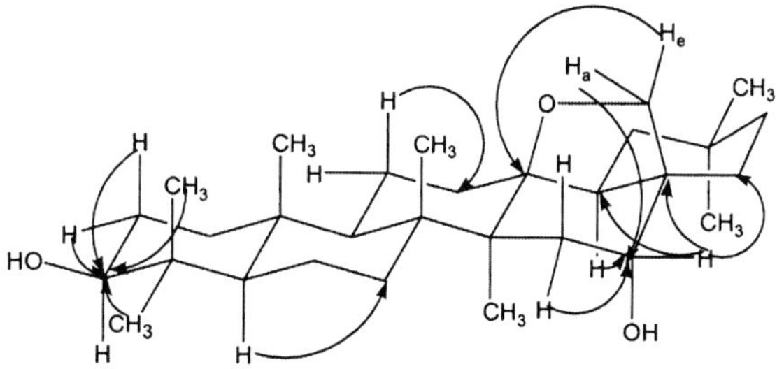

Figure 2. The major HMBC correlations of compound **2**

Norditerpenoids from the Soft Coral *Nephthea chabroli*

Wei-Han Zhang[a]*, Kevin Pan[a] and Chun-Tao Che[b]
[a]Hutchison Medipharma Ltd., Shanghai, China; [b]School of Chinese Medicine, the Chinese University of Hong Kong, Shatin, Hong Kong
**Corresponding email:* weihanz@hmplglobal.com

INTRODUCTION

Marine organisms contain plenty of secondary metabolites and are a rich source of chemical structures with bioactivities [1]. In previous papers [2-4], we have reported the metabolites from *Nephthea chabroli* (phylum *Cnidaria*, class *Anthoza*, order *Alcyonacea*, family *Nephtheidae*). Further research work has led to the isolation of two new norditerpenoids (chabrolone A and chabrolol E). Their structures were elucidated on the basis of spectral data (MS, 1D NMR and 2D NMR) and a comparison with similar structures.

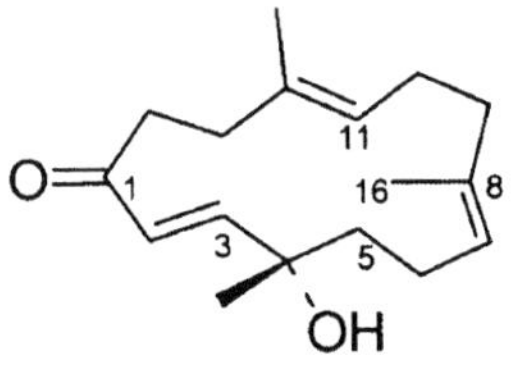

Chabrolone A

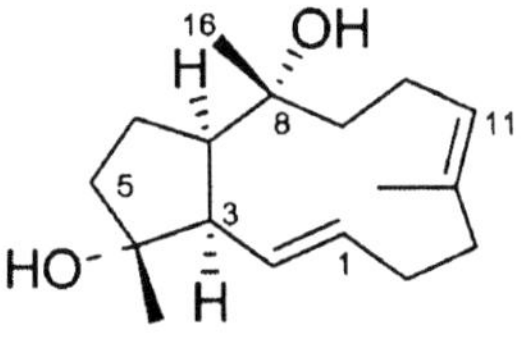

Chabrolol E

MATERIALS AND METHODS

General Experimental Procedures

All solvents used were of analytical grade. Silica gel (230-400 mesh), TLC grade silica gel, and RP-18 silica gel were used for chromatography. NMR spectra were recorded on JEOL JNM-EX-400-FT-NMR spectrometer, the IR absorption spectra on Perkin Elmer 16 PC FT-IR spectrometer and mass spectra on Finnigan TSQ 7000 mass spectrometer. Optical rotations were measured on a Perkin Elmer 241 polarimeter.

Animal Materials

The soft coral, *Nephthea chabroli*, was collected in Xisha Island, the South China Sea and identified by Professor Li Chupu (South China Sea Institute of Oceanography, the Chinese Academy of Sciences, PRC). A voucher specimen was deposited in the Research Center of Organic Natural Products Chemistry, Zhongshan (Sun Yat Sen) University, Guangzhou, China.

Extraction and Isolation

A sample of *Nephthea chabroli* (5 Kg, dry weight) was cut into small species and extracted with EtOH (5 × 5 L) at room temperature. The extract was evaporated to dryness under reduced pressure to obtain a residue that was partitioned between H_2O and $CHCl_3$ to afford a $CHCl_3$ fraction. The $CHCl_3$ fraction (70 g) was taken to dryness in vacuum. Vacuum liquid chromatography of the residue on silica gel using a solvent gradient of increasing polarity (hexane-acetone: 100:8, 100:16, 100:32, 100:64, 0:100) afforded eight fractions (A-H) as judged by TLC. Fraction B was subjected to silica gel chromatography (eluted with gradients of hexane-ethyl acetate mixture) and reversed phase column [RP-18 gel] (eluted with gradients of CH_3OH-H_2O) to afford chabrolone A (21 mg) and chabrolol E (23 mg).

Chabrolone A: viscous liquid. $C_{17}H_{26}O_2$, $[\alpha]^{25}_D$ +2.8° (c 0.029, $CHCl_3$); UV ($CHCl_3$) λ_{max} 245 nm (log ε 2.47); IR (KBr) $_{max}$ 3445, 2985, 1680, 1640 cm^{-1}; ^{13}C and ^1H NMR data, see Tables 1 and 2; EIMS *m/z* 244 (22), 229 (8), 177 (16), 161 (24), 121 (44), 109 (64), 95 (96), 55 (100).

Chabrolol E: viscous liquid. $C_{17}H_{28}O_2$; $[\alpha]^{25}_D$ + 6.5° (c 0.018, $CHCl_3$); UV ($CHCl_3$) λ_{max} 242 nm (log ε 1.4); IR (KBr) $_{max}$ 3450, 2980, 1640 cm^{-1}; ^{13}C and ^1H NMR data, see Tables 1 and 2; EIMS *m/z* 264 (5), 248 (18), 216 (22), 202 (22), 156 (22), 130 (28), 95 (22), 57 (100).

RESULTS AND DISCUSSION

Chabrolone A was isolated as a viscous liquid. The molecular formula $C_{17}H_{26}O_2$ was deduced from the ^{13}C NMR (Table 1) and DEPT spectral data, combined with EIMS *m/z* 244 [M^+], indicating five degrees of unsaturation. The IR spectrum displayed absorption bands at ν_{max} 3445, 2985, 1680, 1640 cm^{-1}, suggesting the presence of hydroxyl, olefinic and α, β-unsaturated ketone functionalities. The UV ($CHCl_3$) absorption at λ_{max} 245 nm (log ε 2.47) indicated an α, β-unsaturation ketone structure.

In the ^1H NMR spectrum, proton resonance signals were observed for a disubstituted double bond at δ_H 6.84 (d, *J* = 16.0 Hz, H-3), 6.30 (d, *J* = 16.0 Hz, H-2), two trisubstituted double bonds at δ_H 4.90 (t, *J* = 4.5 Hz, H-7), 4.80 (t, *J* = 4.0 Hz, H-11) and three methyl groups at δ_H 1.60 (s, H-17), 1.51 (s, H-16), 1.37 (s, H-15). The ^{13}C NMR spectrum showed carbon resonance due to α, β-unsaturated ketone group at δ_C 200.9 (s, C-1), 152.1 (d, C-3), and 125.9 (d, C-2). It also revealed the presence of olefinic carbons at δ_C 134.1 (s, C-8), 126.1 (d, C-7), 132.5 (s, C-12), and 123.1 (d, C-11), as well as an oxygenated quaternary carbon at δ_C 73.1 (s, C-4). Three double bonds and one ketone accounted for four degrees of unsaturation, and the molecule is monocyclic. The connectivity was established by HMBC experiment. Thus, the HMBC correlations between a carbonyl carbon signal at δc 200.9 (C-1) and δ_H 6.84 (d, *J* = 16 Hz, H-3) and 6.30 (d, *J* = 16.0 Hz, H-2) led to the assignment of a partial structure of α, β-unsaturated ketone at C_1-C_2-C_3. In addition, C-3 (δc 152.1) is long-ranged coupled to δ_H 1.37 (s, H-15) and 1.83 (m, H-5); C-4 (δc 73.1) is coupled to δ_H 1.83 (m, H-5), 1.37 (s, H-15), 6.84 (d, *J* = 16 Hz, H-3) and 6.30 (d, *J* = 16.0 Hz, H-2). This information allowed the establishment of the linkage C_3-C_4-C_5. On the other hand, H-H COSY showed the spin system corresponding to H_5-H_6-H_7 linkage. In addition, HMBC cross peaks were observed between δc 126.1 (C-7) and H-16 (δ_H 1.51) / H-5 (δ_H 1.83), between δc 134.1 (C-8) and H-16 (δ_H 1.51) / H-9 (δ_H 1.86), revealing the linkage of C_7-C_8-C_9. The COSY spectrum displayed coupling between H-9 / H-10, H-10 / H-11 and H-11 / H-17. The H-17 long-range coupled to δc 132.5 (C-12), which in turn coupled to signals at δ_H 2.52 (m, H-14) and 2.40 (m, H-13). The linkage of C-14 to C-1 was established by the HMBC observation that δc 200.9 (C-1) correlated with δ_H 2.70 (m, H-14), 2.52 (m, H-14) and 2.40 (m, H-13). All these evidences suggested a cembrane-type 14-membered ring.

The coupling constant between H-2 / H-3 (*J* = 16.0 Hz) suggested a *trans* geometry. The stereochemistry of the other two double bonds was assigned by ^{13}C NMR data. The 14-membered ring of the cembrane skeleton is large enough that transannular effects are small, hence the chemical shift criteria for establishing the stereochemistry of tri-substituted double bonds in acyclic polyisoprenoids can be applied to cembranes. In an acyclic system, the chemical shift for a methyl carbon on an *E*-trisubstituted double bond is 15-16 ppm, while that of a methyl carbon on a *Z*-trisubstituted double bond is 23-24 ppm [5]. The ^{13}C NMR chemical shifts of olefinic methyls in the compound are δc 15.4 (C-16) and 17.0 (C-17), respectively, consistent with *E* stereochemistry. The relative stereochemistry at position 4 was established by DIFNOE experiment. The observed NOE enhancement between H-15 and H-2 suggested the hydroxyl to be α-form. The new structure, chabrolone A, was thus established as shown.

Chabrolol E was obtained as a viscous liquid. Its molecular formula of $C_{17}H_{28}O_2$, deduced from ^{13}C NMR (Table 1) and DEPT spectral data, combined with EIMS *m/z* 264 [M^+], indicated four degrees of unsaturation. The IR spectrum displayed absorption bands at ν_{max} 3450, 2980, 1640 cm^{-1}, suggesting the presence of hydroxyl and olefinic groups. The DEPT spectra displayed signals for three methyls [δ_C 31.7 q, 23.9 q, 19.0 q], six methylenes [δ_C 39.2 t, 36.0 t, 33.9 t, 27.4 t, 23.4 t, 22.9 t], five methines [δ_C 131.7 d, 131.5 d, 127.7 d, 56.2 d, 50.1 d] and four quaternary carbons [δ_C 134.3 s, 81.2 s, 74.8 s]. The ^1H NMR spectrum showed three olefinic proton signals at δ_H 5.55 (ddd, *J* = 16.8, 5.6, 11.2 Hz), 5.38 (dd, *J* = 16.8, 1.5 Hz), 5.34 (m), and three methyl proton signals at δ_H 1.67 (s), 1.14 (s) and 1.12 (s). The ^{13}C NMR spectrum further exhibited signals for two double bonds at δ_C 134.3 (s), 127.7 (d), 131.7 (d), and 131.5 (d). Two double bonds accounted for two degrees of unsaturation, indicating the molecule to be bicyclic.

The DQF H-HCOSY results exhibited couplings between δ_H 5.55 (ddd, *J* = 16.8, 5.6, 11.2 Hz, H-1) and 5.38 (dd, *J* = 16.8, 1.5 Hz, H-2) which, in turn, coupled to a signal at δ_H 2.64 (m, H-3). This information led to the assignment of H_1-H_2-H_3. On the other hand, in HMBC spectrum, H-2 was long-range coupled to an oxygenated carbon at δc 81.2, which was further long range coupled to H-3, H-5 (δ_H 1.74, m), H-6 (δ_H 1.32, m) and H-15 (δ_H 1.12, s). The carbon signal was assigned to C-4. In addition, the couplings between H-3 and H-7, H-7 and H-6, as well as H-6 and H-5 in the H-H COSY spectrum aided the establishment of the following partial cyclopentane structure:

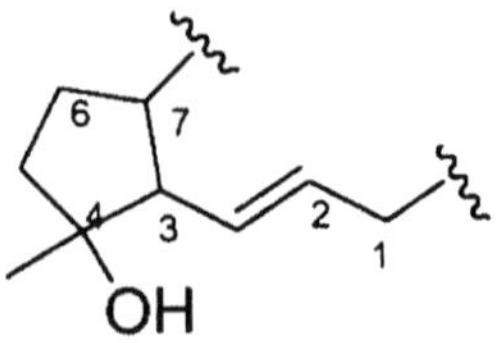

In the HMBC spectrum, H-7 was long-range coupled to the carbon signal at δc 74.8 (C-8), which was further coupled with signals at δ_H 2.64 (m, H-3), 2.36 (m, H-10), 1.14 (s, H-16) and 1.80 (m, H-9), thus establishing the connection of C_8-C_9-C_{10}. Connectivity was then established among H_9-H_{10}-H_{11} through H-H COSY results. The H-17 (δ_H 1.67, s) further long-range coupled to C-12 (δc 134.3). The resonance of C-12 showed long-range coupling with δ_H 2.36 (m, H-10) and 2.02 (m, H-14). In addition, the correlations between δ_C 27.4 (C-14) and δ_H 5.55 (ddd, *J*=16.8, 5.6, 11.2 Hz, H-1) and 5.38 (dd, *J*= 16.8, 1.5 Hz, H-2), led to the connection of C_{14}-C_1-C_2. The ring structure was thus 11-membered.

That the cyclopentane ring fused to the 11-membered ring at C-3 and C-7 positions was deduced by the observation of relevant HMBC correlations. Thus, C-8 (δc 74.8) was coupled to δ_H 2.64 (m, H-3), 1.62 (m, H-6) and 1.96 (m, H-7). On the other hand, C-4 (δc 81.2) was coupled to δ_H 5.38, 2.64 (m, H-3), 1.74 (m, H-5) and 1.96 (m, H-7). These data suggested that chabrolol E possessed the same skeleton as chabrolol B [2].

The relative stereochemistry was established by DIFNOE experiments. NOEs observed between H_3/H_7, H_3/H_{15}, H_7/H_{16} suggested the *syn* stereochemistry between H-3 and H-7, *anti* stereochemistry between H-3 and H-15, and *anti* between H-7 and H-16, respectively. In addition, the olefinic methyl at δ_C 19.0 (C-17) suggested the *E* configuration between C_{11}-C_{12}. The new structure was named chabrolol E. It is a novel norditerpene.

Since a series of norditerpenoids (chabrolone A, chabrolol E, chabrolol A, chabrolol B, chabrolol C) have been identified from *Nephthea chabroli*, a possible biogenetic relationship (oxidation, reduction and cyclization) was proposed as follows:

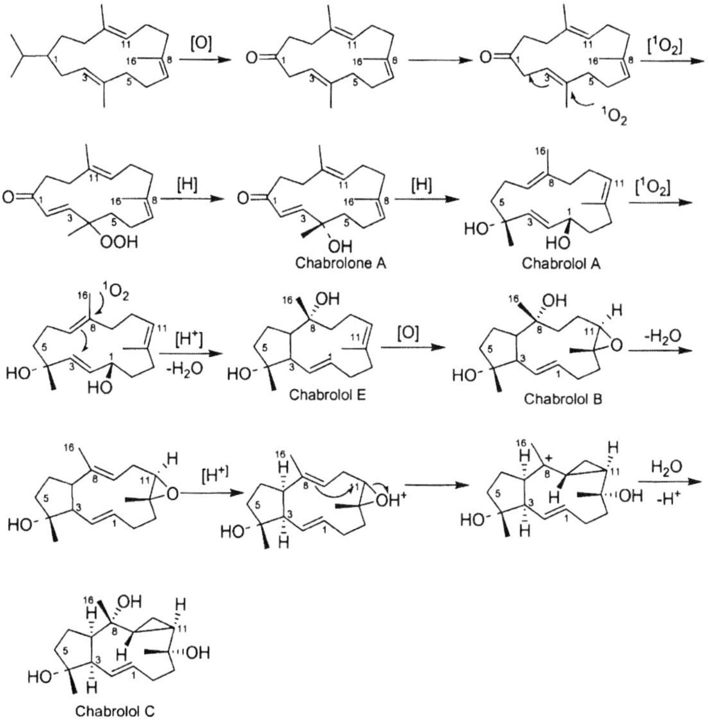

In conclusion, two new norditerpenoids: chabrolone A and chabrolol E, were isolated from the soft coral, *Nephthea chabroli*. Their possible biogenetic pathway was also suggested.

REFERENCES

[1] Faulkner DJ. Marine natural products. Nat Prod Rep 2002; 19(1): 1-48

[2] Zhang WH, Williams ID and Che C T. Chabrolols A, B and C, three new norditerpenes from the soft coral *Nephthea chabroli*. Tetrahedron Lett 2001;42:4681-5

[3] Liu WK, Wong NY, Huang HM, Ho JKC, Zhang WH and Che CT. Growth inhibitory activity of lemnabourside on human prostate cancer cells. Life Sci 2002; 69: 843-53

[4] Zhang WH, Liu WK and Che CT. Polyhydroxylated steroids and other constituents from the soft coral *Nephthea chabroli*. Chem Pharm Bull. 2003; 51(8): 1009-11.

[5] Poet SE and Ravi B N. Three new diterpenes from a soft coral *Nephthea* species. Aust J Chem 1982; 35: 77-83

Table 1. ^{13}C NMR Spectral Data of Chabrolone A and Chabrolol E[a]

Position	Chabrolone A	Chabrolol E
1	200.9 s	131.7 d
2	125.9 d	131.5 d
3	152.1d	50.1 d
4	73.1 s	81.2 s
5	42.4 t	39.2 t
6	22.4 t	23.4 t
7	126.1 d	56.2 d
8	134.1 s	74.8 s
9	38.9 t	33.9 t
10	24.5 t	22.9 t
11	123.1 d	127.7 d
12	132.5 s	134.3 s
13	33.8 t	36.0 t
14	37.4 t	27.4 t
15	29.5 q	23.9 q
16	15.4 q	31.7 q
17	17.0 q	19.0 q

[a] Assignments were based on interpretation of two-dimensional NMR techniques. Values are express in ppm downfield from TMS standard. Spectra were run in $CDCl_3$.

Table 2. ^{1}H NMR Spectral Data of Chabrolone A and Chabrolol E[b]

Position	Chabrolone A	Chabrolol E
1		5.55 (ddd, *J* = 5.6, 11.2, 16.8 Hz)
2	6.30 (d, *J* = 16.0 Hz)	5.38 (dd, *J* = 16.8, 1.2 Hz)
3	6.84 (d, *J* = 16.0 Hz)	2.64 m
4		
5	1.83 m	1.74 m
6	2.05 m, 2.30 m	1.32 m, 1.62 m
7	4.90 (t, *J* = 4.5 Hz)	1.96 m
8		
9	1.86 m, 2.02 m	1.60 m, 1.80 m
10	2.10 m	2.00 m, 2.36 m
11	4.80 (t, *J* = 4 Hz)	5.34 m
12		
13	2.40 m	1.64 m, 2.27 m
14	2.52 m, 2.70 m	2.02 m, 2.60 m
15	1.37 s	1.12 s
16	1.51 s	1.14 s
17	1.60 s	1.67 s

[b] Spectra were recorded in $CDCl_3$ on 400 MHz NMR for ^{1}H, using TMS as internal standard. Proton couplings and one-bond ^{13}C-^{1}H correlation were established by COSY and HETCOR experiments, respectively.

The Metabolites of the Mangrove Fungus *Xylaria* sp (2508)

Zhi-gang She*, Yong-cheng Lin, Zhi-yong Guo, Xiong-yu Wu and Yan Huang
Department of Applied Chemistry, Zhongshan University Guangzhou, 510275, P.R.China
**Corresponding email:* exczgshe@sina.com

INTRODUCTION

A large variety of new bioactive compounds have recently been isolated from marine fungi [1, 2], especially the mangrove fungi. Recently, we embarked on a study of the metabolites of marine fungi including those from mangroves from the South China Sea, and this has yielded a number of interesting compounds [3, 4]. In this paper, the metabolites of the mangrove fungus, strain no. 2508, from the South China Sea and their synthesis are reviewed.

Structure of the Metabolites of the *Xylaria* sp

The mangrove fungus, strain no. 2508, was collected from seeds of an angiosperm tree and identified as *Xylaria* species (Ascomycota). It was found to produce rich secondary metabolites. Over twenty metabolites were isolated from this fungus by Lin et al. (2001). Five unique metabolites, Xyloketals A–E, were isolated from the fermentation broth. Their structures were elucidated by spectroscopic and X-ray diffraction experiments. Xyloketal A is a ketal compound with a *C*3 symmetry while xyloketals B-E are its analogues. Xyloketal A is an inhibitor of acetylcholine esterase. The absolute configuration of xyloketal A and xyloketal D were elucidated by using quantum mechanical calculations of their CD spectra and by comparison with the experimental data (Figure 1) [5].

A novel cyclic peptide containing an allenic ether of a *N*-(*p*-hydroxycinnamoyl) amide, xyloallenolide A, and two aromatic allenic ethers were also isolated. The structures were determined mainly by 2D NMR experiments [6]. The structural feature of these three compounds is the

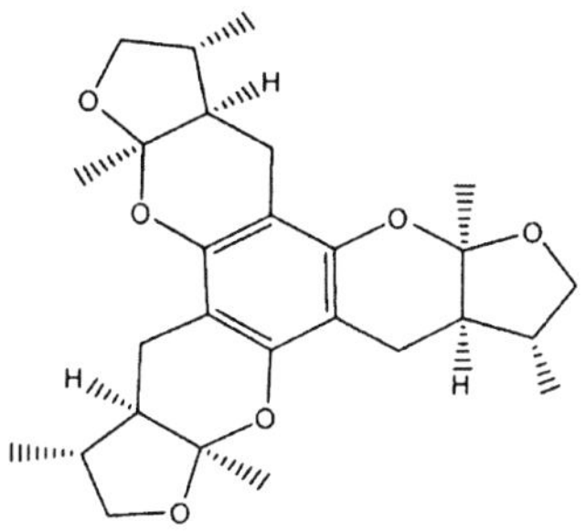

Xyloketal A

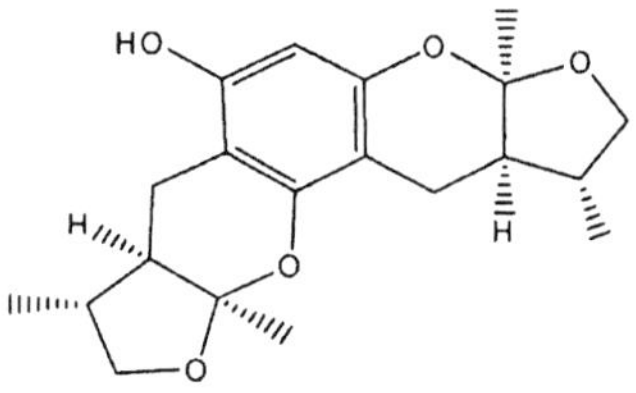

Xyloketal B

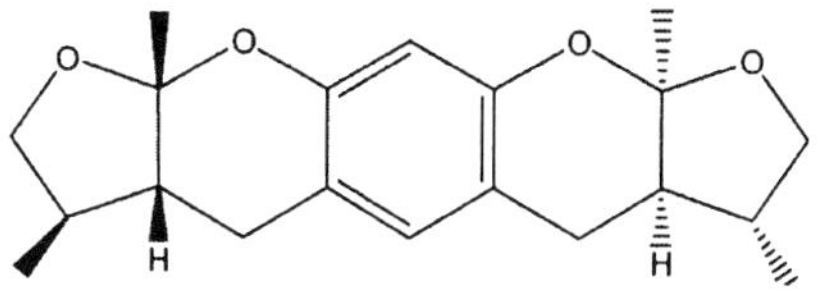

Xyloketal C

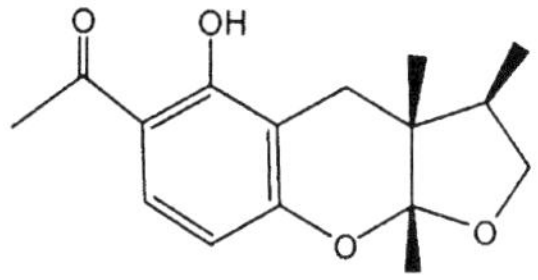

Xyloketal D

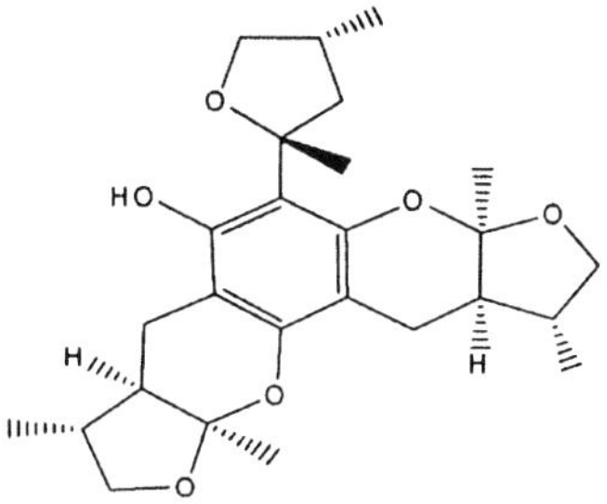

Xyloketal E

Figure 1. Structures of Xyloketals A - E

allenic structure, which is rare in natural products (Figure 2).

Eight known compounds viz. piliformic acid (2-hexylidene-3-methylsuccinic acid), α-glycerol monopalmitate, *p*-hydroxy benzoic acid, protocatechuic acid methyl ester, 4- hydroxy-2-methoxy-acetophe-2-none, ergosterol and 7, 22- *(E)* - diene-3β, 5α, 6β- triol-ergosta, were isolated. The structures were identified by NMR, MS, IR spectroscopic methods and by comparison with the data of authentic compounds [7].

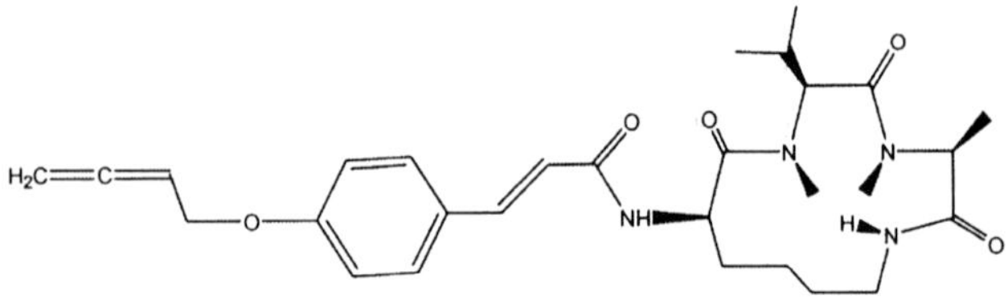

Xyloallenoide A

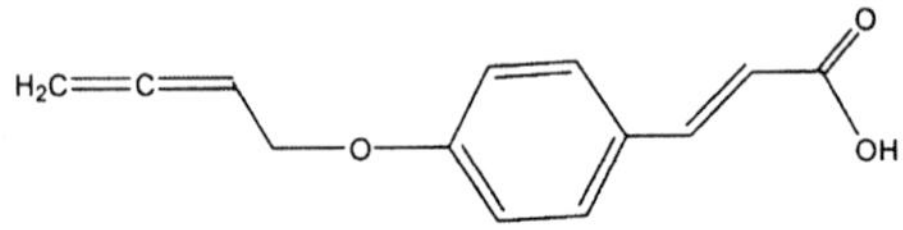

but-2,3-dienyl ether of *p*-hydroxycinnamic acid

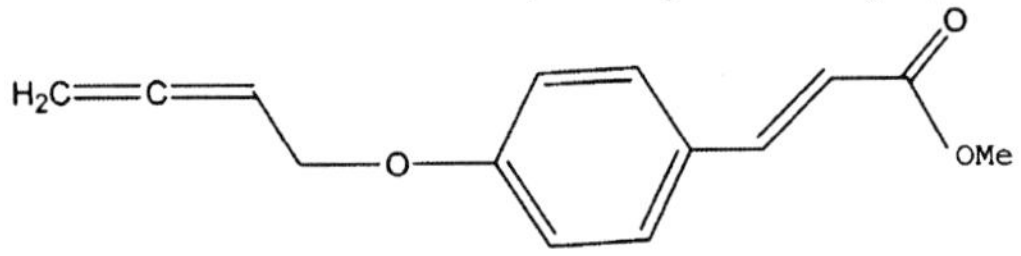

methyl ester of but-2,3-dienyl ether of *p*-hydroxycinnamic acid

Figure 2. Structures of xyloallenolide A and two aromatic allenic ethers

Two polysaccharides, G-11a and G-22a were isolated (8). G-22a exhibited the activity of inhibiting topoisomerase I. Analysis by complete acid hydrolysis and trimethylsilylation indicated that G-22a is composed of rhamnose, mannose and glucose in the ratio of 1:1:2.

Recently a new L-calcium channel blocker, another new xyloketal compound, was obtained and its structure was determined [9].

Synthesis of the metabolites of *Xylaria* sp

Xyloketals have attracted much attention because of their unique structures and bioactivities. Recently, there have been several reports on the synthesis of xyloketal and xyloketal. In these reports, two main methods of ring formation have been described. One is the direct cycloaddition reaction while the other is spontaneous ketal formation after the Michael addition.

Pettigrew et al. (2004) [10] achieved a stereoselective total synthesis of (+/-)-xyloketal D using a [4+2] cycloaddition reaction of an ortho-quinone methide and a dihydrofuran as a key step.

The Mannich base as a precursor for the generation of the *ortho*-quinone methide was prepared from 2, 4-dihydroxyacetophenone, formaldehyde and morpholine (Scheme 1).

The *ortho*-quinone methide was formed by alkylation, oxidation, and photochemical reactions from the Mannich base I of phenols. It is a reaction intermediate (Scheme 2).

Using the same method, two analogues of xyloketal A were synthesized from the Mannich base II and three equivalents of the dihydrofuran (Scheme 3). The 19% yield of this reaction is indeed commendable for this direct process involves nine individual reactions (three alkylation reactions, three elimination reactions, and three subsequent cycloaddition reactions).

In a one-pot multi-step domino reaction, Krohn and Riaz (2004) [11] prepared (+)-xyloketal D by heating optically active 5-hydroxy-4-methyl-3-methylenepentan-2-one (R) in toluene with 2, 4-dihydroxyacetophenone. The reaction was achieved by simply heating the reactants in toluene without any addition of catalysts. Evidently, the acidity of the phenols was sufficient for an autocatalytic process. The first step is the Michael addition of 2, 4-dihydroxyacetophenone to the enone. The reaction that followed is spontaneous condensation to yield the natural product, (+)-xyloketal D (Scheme 4).

With the same method, Krohn et al. (2004) synthesized a series of xyloketals and 5-demethyl-xyloketals by condensation of phloroglucinol or 2, 4-dihydroxy-acetophenone with the enone [12]. Yan and Lin (2004 and 2005) [13,14] synthesized xyloketal C, using phloroglucinol and acrylic acid as starting material (Scheme 5).

Xyloketal F, a new L-calcium channel blocker, was also synthesized by Lin's group (2003) [9].

As stated above, the polyphenol compounds were all used as starting material for synthesizing xyloketals. This possibly provides a new method for the biosynthesis of the xyloketals.

In addition, Alonso et al. (2003) [15] from Spain synthesized a series of substituted perhydrofuro [2, 3-b] pyrans, which are present in xyloketals, from 2-chloromethyl-3- (2-methoxy-ethoxy) propene. The cyclization reaction possibly provides another method for synthesizing xyloketal compounds (Scheme 6).

The mangrove fungus no. 2508 is thus a prolific strain with rich metabolites. We are now investigating new metabolites by changing fermentational conditions and studying biogenetic synthesis of the unique bioactive metabolites.

Scheme 1. Synthesis of Mannich base I

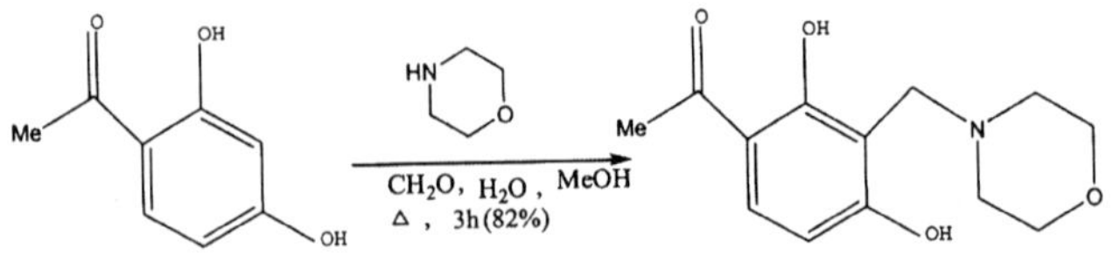

Mannich base I

Scheme 2. Synthesis of (±) xyloketal D

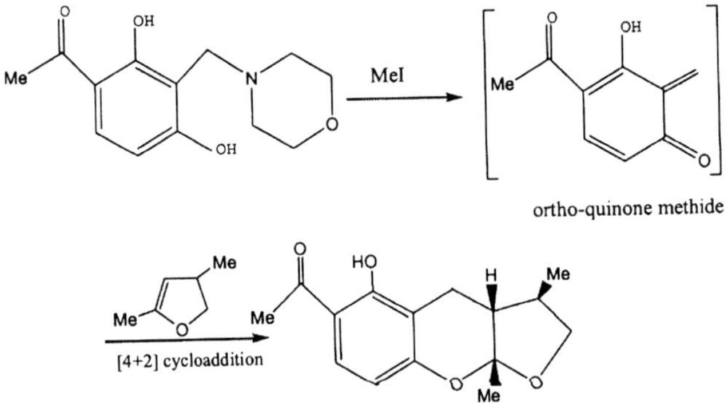

Scheme 3. Synthesis of xyloketal A analogues

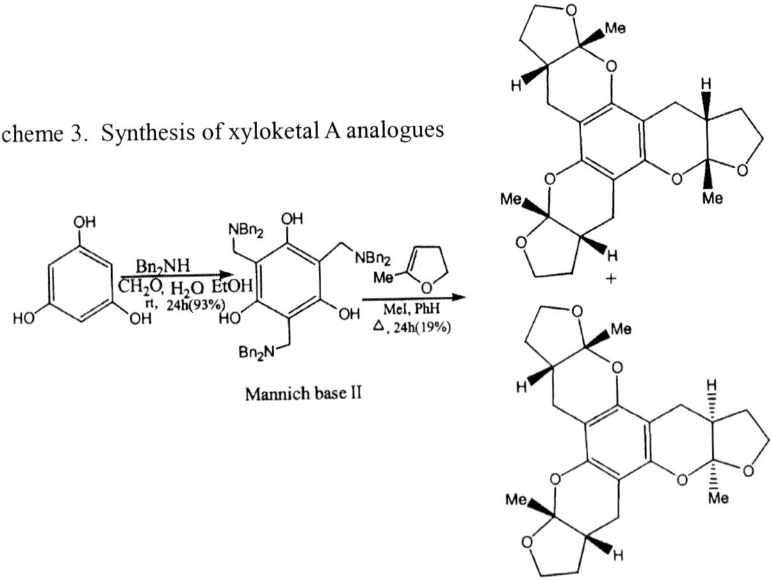

Scheme 4. Synthesis of xyloketal D and its isomer

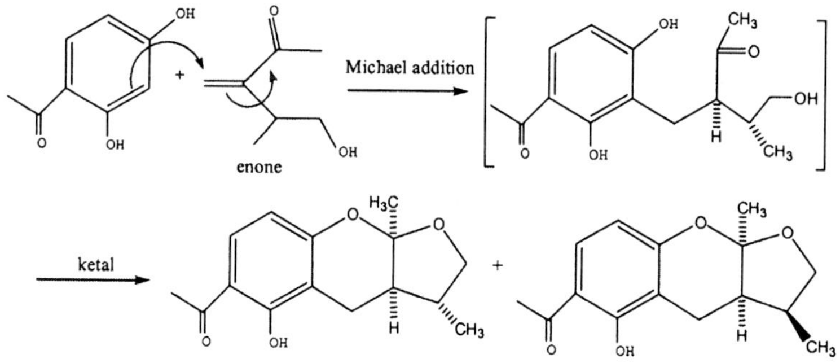

xyloketal D   isomer of the xyloketal D

Scheme 5. Synthesis of xyloketal C

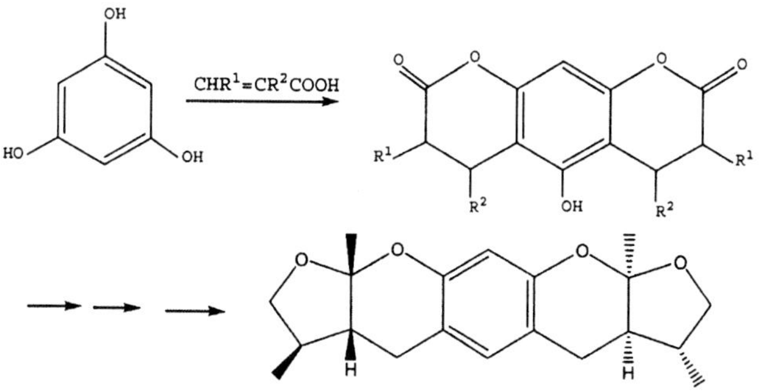

Scheme 6. Synthesis of substituted perhydrofuro [2, 3-b] pyrans

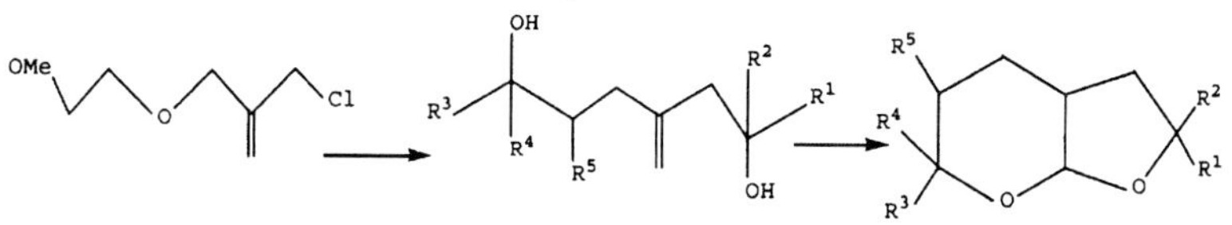

ACKNOWLEDGMENTS

This work was supported by the National Natural Science Foundation of China (20072058), the 863 Foundation of China (2001AA624010 and 2003AA624010), the Natural Science Foundation of Guangdong Province, China (980317), The Science and Technology Plan Foundation of Guangzhou City, China (2004J1-C0061).

REFERENCES

[1] Faulkner D. Marine natural products. *Nat Prod Rep* 1999; 16: 155-98

[2] Fenical W. Chemical Studies of Marine Bacteria: Developing a New Resource. *Chem Rev.* 1993; 93: 1673-83

[3] Lin YC, Shao ZY, Jiang G, Zhou S, Cai J, Vrijmoed, LLP *et al.* Penicillazine, a unique quinolone derivative with 4H-5, 6-dihydro-1,2-oxazine ring system from the marine fungus *Penicillium* sp (strain #386) from the South China Sea. Tetrahedron 2000; *56*: 9607-9

[4] Chen GY, Lin YC, Wen L, Vrijmoed LLP, Jones EBG. Two new metabolites of a marine endophytic fungus (No. 1893) from an estuarine mangrove on the South China Sea coast. Tetrahedron 2003; 59: 4907

[5] Lin YC, Wu XY, Feng S, Jiang GC, Luo JH, Zhou SN, *et al.* Five unique compounds: Xyloketals from mangrove fungus *Xylaria* sp from the South China Sea coast. J of Org Chem 2001; 66: 6252-6

[6] Lin YC, Wu XY, Fen S, Jiang G, Zhou S, Vrijmoed LLP, Jones EBG. A novel N-cinnamoylcyclopeptide containing an allenic ether from the fungus *Xylaria* sp (strain # 2508) from the South China Sea. Tetrahedron Lett. 2001; 42: 449-51

[7] Wu XY, Li ML, Hu GP, Lin YC, Vrijmoed LLP. The Metabolites of the Endophyte Fungus No. 12508 in the Mangrove Tree from the South China Seacoast. Acta cientarium Naturalium Universitatis Unyatseni. 2002; 41(3): 34-6

[8] Guo ZY, She ZG, Chen DM, Lin YC, Lu HN, Wu XY. Studying on the Composition of Polysaccharide G-22a from the Seeds of Marine Endophytic Fungus No. 2508 from the South China Sea. Acta Scientarium Naturalium Universitatis Unyatseni. 2003; 42(4): 127-8

[9] Lin YC, She ZG, Li HJ. The preparation and application of a compound: Xyloketal F .PCT/CNO3/00877. Application date :2003.10.21

[10] Pettigrew JD, Bexrud JA, Freeman RP, Wilson PD. Total Synthesis of (±)-Xyloketal D and Model Studies towards the total Synthesis of (-)-Xyloketal A. Herocycles 2004; 62: 445-51

[11] Krohn K, Riaz M. Total synthesis of (+)-xyloketal D, a secondary metabolite from the mangrove fungus *Xylaria* sp. Tetrahedron Lett 2004; 45: 293-4

[12] Krohn K, Riaz M, Flörke U. Synthesis of Xyloketals, Natural Products from the Mangrove Fungus *Xylaria* sp. Eur. J. Org. Chem. 2004; 1261-70

[13] Yan H, Lin YC. Synthesis of dihydrocoumarins catalyzed by acidic ion-exchange resin. Chin J Org Chem 2004; 24 (11): 1451-3

[14] Yan H, Lin YC. Synthesis of cyclic dienol ether and its properties. Chin J Org .Chem 2005; 25 (7): 835-7

[15] Alonso F, Lorenzo E, Melendez J, Yus M. Straight and versatile synthesis of substituted perhydrofuro[2,3-b] pyrans from 2-chloromethyl-3-(2-methoxyethoxy) propene. Tetrahedron 2003; 59 (28): 5199-208

Phytochemical Evaluation of Polyherbal Formulations Using HPTLC

Milind S Bagul and M Rajani*
B. V. Patel Pharmaceutical Education and Research Development (PERD) Centre, Thaltej-Gandhinagar Highway, Thaltej, Ahmedabad – 380 054, India
**Corresponding email: rajanivenkat@hotmail.com*

INTRODUCTION

Traditional systems of medicine have been steadily gaining interest and acceptance all over the world, even among the practitioners of modern medicine. Consequently plant materials and herbal drugs derived from them represent a substantial proportion of the current global market. In this scenario, there is a need to ensure quality of herbal preparations. The only way to ensure that herbal drugs and preparations made with them achieve optimum and consistent quality is to create and maintain a comprehensive quality assurance system. Traditional systems of medicine have been steadily gaining interest and acceptance even among the practitioners of modern medicine. Consequently plant materials and herbal drugs derived from them represent a substantial proportion of the current global market. In this scenario, there is a need to ensure the quality of herbal preparations. In this context the only way to ensure that herbal drugs and preparations made with them achieve optimum and consistent quality is to create and maintain a comprehensive quality assurance system.

Most of the herbal formulations, especially the classical formulations, are polyherbal. Further, the unique processing methods followed for the manufacturing of these drugs turn the herbal ingredients into very complex mixtures, from which separation, identification and estimation of chemical components is very difficult [1, 2]. The recent advancement in chromatographic techniques made it possible to separate and quantify the chemical constituents in a mixture with comparatively little clean-up requirement using HPTLC [3, 4].

In the present study systematic phytochemical evaluation has been carried out for classical formulations *viz., Chandraprabha vati*, *Triphala guggulu*, *Phalatrikadi kvatha churna*, *Prabhakara vati*, *Sharbat-e-*

Dinar, *Itrifal Mulaiyin,* using HPTLC by adapting multi-marker based standardization including establishment of fingerprint profiles and developing methods for quantification of chemical / biological markers.

MATERIALS AND METHODS

Drug samples

One authentic sample each of *Chandraprabha vati*, *Phalatrikadi kvatha churna*, *Prabhakara vati* was prepared by Ayurveda practitioner specialized in the preparation of classical formulations (*Rasashastra*). This sample was denoted as PS. Two commercial samples of the above four formulations were procured from local market which were denoted as sample MS1 and sample MS2. *Triphala guggulu*, *Sharbat-e-Dinar* and *Itrifal Mulaiyin* were procured from local market.

Chemicals

All chemicals used in the experiments were of analytical grade. Gallic acid was a gift sample from Tetrahedron Ltd., India. Ellagic acid, berberine and β-sitosterol were purchased from Natural Remedies Pvt. Ltd. Z-Guggulsterone and kutkin (a 70:30 mixture of picroside–I & II) were purchased from Regional Research Laboratory Jammu, India. Emodine was purchased from Sigma-Aldrich, Germany. Piperine was isolated from *Piper nigrum* by the reported method [5]. The identity and purity of the piperine was confirmed by determining their melting point and by comparing IR and UV spectral data with the reported values.

TLC conditions

Plate -Precoated silica gel 60 F_{254} TLC plate (E. Merck) (0.2 mm thickness)
Spotter - CAMAG Linomat IV Automatic Sample Spotter
Developing chamber - CAMAG glass twin trough chamber (20 x 10 cm)
Scanner - CAMAG TLC Scanner 3 and CATS 4 software
Experimental conditions - temperature 25 ± 2°C, relative humidity 40 %

Preparation of extracts

Chandraprabha vati, *Triphala guggulu* and *Prabhakara vati*

Twenty pills (*vatis*) of each of the samples were weighed and powdered. 2 gm each of the powdered samples were extracted under reflux for 15 min with methanol (4 X 25 ml). The extracts were filtered, pooled and concentrated under reduced pressure to 50 ml in a volumetric flask.

Phalatrikadi kvatha churna

1.5 gm of *Phalatrikadi kvatha churna* was extracted with methanol (4 X 25 ml). The extracts were filtered, pooled and concentrated under reduced pressure to 25 ml in a volumetric flask.

Itrifal Mulaiyin and *Sharbat-e-Dinar*

Itrifal Mulaiyin (2 gm) and *Sharbat-e-Dinar* (2 ml) were extracted with methanol (4 X 25 ml). The extracts were filtered, pooled and concentrated under reduced pressure to 25 ml in a volumetric flask.

Preparation of standard solutions

Standard solutions of gallic acid

A stock solution of gallic acid (100 g/ml) was prepared by dissolving 10 mg of accurately weighed gallic acid in methanol and making up the volume to 100 ml with methanol. The aliquots (1.5 to 7.5 ml) of stock solution were transferred to 10 ml volumetric flasks and the volume of each was adjusted to 10 ml with methanol to obtain standard solutions containing 150 to 750 μg/ml.

Standard solutions of ellagic acid

A stock solution of ellagic acid was prepared by dissolving 10 mg of accurately weighed ellagic acid in methanol and making up the volume to 100 ml with methanol. From this stock solution standard solutions of 15 to 35 mg/ml were prepared by transferring aliquots (7.5 to 17.5 ml) of stock solution to 50 ml volumetric flasks and adjusting the volume with methanol.

Standard solutions of piperine

A stock solution of piperine (50 μg/ml) was prepared by dissolving 5 mg of accurately weighed piperine in methanol and making up the volume of the solution to 100 ml with methanol. The aliquots (0.8 to 4.8 ml) of stock solution were transferred to 10 ml volumetric flasks and the volume of each was adjusted to 10 ml with methanol to obtain standard solutions containing 4 to 24 μg/ml piperine, respectively.

Standard solution of berberine

A stock solution of berberine (50 μg/ml) was prepared by dissolving 5 mg of accurately weighed berberine in methanol and making up the volume of the solution to 100 ml with methanol. The aliquots (0.8 to 4.8 ml) of stock solution were transferred to 10 ml volumetric flasks and the volume of each was adjusted to 10 ml with methanol to obtain standard solutions containing 4 to 24 μg/ml berberine, respectively.

Standard solutions of Z-guggulsterone

A stock solution of Z-guggulsterone (100 mg/ml) was prepared by dissolving 10 mg of accurately weighed Z-guggulsterone in chloroform and making up the volume to 100 ml with methanol. The aliquots (1 to 6 ml) of stock solution were transferred to 10 ml volumetric flasks and the volume of each was adjusted to 10 ml with chloroform to obtain standard solutions containing 100 to 600μg/ml.

Standard solutions of β-sitosterol

A stock solution of β-sitosterol (50 μg/ml) was prepared by dissolving 5 mg of accurately weighed β-sitosterol in methanol and making up the volume of the solution to 100 ml with methanol. The aliquots (1.6 to 9.6 ml) of stock solution were transferred to 10 ml volumetric flasks and the volume of each was adjusted to 10 ml with methanol to obtain standard solutions containing 8 to 48 μg/ml β-sitosterol, respectively.

Standard solutions of kutkin

A stock solution of kutkin was prepared by dissolving 5 mg of accurately weighed kutkin in methanol and making up the volume to 25 ml with methanol. From this stock solution standard solutions of 20 to 100 mg/ml were prepared by transferring aliquots (1 to 5 ml) of stock solution to 10 ml volumetric flasks and adjusting the volume to10 ml with methanol. Kutkin standard used was a mixture of picroside–I & II (70:30).

Standard solutions of emodine

A stock solution of emodine (40 mg/ml) was prepared by dissolving 4 mg of accurately weighed emodine in methanol and making up the volume to 100 ml with methanol. From this stock solution standard solutions of 16 to 80 mg/ml were prepared by transferring aliquots (7.5 to 17.5 ml) of stock solution to 50 ml volumetric flasks and adjusting the volume with methanol.

Preparation of calibration curve

Calibration curve for gallic acid

10 mml of each of the standard solutions (150 ng to 750 ng per respective spot) were applied in triplicate on a TLC plate. The plate was developed in a solvent system of toluene/ethyl acetate/formic acid/methanol (3:3:0.8:0.2 v/v) in a chamber up to a distance of 8 cm. After development, the plates were dried in air and scanned at 280 nm. The peak areas were recorded. Calibration curve of gallic acid was prepared by plotting peak areas *vs* concentration.

Calibration curve for ellagic acid

10 mml of the standard solutions (160 to 960 ng per respective spot) were applied in triplicate on a TLC plate. The plate was developed in a solvent system of toluene/ ethyl acetate/ formic acid/ methanol (3:3:0.8:0.2 v/v) in a chamber up to a distance of 8 cm. After development, the plates were dried in air and scanned at 280 nm. The peak areas were recorded. Calibration curve of ellagic acid was prepared by plotting peak areas *vs* concentration of ellagic acid applied.

Calibration curve for piperine

10 μl each of the standard solutions (40 ng to 240 ng per respective spot) were applied in triplicate on a TLC plate. The plate was developed in a solvent system of dichloromethane/ ethyl acetate (7.5:1.0 v/v) in a chamber up to a distance of 8 cm. After development, the plates were dried in air and scanned at 337 nm. The peak areas were recorded. Calibration curve of piperine was obtained by plotting peak areas *vs* concentration of piperine applied.

Calibration curve for berberine

10 μl each of the standard solutions (40 ng to 240 ng per respective spot) were applied in triplicate on a TLC. The plate was developed in a solvent system of toluene/ ethyl acetate/ formic acid/methanol (3:3:0.8:0.2 v/v) in a chamber up to a distance of 8 cm. After development, the plates were dried in air and scanned at 348 nm. The peak areas were recorded.

Calibration curve of berberine was obtained by plotting peak areas *vs* concentration of berberine applied.

Calibration curve for Z-guggulsterone

10 mml of each of the standard solutions (100 ng to 600 ng per respective spot) were applied in triplicate on a TLC plate. The plate was developed in a solvent system of petroleum ether/ ethyl acetate (6:4 v/v) in a chamber up to a distance of 8 cm. After development, the plates were dried in air and scanned at 248 nm. The peak areas were recorded. Calibration curve of Z-guggulsterone was prepared by plotting peak areas *vs* concentration.

Calibration curve for β-sitosterol

10 μl each of the standard solutions (80 ng, to 480 ng per respective spot) were applied in triplicate on a TLC plate. The plate was developed in a solvent system of toluene/methanol (9 :1 v/v) in a chamber up to a distance of 8 cm. After development, the plates were dried in air and derivatized with anisaldehyde sulphuric acid reagent and kept in oven at 110°C for 5 min and scanned at 524 nm. The peak areas were recorded. Calibration curve of β-sitosterol was obtained by plotting peak areas *vs* concentration of β-sitosterol applied.

Calibration curve for picroside–I (using kutkin, a 70:30 mixture of picroside–I & II)

10 mml of the standard solutions (200 to 1400 ng per respective spot) were applied in triplicate on a TLC plate. The plate was developed in a solvent system of ethyl acetate/formic acid/ methanol (6:0.6:0.4) in a chamber up to a distance of 8 cm. After development, the plates were dried in air and scanned at 280 nm. The peak areas were recorded. Since kutkin (a 70:30 mixture of picroside–I & II) was used, the peaks were identified from their relative percentage of peak areas. Calibration curve of picroside–I was prepared by plotting peak areas *vs* concentration applied.

Calibration curve for emodine

10 mml of the standard solutions (160 to 800 ng per respective spot) were applied in triplicate on TLC plate. The plate was developed in

a solvent system of toluene/ethyl acetate/formic acid/methanol (3:3:0.8:0.2) up to a distance of 8 cm. After development, the plates were dried in air and scanned at 280 nm. The peak areas were recorded. Calibration curve of emodine was prepared by plotting peak areas *vs* concentration of emodine applied.

Fingerprint profile of extracts

TLC Fingerprinting profile of methanolic extract of Triphala guggulu and Chandraprabha vati

10 μl of suitably diluted stock solutions of methanolic extract from all the samples were applied on to precoated silica gel plates. The plates were developed with two different mobile phases separately (1) toluene/ methanol (9:1), (2) petroleum ether (60-80°C)/ ethyl acetate (3: 1). After developing the plates were dried at room temperature and scanned at 254 nm and 366 nm, chromatograms and absorption spectra of the resolved bands were recorded. The band properties such as colour, fluorescent/non-fluorescent nature, R_f--and $\lambda_{max,}$ shoulder inflections were also noted. The relative concentrations were determined by densitometric scanning. Plates were derivatized in anisaldehyde sulphuric acid reagent and $R_{f,}$- colours of the bands were noted.

Fingerprint profile of formulation with its ingredients

TLC fingerprint profile of methanolic extract of Prabhakara vati and T. arjuna stem bark along with gallic acid and ellagic acid standards

10 μl of methanolic extract of the three samples and *T. arjuna* stem bark were applied on precoated TLC plate with 5 μl each of standard solutions of gallic acid and ellagic acid. The plate was developed in a solvent system of toluene/ethyl acetate/formic acid/methanol (3:3: 0.8:0.2) in a chamber up to a distance of 8 cm. After development, the plate was dried in air and scanned at 280 nm.

Fingerprint profile with marker compounds

Methanolic extract of Triphala guggulu and Chandraprabha vati

Solvent systems suitable for resolving the different compounds were optimized after experimenting with different solvent systems.

Gallic acid and ellagic acid were resolved in the solvent system of toluene/ethyl acetate/ formic acid/methanol (3:3:0.8:0.2), Z-guggulsterone and piperine were resolved in the solvent system of petroleum ether (60-80°C)/ethyl acetate (6:4).

Itrifal Mulaiyin and Sharbat-e-Dinar

10 μl extract of *Itrifal Mulaiyin* and *Sharbat-e-Dinar*, and 4 μl of standard solutions of gallic acid and emodine were applied on precoated TLC plate. The plate was developed in a solvent system of toluene/ethyl acetate/formic acid/methanol (3:3:0.8:0.2) up to a distance of 8 cm. After development, the plate was dried in air and scanned at 280 nm. The resolved bands were evaluated for their spectral details, and the relative concentrations were determined by densitometry. R_f and relative percentage of resolved bands were noted.

Quantification of marker compounds

Preparation of sample solution

For the estimation of gallic acid, Z-gugglesterone, berberine, piperine

1 gm of sample powder was extracted with methanol (4 X 25 ml). The extract was filtered, filtrates were combined and concentrated to 25 ml.

For the estimation of ellagic acid

To 1 gm of powder 10 ml of water and 25 ml of methanolic hydrochloric acid were added and refluxed for 2 hr. The extract was filtered and concentrated to 25 ml.

For the estimation of ß-sitosterol

Two gm of powder extracted with petroleum ether (4 X 25 ml). The extract was filtered, filtrates were combined and concentrated to 25 ml.

For the estimation of kutkin – sample preparation is as given above for *Phalatrikadi kvatha churna* in.

For the estimation of emodine – sample preparation is as given above for *Itrifal Mulaiyin* and *Sharbat-e-Dinar.*

Estimation of marker compounds

For the quantification of gallic acid method adopted by us was adopted [6].

Simultaneous estimation of Z-guggulsterone and piperine from Chandraprabha vati

10 μl each of sample solutions were applied in triplicate on a TLC plate. The plate was developed and scanned as mentioned in the calibration curve for Z-guggulsterone.

Simultaneous estimation of gallic acid and berberine from Chandraprabha vati

10 μl each of sample solutions were applied in triplicate on a TLC plate. The plate was developed and scanned as mentioned in the calibration curve for gallic acid.

Estimation of ellagic acid from Chandraprabha vati

10 μl each of sample solutions were applied in triplicate on a TLC plate. The plate was developed and scanned as mentioned in the calibration curve for ellagic acid.

Estimation of β-sitosterol from Chandraprabha vati

15 μl of each of sample solutions were applied in triplicate on a TLC plate. The plate was developed and scanned as mentioned in the calibration curve for Z-guggulsterone.

Simultaneous estimation of gallic acid and emodine from *Itrifal Mulaiyin* and *Sharbat-e-Dinar*

10 μl each of sample solutions were applied in triplicate on a TLC plate with 4 μl of standard solutions of gallic acid and emodine. The plate was developed and scanned as mentioned in the calibration curve for emodine.

Simultaneous estimation of gallic acid and ellagic acid from Prabhakara vati and Phalatrikadi kvatha churna

10 μl each of sample solutions were applied in triplicate on a TLC plate. The plate was developed and scanned at 280 nm. The peak areas and absorption spectra were recorded. The purity of gallic acid, and ellagic

acid bands in the sample extracts was checked by recording the absorption spectra at start, middle and end position of the bands. The amount of gallic acid and ellagic acid in different samples was calculated using the respective calibration curve.

Estimation of picroside–I from Phalatrikadi kvatha churna

10 μl each of sample solutions were applied in triplicate on a TLC plate. The plate was developed and scanned as mentioned in the calibration curve for picroside–I. The peak areas and absorption spectra were recorded. The purity of marker compound in the sample extracts was checked by recording the absorption spectra at start, middle and end position of the bands. The amount of compound in different samples was calculated using the respective calibration curve.

RESULTS AND DISCUSSION

Some important polyherbal formulations of Ayurveda and Unani systems of medicine which are widely used for the treatment of various disease and disorders were selected for the present study (Tables 1 to 6). Quality control parameters mainly include qualitative, semi-quantitative and quantitative analysis were developed for the selected formulations. TLC fingerprint profile for the methanolic extract of the formulations *Chandraprabha Vati* and *Triphala guggulu* was carried out. Fingerprint pattern of the prepared samples (PS) was compared with market samples (MS1 and MS2). TLC fingerprint profile of the prepared formulation (*Prabhakar vati*) with its ingredient (*T. arjuna* stem bark) was carried and compared with market samples. TLC fingerprint profile of the formulations (*Chandraprabha Vati* and *Triphala guggulu*) with marker/biomarker compounds *viz*., Z-guggulsterone, piperine, gallic acid, ellagic acid, berberine and *β*-sitosterol was carried out. Methods were developed and validated for the TLC densitometric quantification of above mentioned marker/biomarker compounds including picroside–I.

TLC Fingerprint profiles of the formulations

Complete TLC fingerprint profiles were established for *Chandraprabha Vati* and *Triphala guggulu*. For establishing fingerprint profiles, the methanolic extract of the formulations was used. Chromatogram and absorption spectra of the resolved bands were recorded under UV 254 nm and 366 nm. Further, the plates were derivatized with anisaldehyde sulphuric acid reagent. From this R_f, λ_{max}, shoulder inflections if any, relative percentage and colour of the resolved bands were noted. Solvent systems were optimized to achieve best resolution of different components of the extracts. Two market samples of *Chandraprabha Vati* were also worked out in a similar way and compared. The details of the study are given in Tables 7 & 8. In both the solvent system tried, we find out variations in the TLC fingerprint pattern of the prepared samples and market samples. (Figures 1, 2 & 7).

TLC Fingerprint profile of the formulation with its ingredients

Herbo-mineral preparation *Prabhakara vati* (Table 4) is widely used in cardiac disorders [7]. As per the classical texts, decoction of *Terminalia arjuna* stem bark is used in the preparation of the formulation. Methanolic extracts of prepared sample showed ten bands when scanned at 280 nm, of which four bands at corresponding R_f were present in the track of *T. arjuna* stem bark extract, the spectra of which overlap with those of the bands in the extract of *T. arjuna* stem bark. These four bands were absent in MS1 and MS2, which showed the presence of some other components. Further in MS2 a band corresponding to gallic acid at R_f 0.43 was found to be present. However, the spectral comparison revealed that it was not gallic acid (Figure 3).

TLC Fingerprint profile with marker compounds

Presence of important marker compounds indicate the presence of the respective raw materials in the formulation. Spectral comparison is a further confirmation. Multiple marker based evaluation ensures the quality with respect to the ingredients containing these marker compounds. However it is practically impossible to have marker compounds representing all the ingredients of a polyherbal formulation. Hence, in the present study marker compounds representative of important/major ingredients of the formulations were used.

Triphala guggulu and Chandraprabha vati

The different ingredients of the formulation contain important marker compounds *viz.* Z-guggulsterone, berberine, piperine, gallic acid, ellagic acid. These compounds are biomarkers since they have been shown to have several biological activities. Suitable extraction procedures were adapted to effect complete extraction of the compounds from the samples. The presence of the markers in the sample extracts was ascertained by co-chromatography and comparison of the R_f and absorption spectra with

Table 1. Composition of *Chandraprabha* vati

Ingredients	Botanical source*	Quantity
Chandra prabha	*Cinnamomum camphora* (Karpura)	3 gm
Vacha	*Acorus calamus* (Rz)	3 gm
Musta	*Cyprus rotundus* (Rz)	3 gm
Bhuinimba	*Phyllanthus amarus* (W Pl)	3 gm
Amrta (gaduchi)	*Tinospora cordifolia* (St)	3 gm
Daruka (Davadaru	*Cedrus deodar* (Ht.wd.)	3 gm
Haridra	*Curcuma longa* (Rz)	3 gm
Ativisa	*Aconitum heterophyllum* (Rt)	3 gm
Daru haridra (Darvi)	*Berberis aristata* (St)	3 gm
Pipppali Mula	*Piper longum* (Rt)	3 gm
Chitraka	*Plumbago zeylanica* (Rt)	3 gm
Dhnayaka	*Coriandrum sativam* (Fr)	3 gm
Haritaki	*Terminalia chebula* (Fr P)	3 gm
Bhibhitaka	*Terminalia bellirica* (Fr P)	3 gm
Amlaki	*Emblica officinalis* (Fr P)	3 gm
Cavya	*Piper chaba* (St)	3 gm
Vidang	*Embelia ribes* (Fr)	3 gm
Gaja pippali	*Saussaria lappa* (Fr)	3 gm
Sunthi	*Zingiber officinalis* (Rz)	3 gm
Marica	*Piper nigrum* (Fr)	3 gm
Pippali	*Piper longum* (Fr)	3 gm
Swarna Masika bhasma	Iron pyrite	3 gm
Yava khsara	Ammonium chloride	3 gm
Sarji khsara	Natural Sodium Carbonate	3 gm
Saindhava lavana	Rock Salt	3 gm
Sauvarchala lavan	Black Salt	3 gm
Vida lavana	Ammonium chloride	3 gm
Trivrt	*Ipomea turpethum* (Rt)	12 gm
Danti	*Balliospermum montanum* (Rt)	12 gm
Patraka (tej patra)	*Cinnamomum tamala* (Lf)	12 gm
Tvak	*Cinnamomum zeylanicum* (St bk)	12 gm
Ela	*Elettaria cardamom* (Sd)	12 gm
Vamsalochana	*Bambusa arundinacea*	12 gm
Hata loha (lauha bhasma)	Iron	12 gm
Sita (sarkara)	Sugar	48 gm
Silajatu	*Shilajit*	96 gm
Guggulu	*Commophora mukul* (oleogum resin)	96 gm

***Part used:** Fr, fruit; Fr P, fruit pulp; Ht wd, heart wood; Lf, leaf; Rt, root; Rz, rhizome; Sd, seed; St, stem; St bk, stem bark; W Pl, whole plant.
Uses - for various disorders like anemia, pain, indigestion, renal calculi.

Table 2. Composition of *Triphala guggulu*

Ingredient	**Botanical name**	**Part used**	**Quantity (gm)**
Haritaki	*Terminalia chebula*	Fruit pulp	48
Bibhitaka	*Terminalia bellirica*	Fruit pulp	48
Amalaki	*Emblica officinalis*	Fruit pulp	48
Pippali	*Piper longum*	Fruit	48
Guggulu	*Commiphora mukul*	Oleogum resin	240

Uses – Prescribed for fistula-in-ano, and also used as anti-dropsy/anti-odematous

Table 3. Composition of *Phaltrikadi kvatha churna*

Ingredient	**Botanical name**	**Part used**	**Quantity**
Haritaki	*Terminalia chebula*	Fruit pulp	1 part
Bibhitaka	*Terminalia bellirica*	Fruit pulp	1 part
Amlaki	*Embelica officinalis*	Fruit pulp	1 part
Patola	*Tricosanthes dioica*	Whole plant	1 part
Tikta	*Picrorhiza kurroa*	Rhizome	1 part

Uses – used as antipyretic and antiemetic

Table 4. Composition of *Prabhakara vati*

Ingredient	**Description**	**Quantity**
Maksika	Swarna maksika bhasma	1 part
Lauha	Calcinied iron	1 part
Abhra	Abhraka bhasma	1 part
Tugaksiri	Siliceous encrustations from culms of *Bambusa arundinacea*	1 part
Silajatu	Exudates from rock crevices	1 part
Partha vari	*Terminalia arjuna* stem bark	For *bhavana*

Uses – used in cardiac disorders

Table 5. Composition of *Itrifal Mulaiyin*

Ingredient	Botanical name	Quantity
Post-e-Halela Kabli	*Terminalia chebula*	10 g
Post-e-Balela	*Terminalia bellerica*	10 g
Halela Siyah	*Terminalia chebula*	10 g
Aamla	*Emblica officinalis*	10 g
Turbud	*Operculina turpethum*	10 g
Badiyan	*Foeniculum vulgare*	25 g
Mastagi	*Pistacia lentiscus*	25 g
Ustukhuddus	*Lavandula stoechas*	25 g
Saqmonia	*Convolvulus scammonia*	25 g
Rewand Chini	*Rheum emodi*	25 g
Asal or Qand Safaid	Honey or Sugar	525 g

Uses – Widely used for constipation, chronic headache, coryza

Table 6. Composition of *Sharbat-e-Dinar*

Ingredient	Botonical name	Quantity
Post-e-Bekh-e-Kasni	*Cichorium intybus*	170 g
Tukhm-e-Kasoos	*Cuscuta refluxa*	100 g
Tukhm-e-Kasni	*Cichorium intybus*	85 g
Ghuncha-e-Gul-e-Surkh	*Rosa damascene*	85 g
Rewand Chini	*Rheum emodi*	60 g
Gul-e-Nilofar	*Nymphea alba*	45 g
Gaozaban	*Borago officinalis*	45 g
Aab	Water	Q.S
Qand Safaid	Sugar	1.2 kg

Uses – Widely used for hepatitis, uteritis, obstructive jaundice, dropsy and constipation.

Table 7. TLC Fingerprint profile of two different samples of *Triphala guggulu* (solvent system-1)

A. Scanned at 254 nm

MS1			MS2		
R_f value	λ_{max}	Relative %	R_f value	λ_{max}	Relative %
0.08	--	19.83	0.09	--	11.41
0.11	327	4.11	0.11	328	7.11
0.15	--	17.12	0.16	327	14.28
0.21	325	12.86	0.21	325	14.64
0.26	341	2.60	--	--	--
0.29	--	4.12	0.29	252	8.94
0.35	342	6.74	0.36	--	7.70
0.38	339	1.97	0.39	253	3.33
0.42	--	1.74	0.43	328	0.94
0.47	254	6.64	--	--	--
0.51	253	3.90	0.50	--	2.55
--	--	--	0.53	253	0.82
0.56	251	12.03	0.58	254	14.43
--	--	--	0.64	251	4.28
0.69	432	3.57	0.72	200	1.95
--	--	--	0.79	200	5.88
--	--	--	0.87	200	1.74
0.98	226	2.77	--	--	--

B. Scanned at 366 nm

MS1				MS2			
R_f value	Colour of the band	λ_{max}	Relative %	R_f value	Colour of the band*	λ_{max}	Relative %
--	--	--	--	0.08	Blue	287	1.64
0.11	Bf	326	5.56	0.11	Blue	328	4.54
0.17	Bf	--	4.01	0.17	Green	327	3.92
0.21	Gf	326	9.45	--	--	--	--
--	--	--	--	0.24	Blue	299	3.73
0.28	Bf	252	6.74	0.28	Blue	253	5.19
0.35	Gf	249	38.33	0.36	Green	210	33.57
0.44	Bf	254	4.12	0.45	Blue	252	3.54
0.53	Bf	253	15.78	0.54	Blue	256	20.93
0.57	Bf	257	13.40	0.59	Blue	254	17.83
0.75	Gf	385	2.61	0.74	Green	200,320	3.18

*fluorescent bands

C. After derivatization with anisaldehyde sulphuric acid reagent

R_f value	**Colour of the band**
0.10	Green
0.13	Violet
0.22	Green
0.29	Pink
0.40	Purple
0.42	Violet
0.48	Yellow
0.53	Pink
0.58	Pink
0.72	Blue
0.80	Purple
0.92	Light blue
0.94	Violet
0.98	Light blue

Table 8. Fingerprint profile of two different samples of *Triphala guggulu* (solvent system-2)

A. Scanned at 254 nm

MS1			**MS2**		
R_f value	**λ_{max}**	**Relative %**	**R_f value**	**λ_{max}**	**Relative %**
0.09	302,219*	5.59	0.09	305,219*	6.84
0.17	250,331	2.06	0.17	250	3.33
--	--	--	0.21	289,250*	1.31
0.32	339	6.71	0.29	339,250*	7.36
0.38	252,335*	10.48	0.36	252,335*	7.65
0.49	251	10.22	0.47	251	19.14
0.68	266	13.57	0.65	260	9.04
0.79	265	2.43	0.76	267	12.17
0.84	290,409*		0.89	251	4.38
0.90	200,249*	3.82	--	--	--

* Shoulder inflection

that of the standards of the marker compounds (Figures 4, 5, 8 & 9, Tables 7, 8, 9 & 10).

B. Scanned at 366 nm

MS1				MS2			
R_f	Colour	λ_{max}	Relative %	R_f	Colour	λ_{max}	Relative %
0.08	318,220*	Bf	2.81	0.08	328,244*	Bf	3.40
--	--	--	--	0.11	--	Bf	4.34
0.14	419,332*	Bf	3.57	0.14	--	Bf	1.07
--	--	--	--	0.18	251	--	1.86
0.27	200,337*	Bf	3.17	0.24	200,334*	Bf	4.70
0.32	339	Gf	29.25	0.31	339,250*	Gf	13.56
0.41	228,340*	Bf	26.36	0.39	246,340*	Bf	28.38
0.50	252	Bf	2.10	0.47	251	Bf	2.26
0.59	227,393*	Gf	24.73	0.56	200,364*	Gf	29.35
--	--	--	--	0.75	268	Bf	2.38
0.87	200,293*	Gf	5.18	0.86	200,343*	Gf	8.70

* Shoulder inflection

C. After derivatization with anisaldehyde sulphuric acid reagent

R_f value	Colour of the band
0.16	Light pink
0.26	Brown
0.63	Pink
0.69	Light pink
0.78	Pinkish brown
0.80	Pinkish brown
0.90	Light pink
0.95	Light pink

Itrifal Mulaiyin and Sharbat-e-Dinar

Itrifal Mulaiyin and *Sharbat-e-Dinar* are prescribed for various diseases (Tables 5 & 6). *Itrifal Mulaiyin* consists of 11 ingredients including *Triphala* (equal mixture of fruit pulp of *Terminalia bellerica*, *T. chebula* and *Emblica officinalis*) and *Rewand chini* (*Rheum emodi* rhizome), while *Sharbat-e-Dinar* contains 9 ingredients including *Gul-e-Nilofar* (*Nymphea alba* flower) and *Rheum emodi* rhizome (Tables 5 & 6). Gallic acid was reported from *Triphala* and *Gul-e-Nilofar*, emodine was reported from *Rewand chini* [8]. Efforts were made to simplify the sample preparation step and resolve gallic acid and emodine in the same solvent system. *Itrifal Mulaiyin* was found to contain gallic acid and emodine. In

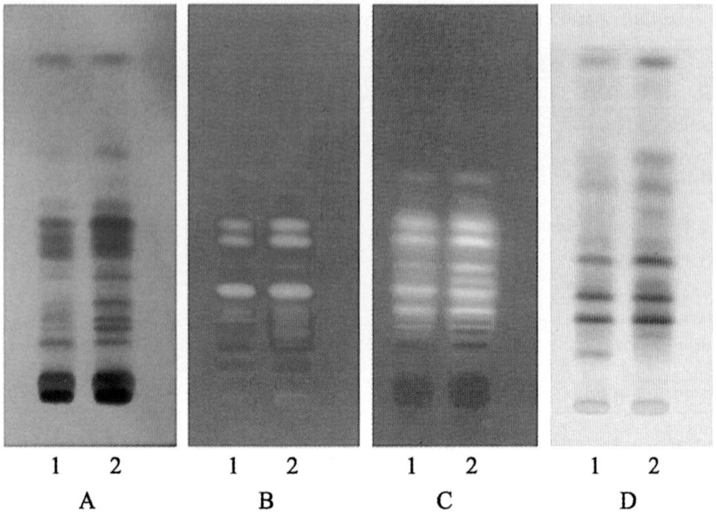

Figure 1. TLC Fingerprint profile of two different samples of *Triphala guggulu* in solvent system-1.

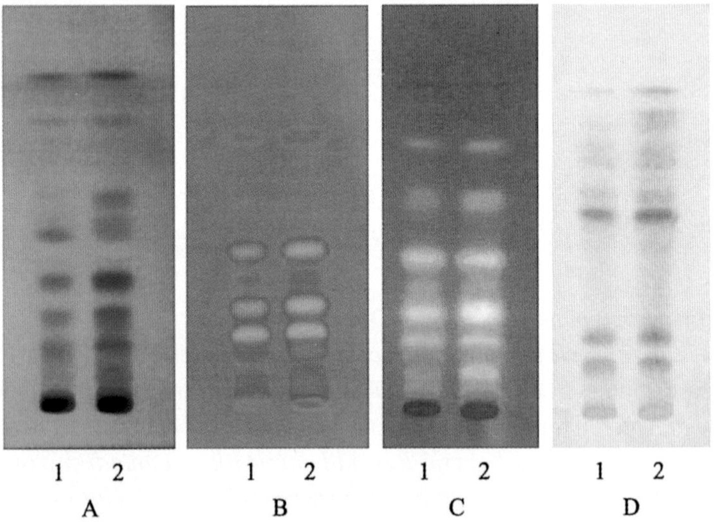

Figure 2. TLC Fingerprint profile of two different samples of *Triphala guggulu* in solvent system-2. (**A**) Under UV 254 nm, (**B**) Under UV 366 nm, (**C**) Under UV 366 nm after derivatization with anisaldehyde sulphuric acid reagent, (**D**) In visible light after derivatization with anisaldehyde sulphuric acid reagent. 1 – MS1, 2 – MS2.

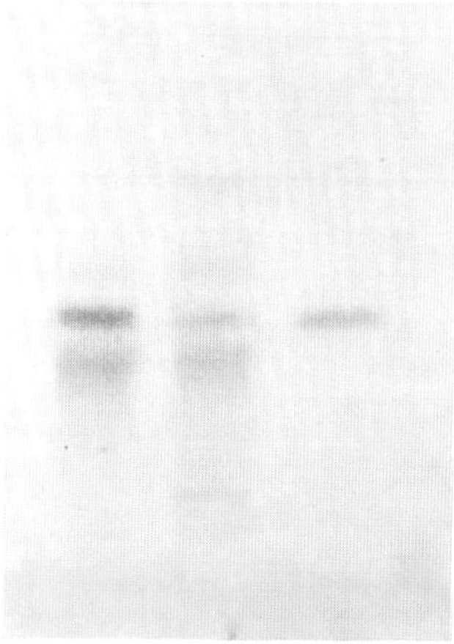

PS TA G

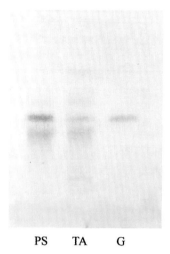

Figure 3. TLC Fingerprint profile of *Prabhakara vati* with *Terminalia ajuna* extract and gallic acid. PS- prepared sample, TA- Extract of *Terminalia ajuna* stem bark, G- Gallic acid standard.

Sharbat-e-Dinar gallic acid was found but emodine was not detected (Figures 6 & 11). This indicates either the absence or poor quality of *Rewand chini* used in the formulation.

Quantification of marker compounds

Efforts were made to resolve more than one marker in the same solvent system which enabled simultaneous estimation of some of the marker compounds.

Chandraprabha vati

It was possible to resolve Z-guggulsterone and pierine from methanolic extract of CPV in the same solvent system and gallic acid and berberine in another solvent system and these were quantified simultaneously. Though gallic acid and ellagic acid resolved in the same solvent system, gallic acid was present in free form in the sample and was extracted in methanol without hydrolysis, while hydrolysis step was essential for ellagic acid extraction. Hence ellagic acid was estimated after hydrolysis.

Six marker compounds were quantified in all the three samples and compared (Table 11A). The amount of Z-guggulsterone varied from 0.05

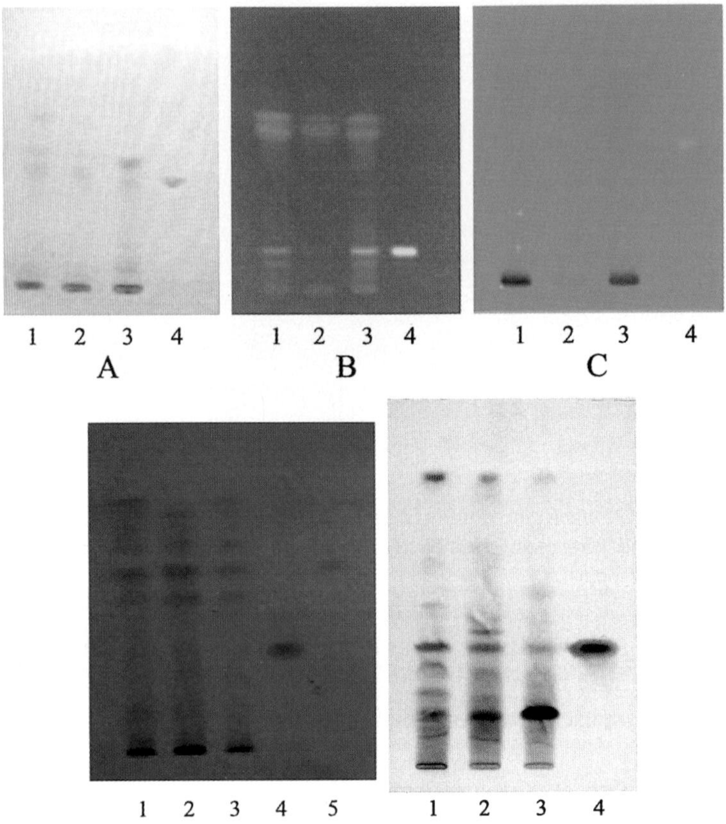

Figure 4. TLC Fingerprint profile of *Chandraprabha vati* with marker compounds, (**A**) with berberine (**B**) with gallic acid (**C**) with ellagic acid (**D**) with piperine and Z-guggulsterone (**E**) with β-sitosterol; 1, 2, 3 – sample solutions; 4,5 – standards.

to 0.3% in different samples and it was found to fall within the range expected. Expected Z-guggulsterone content from the formulation was 0.092 – 0.18%, as calculated from the reported value of *Z*-guggulsterone content in Guggul (0.40 – 1.5%) (Table 11A). *Z*-Guggulsterone, piperine and berberine are specific markers, the presence of which in the samples confirms the presence of the respective raw materials in the formulation. The presence and quantity of Z-guggulsterone is important since guggul is major ingredient of the formulation. Gallic acid and ellagic acid are general

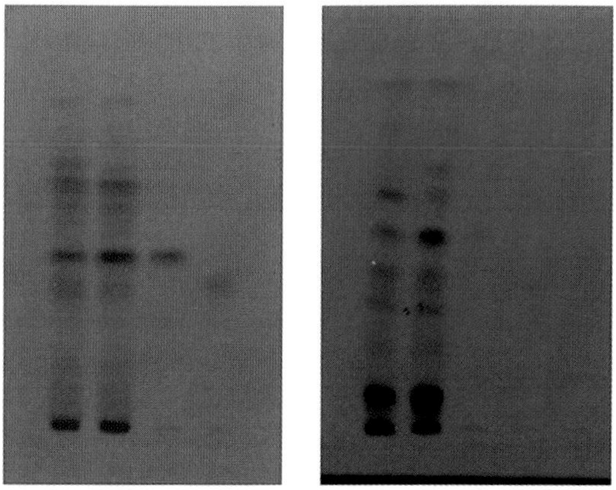

Figure 5. TLC Fingerprint profile of two different samples of *Triphala guggulu* with marker compounds, (**A**) with piperine and Z-guggulsterone, (**B**) with gallic acid and ellagic acid.
1 – MS1; 2 – MS2; G – Gallic acid; E – Ellagic acid; P – Piperine; Z – Z-Guggulsterone.

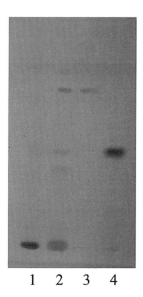

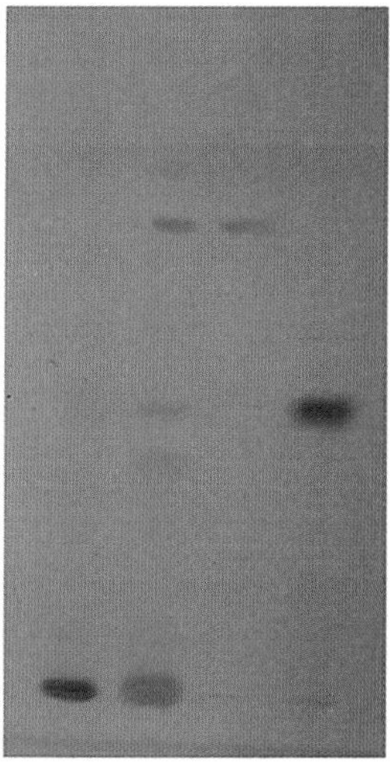

Figure 6. TLC fingerprint profile of methanolic extract of *Sharbat-e-dinar* and *Itrifal muliyan* with gallic acid and emodine, 1. Sample solution of *Sharbat-e-dinar*, 2. Sample solution of *Itrifal muliyan*, 3. Emodine standard, 4. Gallic acid standard.

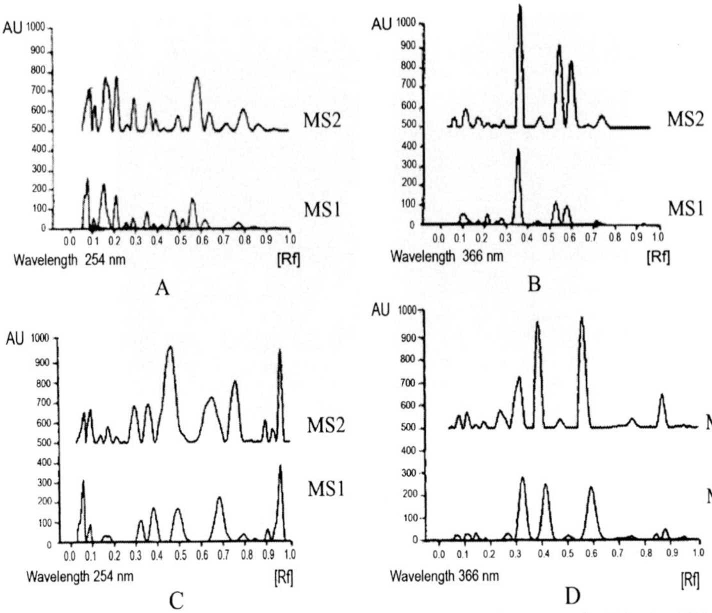

Figure 7. TLC densitometric chromatogram of *Triphala guggulu* (**A**). At 254 nm in solvent system-1 (**B**) at 366 nm in solvent system-1 (**C**) At 254 nm in solvent system-2. (**D**) at 366 nm in solvent system-2.

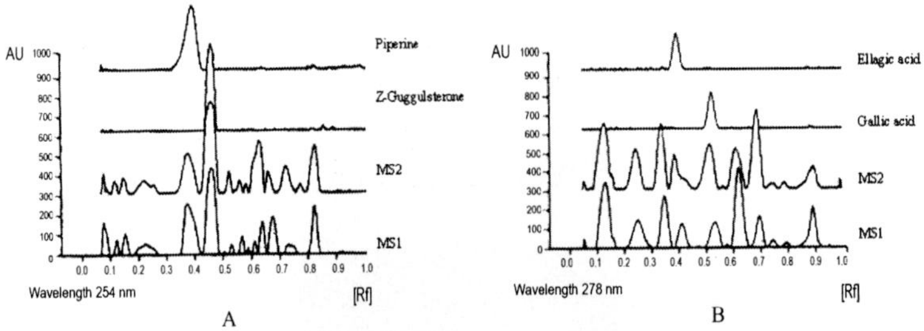

Figure 8. TLC densitometric chromatogram of *Triphala guggulu* with marker compounds. (**A**) with piperine and Z-guggulsterone, (**B**) with gallic acid and ellagic acid.

markers since they are present in many ingredients of the formulation. However, the amount of these markers in the formulation can serve as an extra parameter in quality control.

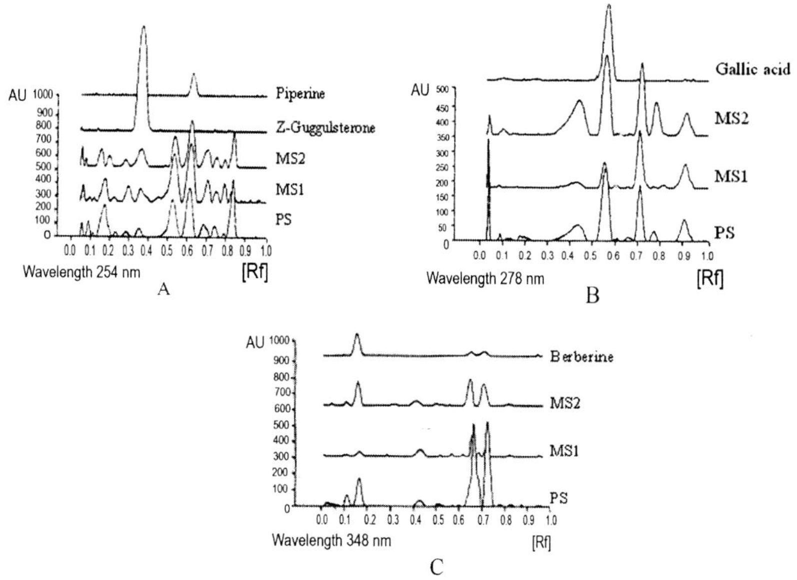

Figure 9. TLC densitometric chromatogram of fingerprint profile of *Chandraprabha vati* with marker compounds. (**A**) with piperine and Z-guggulsterone, (**B**) with gallic acid, (**C**) with berberine.

Itrifal Mulaiyin and *Sharbat-e-Dinar*

Itrifal Mulaiyin was found to contain 0.13% gallic acid and 0.015% emodine. In *Sharbat-e-Dinar* gallic acid was found to be 0.79 mg/ml, but emodine was not detected. This indicates either the absence or poor quality of *Rewand chini* used in the formulation (Table 11B).

Prabhakara vati

The amount of galic acid and ellagic acid in the authentic sample was 0.115% and 0.0164%, whereas these markers were not detected in the market samples (Table 11C). The fingerprint analysis and the absence of marker compounds in the market samples indicate that decoction of *T. arjuna* stem bark was not seemed to have been used in the preparation of these market samples. *T. arjuna* stem bark was established for its activity in cardiac disorders [9, 10]. Since *Prabhakara vati* is prescribed for cardiac disorders, during its preparation bhavana with *T. arjuna* stem bark decoction is an important step for the efficacy of the preparation. Hence, the phytochemical evaluation to ascertain the presence of the

Table 9. TLC of methanolic extract of *Triphala guggul* with marker compounds under UV at 254 nm

Z-Guggulsterone & Piperine		Gallic acid & Ellagic acid	
MS1	**MS2**	**MS1**	**MS2**
0.05	0.05	0.08	0.08
0.13	0.13	0.13	0.12
0.25	0.24	0.15	0.15
0.35	0.34	0.23	0.22
0.40 (Piperine)	0.39	-	0.26
0.53 (Z-Guggulsterone)	0.52	0.37 (Ellagic acid)	0.38
-	0.56	0.46 (Gallic acid)	0.46
0.62	0.60	0.53	0.52
0.69	0.68	0.56	0.56
0.74	0.74	0.58	0.58
-	0.78	0.63	0.63
0.88	0.88	0.65	0.65
-	0.98	0.72	0.72
		-	-
		0.82	0.82

Table 10. TLC of methanolic extract of *Chandraprabha vati* with marker compounds

A. With piperine and Z-guggulsterone standards

PS	MS1	MS2
0.06	0.06	0.06
0.08	0.09	0.10
0.16	0.11	0.12
0.20	0.18	0.18
0.28	0.23	0.22
0.36 (Piperine)	0.28	0.30
0.43	0.36	0.36
0.53	0.38	0.53
0.62 (Z-Guggulsterone)	0.52	0.62
0.70	0.62	0.70
0.75	0.68	0.75
0.79	0.69	0.79
0.84	0.74	0.83
	0.74	0.92
	0.79	

B. With gallic acid and ellagic acid

PS	MS1	MS2
0.06	0.05	0.05
0.09	0.09	0.10
-	0.20	0.21
0.33	-	0.34
0.42 (Ellagic acid)	0.41	0.42
0.46 (Gallic acid)	0.46	0.46
0.55	--	0. 55
0.58	0.58	--
0.63	--	--
0.78	0.76	0.78

C. With β-sitosterol (After derivatization with anisaldehyde-sulphuric acid reagent)

PS		MS1		MS2	
R_f	**Colour of the band**	**R_f**	**Colour of the band**	**R_f**	**Colour of the band**
0.16	Brown	0.16	Brown	0.16	Brown
0.19	Light pink	0.19	Light pink	-	-
0.23	Light pink	0.23	Light pink	0.23	Light pink
0.26	Light pink	0.26	Light pink	0.26	Light pink
-	-	0.28	Light pink	-	-
-	-	0.30	Light pink	0.30	Light pink
-	-	0.33	Light pink	-	-
0.36	Dark pink (β-Sitosterol)	0.36	Dark pink	0.36	Dark pink
0.41	Light pink	0.41	Light pink	0.41	Light pink
0.55	Dark pink	0.50	Dark pink	0.50	Dark pink
0.62	Light pink	0.62	Light pink	0.59	Light pink
0.67	Light pink	0.67	Light pink	0.67	Light pink
0.75	Light pink	-	-	-	-
0.85	Light pink	0.85	Light pink	0.85	Light pink
0.93	Grey	0.93	Grey	0.93	Grey

components of *T. arjuna* stem bark in the preparation is an essential step in quality control.

Phalatrikadi kvatha churna

Phalatrikadi kvatha churna contains *Picrorhiza kurroa* rhizome, *Triphala* and *Tricosanthes dioica* whole plant (Table 3) and it is widely used in ayurveda as antipyretic and antiemetic [11].

Table 11. Estimation of marker compounds

A. *Chandraprabha vati*

Markers	MS1	MS2	PS
Z-Guggulsterone	0.147 ± 0.001	0.212 ± 0.0072	0.177 ± 0.007
Gallic acid	0.191 ± 0.002	0.132 ± 0.003	0.150 ± 0.007
Ellagic acid	0.196 ± 0.002	0.079 ± 0.003	0.155 ± 0.00
Piperine	0.037 ± 0.002	0.042 ± 0.001	0.028 ± 0.001
Berberine	0.0055 ± 0.0004	0.0028 ±0.005	0.0071 ± 0.0001
β-Sitosterol	0.0232 ± 0.0007	0.0206 ± 0.0002	0.0252 ± 0.0005

B. *Itrifal Mulaiyin* and *Sharbat-e-Dinar*

Markers	***Itrifal Mulaiyin*** (% w/w)	***Sharbat-e-Dinar*** (mg/ml)
Gallic acid	0.13 ± 0.011	0.79 ± 0.001
Emodine	0.015 ± 0.0008	Not detected

C. *Prabhakara vati*

Samples	Gallic acid	Ellagic acid
PS	0.115	0.0164
MS1	-	-
MS2	-	-

\- not detected

D. *Phalatrikadi kvatha churna*

Markers	PS	MS1	MS2
Gallic acid	0.780 ± 0.059	0.368 ± 0.0084	0.348 ± 0.0023
Ellagic acid	0.121 ± 0.0056	0.085 ± 0.0049	0.065 ± 0.049
Picroside–I	0.165 ± 0.001	0.037 ± 0.007	0.039 ± 0.018

Efforts were made to simplify the sample preparation step and resolve more than one marker in the same solvent system, which enabled simultaneous estimation of gallic acid and ellagic acid. The amount of gallic acid in the authentic sample was 0.7%, whereas in market samples it was 0.35 - 0.37%. Ellagic acid was found to be 0.12% in the authentic sample,

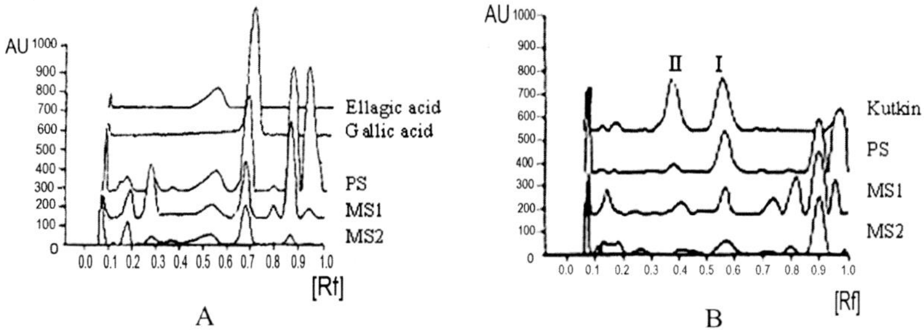

Figure 10. TLC densitometric chromatogram of fingerprint profile of *Phalatrikadi kvatha churna* with marker compounds. (**A**) with gallic acid and ellagic acid

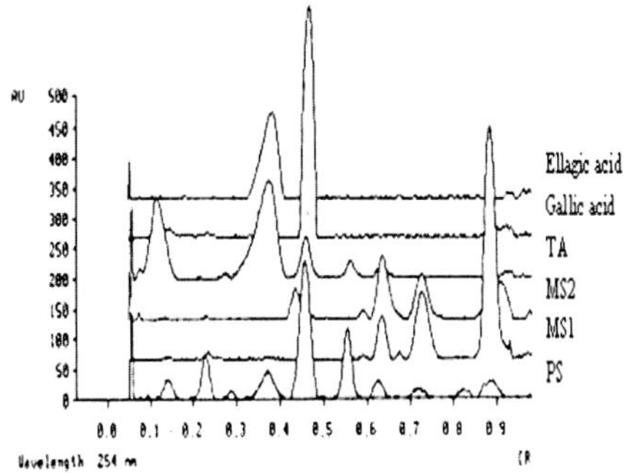

Figure 11. TLC densitometric chromatogram of fingerprint profile of *Prabhakara vati.* PS –

whereas it varied from 0.065 - 0.085 in the market samples. In authentic sample picroside–I was found to be 0.165%, but in market samples it was found to be in very less amount i.e. 0.037 - 0.039% (Table 11D, Figure 10).

CONCLUSION

Present study of qualitative and multiple-marker based semi-quantitative and quantitative phytochemical evaluation of several classical formulations of Indian Systems of Medicine is an effort to develop and establish methods to assess and maintain quality standards of polyherbal preparations. The general methodology evolved is applicable to any herbal preparations, keeping in view the well accepted synergistic and additive role of different chemical constituents of the herbal ingredients of the preparations.

ACKNOWLEDGEMENTS

The authors thank Prof. Harish Padh, Director, B. V. Patel PERD Centre, for providing the facilities and Industries Commissionerate, Govt. of Gujarat, for the financial aid towards instrumentation facility.

REFERENCES

[1] Forni GP. Thin layer chromatography and high performance liquid chromatography for analysis of extracts. Fitoterapia 1980; 51: 13-33

[2] Ravishankara M, Shrivastava NN, Padh H, Rajani M. HPTLC method for the estimation of alkaloids from *Cinchona officinalis* stem bark and its marketed formulations. Planta Med 2001; 67: 294-6

[3] Rajani M, Ravishankara, MN, Shrivastava N, Padh H. HPTLC-aided phytochemical fingerprinting analysis as a tool for evaluation of herbal drugs. A case study of Ushaq (Ammoniacum gum). J Planar Chromatogr 2001; 14: 34-41

[4] Quality Standards of Indian Medicinal Plants, vol-1, Indian Council of Medical Research, New Delhi; 2003

[5] Stahl E, Schild W. Pharmazeutische Biologie. 4. Drogenanalyse II : Inhaltsstoffe und Isolierungen, Gustav Fischer Verlag, Stutgart, New York; 1981, p. 410

[6] Bagul MS, Ravishankara MN, Padh H, Rajani M. Phytochemical evaluation and free radical scavenging properties of rhizome of *Bergenia ciliata* (Haw) sternb. forma *ligulata* Yeo. J Nat Rem 2003; 3: 83-7

[7] The Ayurvedic Formulary of India, Government of India, Ministry of Health and Family Welfare, Department of ISM &H, New Delhi, Part-II, 2000; p. 175

[8] *National Formulary of Unani Medicine*, Part-I. Government of India, Ministry of Health and Family Welfare, Department of Health, New Delhi; 2000. p. 94, 222.

[9] Bharani A., Ganguli A., Mathur LK, Jamra Y, Raman PG. Efficacy of *Terminalia arjuna* in chronic stable angina: A double blind placebo controlled, crossover study comparing *T. arjuna* with isosorbide mononitrate. Indian Heart J 2002; 54: 170-5

[10] Verma SK, Bordia A. Effect of *Terminalia arjuna* bark (Arjun chal) in patients of congestive heart failure and hypertension. J Res Edu Indian Med 1988; 7: 31-6

Application of Chromatographic Fingerprint to Quality Control for *Clematis chinensis*

Zhi Zeng[a,b] * and Jiuwei Teng[c]
[a]Department of Chemistry, South China Normal University, Guangdong, Guangzhou 510631, P. R. China; [b]Department of Biotechnology, Faculty of Bioresources Science, Akita Prefactural University, Akita 010-0195, Japan; [c]China National Analytical Center, Guangzhou 510070, P. R. China

INTRODUCTION

Chinese herbal medicines were traditionally identified mainly on their visual appearance either with the naked eye or microscope. Their specifications consisted of verbal descriptions, drawings, or photographs. Other approaches based on human senses such as taste and smell have also been used for quality control for Chinese herbal medicines by highly experienced individuals. However, such methods were far from objective and depended more on personal skill than scientific methods. On the other hand, using one of the powerful analytical techniques, the chromatographic fingerprint has recently been developed and used for the accurate identification and quality control of herbal medicines [1]. Fingerprint analysis has been introduced and accepted by WHO as a strategy for the assessment of herbal medicines [2]. Recently, fingerprint has also been required by the Drug Administration Bureau of China to standardize injections made from herbal medicines and corresponding raw materials [3]. The use of fingerprints in herbs tends to focus on identifying major constituents and assessing the compositional stability of the raw materials. Fingerprint analysis by high performance liquid chromatography (HPLC) and gas chromatography (GC) has been reported to be used on botanical medicines and their raw materials for similar purposes [4 ~ 8].

Pyrolysis-gas chromatography (Py-GC) is a powerful tool for the investigation of plant materials. This feature has been exploited in a number of applications of Py-GC to the study of plant forage [9] and cigarettes [10]. Pyrolysis-gas chromatography (Py-GC) is a technique combining pyrolysis with gas chromatography. It consists of thermal pyrolysis of biomolecules by heating them for a few seconds at a high temperature.

The products of the pyrolysis are small volatile molecules, which can be analyzed by GC thus providing information on the original sample. The most important is that the volatile composition released thermally from plant materials is analyzed by GC after interfacing automatically into GC instrument. Analytical pyrolysis has certain advantages over other conventional degradation techniques in that it can be easily interfaced to a GC instrument and automated; it needs very small sample amounts and practically minimum preparation, and is capable of providing information on the composition of different classes of substances. If plant material was pyrolyzed, the polymeric lignocellulose's matrix decomposes and typical volatile or semi-volatile pyrolysis products are formed. Further on, low molecular weight components present in plant materials are thermally released and eventually partly degraded during the process.

Clematis chinensis Osbeck is a liana distributed widely in China. Its root has been a source of the Chinese crude drug, "Wei-Ling-Xian", which has been recorded in Chinese Pharmacopoeia (2000 edition). As a commonly used Chinese herbal medicine, it has been used as an analgesic, diuretic and anti-inflammatory agent [11]. Although a number of chemical constituents have been isolated from this plant, the identities of the bioactive compounds remain unknown [12]. The aim of this study was to develop a characteristic fingerprint of *Clematis chinensis Osbeck* by using Py-GC to identify the raw herb.

MATERIALS AND METHODS

A Shimadzu GC-16A gas chromatograph with FID detector and PRY-2A pyrolysis system (Shimadzu) was used in the present study. Silicone capillary column 25 m × ϕ 0.32 mm ZYT-1 was purchased from Lanzhou Institute of Chemistry and Physics, the Chinese Academy of Sciences, China. Eight specimens of *Clematis chinensis Osbeck* from different producing regions used in this experiment (Table 1) had been authenticated by authors. Voucher specimens were deposited at the Laboratory of the Department of Chemistry, South China Normal University.

2 g of dry sample of *Clematis chinensis Osbeck* from each source was ground into powder. The samples were kept in small vials before analysis. Column temperature was initially set at 60 °C, held for 5 min, and then increased linearly at 4 °C min^{-1} to 250 °C. 10 mg of powdered

Table 1. Source of *Clematis chinensis Osbeck* samples

Sample No.	1	2	3	4	5	6	7	8
Origin	Liao-ning-1	Jiang-su	Shan-xi	Hu-nan	Guang-dong	Guang-xi	Liao-ning-2	Liao-ning-3

sample was directly placed into the pyrolyzer at 400 °C for 30 seconds at a flow rate of 35 ml min^{-1} carrier gas. Split ratio was 100: 1.

RESULTS AND DISCUSSION

Pyrogram of sample No. 2 of *Clematis chinensis Osbeck* obtained under the above conditions is shown in Figure 1. The procedure for raw data collection is basically the same as that described in our previous paper [5 ~ 8]. Peaks with relative peak areas (ratio of each peak area to the total area) larger than 1 % were selected for the fingerprint analysis. There were 34 corresponding peaks, i.e. fingerprints, for all of the eight

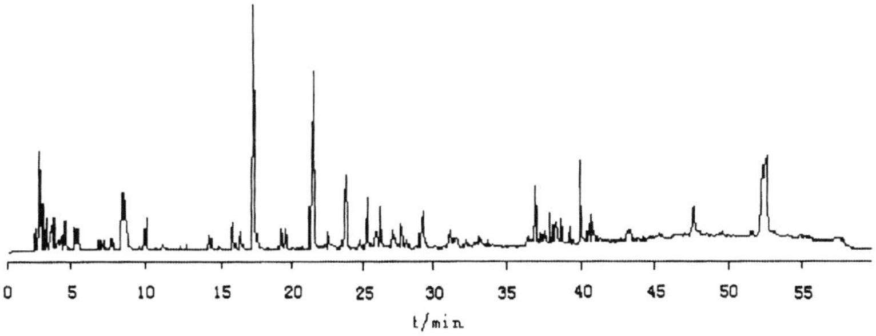

Figure 1. Pyrolysis-gas chromatography (Py-GC) fingerprint of sample No.2

samples in this study. A reference peak was also selected for the calculation of the relative retention values, α. This internal standard should be a pyrolyzate that is common to all samples and have a stable and substantial intensity. Peak 11 was chosen for such purpose and its α is 1. The α and Sr values [5, 7] of pyrolysis-gas chromatographic (Py-GC) fingerprints of eight *Clematis chinensis Osbeck* from different sources are listed in Table 2 and the pyrolysis-gas chromatogram of the eight crude materials are shown in Figure 2.

Spectral contrast angle (SCA) method [13] was based on the concept that a spectrum with N peaks can be expressed as a N-dimension vector. For example, a chromatogram, A, in this paper may be represented by vector (A_1, A_2, …, A_N), where A_i is the relative area of the *i*th peak. The similarity of two such chromatograms, A and B, then could be measured by the contrast angle θ of their two corresponding vectors, which must be of the same dimension:

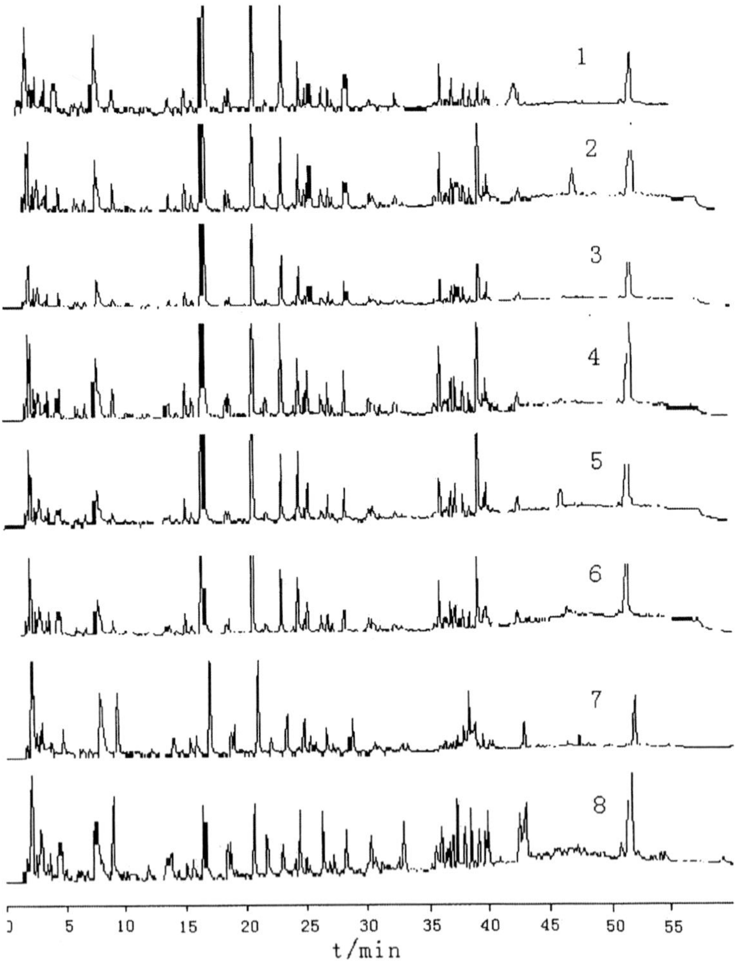

Figure 2. Pyrolysis-gas chromatographic (Py-GC) fingerprints of *Clematis chinensis Osbeck* from different producing regions

Table 2. The α and Sr values of Py-GC fingerprints of 8 samples of *Clematis chinensis Osbeck*

α	Sr							
	1	2	3	4	5	6	7	8
0.022	1.97	1.17	1.14	0.94	0.90	0.96	2.53	1.42
0.070	1.59	1.13	1.08	0.82	0.85	1.05	1.29	0.89
0.112	1.44	1.15	1.16	0.99	1.02	1.17	2.58	2.12
0.280	6.98	4.00	3.21	2.54	2.42	3.83	4.77	6.14
0.366	2.56	1.64	1.16	0.99	0.86	1.92	5.86	7.60
0.523	0.39	0.53	0.50	0.25	0.19	0.48	0.82	0.92
0.615	1.41	1.31	0.93	0.77	0.61	0.90	1.66	2.07
0.692	1.99	1.79	2.04	1.83	1.72	2.09	1.76	0.96
0.723	0.94	0.84	0.82	0.66	0.66	1.01	1.57	1.16
0.772	20.61	11.94	12.61	11.86	10.94	14.40	10.23	3.93
0.879	1.37	1.27	1.15	1.03	0.95	1.28	2.32	3.84
0.896	1.57	1.17	1.31	1.19	1.08	1.39	2.46	2.22
1.000	10.27	9.61	10.58	10.33	10.87	10.90	9.58	7.61
1.062	1.14	1.41	1.39	1.23	1.13	1.42	2.20	4.55
1.125	8.00	4.20	5.48	5.06	4.21	6.03	3.92	2.46
1.183	0.49	0.68	0.77	0.72	0.60	0.63	0.95	1.32
1.206	2.45	3.35	4.16	4.01	4.32	3.13	3.26	3.53
1.237	1.18	1.10	1.16	1.11	0.97	1.25	1.35	1.26
1.252	1.43	2.53	2.64	2.32	2.96	2.61	1.15	0.77
1.304	1.22	1.08	1.07	0.94	0.72	1.05	1.77	3.02
1.336	1.40	1.53	1.63	1.46	1.56	1.61	1.04	1.19
1.416	2.53	2.09	2.09	1.98	1.88	2.18	2.88	2.99
1.525	1.09	1.18	1.32	1.36	2.26	1.19	1.19	2.12
1.672	0.97	0.63	0.28	0.77	0.71	0.70	1.84	4.56
1.838	2.64	3.09	3.27	3.48	3.17	3.19	1.49	1.78
1.895	1.96	1.97	1.91	2.15	1.73	1.94	1.55	1.86
1.914	0.73	2.16	2.52	2.53	2.54	2.07	3.12	3.65
1.938	1.59	1.83	2.13	2.32	1.59	1.97	5.67	2.39
1.970	0.94	1.44	1.64	1.79	1.58	1.27	3.48	4.11
2.008	1.88	6.03	6.42	6.40	7.53	6.39	2.36	2.89
2.050	0.89	2.10	1.94	2.03	2.08	1.67	1.15	2.21
2.190	5.76	1.76	2.11	2.13	2.01	1.61	2.95	1.81
2.419	0.77	6.10	3.64	5.09	7.02	1.23	1.67	1.93
2.700	7.85	16.22	14.75	16.92	16.37	15.47	7.60	8.75

$$\cos\theta = \Sigma A_i B_i \ / \sqrt{\Sigma A_i^2 \times \Sigma B_i^2} \quad (1)$$

Where $\theta = 0°$, A and B are identical while 90°($\pi/2$) implies they are totally different. The similarity index defined in reference [13] is actually the measurement of difference. The following modified equation measures the similarity index (SI):

$$SI = 100 - \sqrt{\Sigma \{[(i_a - i_b)/(i_a + i_b)] / N\}} \times 100 \quad (2)$$

Big Eight Peaks (BEP) is a simple empirical approach for fingerprint analysis [5]. In its simplest form of spectral comparison, instead of comparing every corresponding peak, only the strongest eight peaks were compared. Those BEPs, however, were often different from spectrum to spectrum. In the case of comparison of multiple spectra, the top eight of the BEPs shared by most spectra were employed in the spectral comparison. In this study, spectral comparison had been carried out by three approaches: (a) comparison with all peaks whose relative peak areas are larger than 1 %, i.e. 43 peaks; (b) comparison with all BEPs; (c) comparison with top eight most shared BEPs. The numbers of peaks compared were 34, 14 and 8. Tables 3, 4 and 5 show the SIs calculated from equation (2) by methods (a), (b) and (c), respectively. Similarly, SCAs data have been shown in Tables 6, 7 and 8. All the tables are diagonally symmetrical, due to the symmetry of both spectral comparison techniques. That is, A compared to B is identical to B compared to A for both SI and SCA. So only half of the data were needed. The values highlighted in bold, 100 in SI tables and 0 in SCA tables, resulted from self-comparison and are identical.

The pattern was very consistent, as shown in the six tables. There were two distinguished groups of samples. Group I consisted of samples 2, 3, 4, 5, and 6 and were very similar to one another, while group II included samples 1, 7, and 8 and were quite different from all other samples. One interesting phenomenon was that the number of peaks selected for spectral comparison had very limited effects on both SIs and SCAs in this

Table 3. Similarity Indexes of 8 samples, compared with all 43 peaks

SI	1	2	3	4	5	6	7	8
1	**100.00**	73.45	71.85	70.24	67.14	77.60	72.18	62.59
2	73.45	**100.00**	89.64	88.14	84.43	86.76	70.45	65.85
3	71.85	89.64	**100.00**	88.20	84.18	86.57	69.17	64.26
4	70.24	88.14	88.20	**100.00**	91.36	84.15	67.07	62.51
5	67.14	84.43	84.18	91.36	**100.00**	80.18	62.92	59.73
6	77.60	86.76	86.57	84.15	80.18	**100.00**	71.55	65.25
7	72.18	70.45	69.17	67.07	62.92	71.55	**100.00**	78.99
8	62.59	65.85	64.26	62.51	59.73	65.25	78.99	**100.00**

Table 4. Similarity Indexes of 8 samples, compared with all 14 BEPs

SI	1	2	3	4	5	6	7	8
1	**100.00**	65.54	66.35	62.83	60.51	73.26	67.07	58.74
2	65.54	**100.00**	89.32	88.62	87.45	81.05	65.70	63.35
3	66.35	89.32	**100.00**	93.26	87.84	83.59	66.73	62.59
4	62.83	88.62	93.26	**100.00**	91.83	79.08	64.24	60.31
5	60.51	87.45	87.84	91.83	**100.00**	75.79	60.37	58.15
6	73.26	81.05	83.59	79.08	75.79	**100.00**	68.22	63.52
7	67.07	65.70	66.73	64.24	60.37	68.22	**100.00**	78.01
8	58.74	63.35	62.59	60.31	58.15	63.52	78.01	**100.00**

Table 5. Similarity Indexes, compared with the top eight most-shared BEPs

SI	1	2	3	4	5	6	7	8
1	**100.00**	71.08	70.40	67.40	64.73	73.43	78.27	66.70
2	71.08	**100.00**	92.24	90.47	89.12	92.26	76.04	71.11
3	70.40	92.24	**100.00**	94.73	91.96	93.35	73.78	67.45
4	67.40	90.47	94.73	**100.00**	94.79	90.14	71.62	66.10
5	64.73	89.12	91.96	94.79	**100.00**	87.05	71.06	66.64
6	73.43	92.26	93.35	90.14	87.05	**100.00**	73.77	66.61
7	78.27	76.04	73.78	71.62	71.06	73.77	**100.00**	80.51
8	66.70	71.11	67.45	66.10	66.64	66.61	80.51	**100.00**

Table 6. Spectral Contrast Angles, compared with all 43 peaks

SCA	1	2	3	4	5	6	7	8
1	**0.00**	32.61	28.83	33.47	36.80	25.86	27.90	46.34
2	32.61	**0.00**	8.16	5.77	7.56	12.82	29.52	37.28
3	28.83	8.16	**0.00**	6.26	10.22	7.34	27.17	37.97
4	33.47	5.77	6.26	**0.00**	6.61	11.61	30.32	38.70
5	36.80	7.56	10.22	6.61	**0.00**	16.30	32.85	39.55
6	25.86	12.82	7.34	11.61	16.30	**0.00**	26.92	38.65
7	27.90	29.52	27.17	30.32	32.85	26.92	**0.00**	25.17
8	46.34	37.28	37.97	38.70	39.55	38.65	25.17	**0.00**

Table 7. Spectral Contrast Angles, compared with all 14 BEPs

SCA	1	2	3	4	5	6	7	8
1	**0.00**	33.10	28.99	33.81	37.08	26.07	26.71	44.98
2	33.10	**0.00**	8.26	5.50	6.89	13.10	28.79	34.56
3	28.99	8.26	**0.00**	6.06	10.05	7.31	26.06	35.17
4	33.81	5.50	6.06	**0.00**	6.23	11.76	29.28	35.97
5	37.08	6.89	10.05	6.23	**0.00**	16.43	31.80	36.70
6	26.07	13.10	7.31	11.76	16.43	**0.00**	25.60	35.91
7	26.71	28.79	26.06	29.28	31.80	25.60	**0.00**	24.61
8	44.98	34.56	35.17	35.97	36.70	35.91	24.61	**0.00**

Table 8. Spectral Contrast Angles, compared with top 8 most-shared BEPs

SCA	1	2	3	4	5	6	7	8
1	**0.00**	30.52	27.69	32.19	34.41	25.03	18.11	38.71
2	30.52	**0.00**	6.05	4.48	6.62	6.75	19.66	22.12
3	27.69	6.05	**0.00**	5.40	6.94	4.01	16.73	23.29
4	32.19	4.48	5.40	**0.00**	4.28	7.64	21.16	23.78
5	34.41	6.62	6.94	4.28	**0.00**	9.92	22.06	22.96
6	25.03	6.75	4.01	7.64	9.92	**0.00**	15.88	24.83
7	18.11	19.66	16.73	21.16	22.06	15.88	**0.00**	22.92
8	38.71	22.12	23.29	23.78	22.96	24.83	22.92	**0.00**

study, as shown in Table 9.

Table 9. Comparisons Between Methods and Samples

Method	a		b		c	
Group	G I	G II	G I	G II	G I	G II
SI	86.36±3.23	68.49±5.30	85.78±5.70	64.75±4.98	91.61±2.46	70.93±4.45
SCA	9.27±3.41	33.11±5.89	9.16±3.60	31.96±5.36	6.21±1.79	24.56±6.10

Notes: **1.** Data are shown as ***mean ± standard deviation***.

2. G I indicated comparison within Group I, i.e., samples 2, 3, 4, 5, and 6, while G II indicated comparison within Group II, i.e., samples 1, 7, and 8, compared with all other samples.

The equation (3) is for the calculation of standard deviation S, where n is the number of samples, X_i is the SI or SCA of the *i*th sample, and $\overline{X}$ is the average of the SIs or SCA.

$$S = \sqrt{\frac{1}{(n-1)}\sum_{i=1}^{n}(X_i - \bar{X})^2} \qquad (\bar{X} = \frac{1}{n}\sum_{i=1}^{n} X_i \ ) \qquad (3)$$

Relative standard deviation (RSD) of peak area had been determined as 17 % in these experiments. The fluctuations of SI and SCA values caused by choosing a different number of peaks for comparison were smaller than that in the experiment, except for SCA measurements by method c. Apparently the SCA method was more sensitive than SI in spectral comparison. As demonstrated in Table 9, the same difference in samples had caused SCA values to change from 9.27 to 33.11, compared to changes in SI values from 86.36 to 68.49. The different behaviors of these two techniques result directly from their definitions.

CONCLUSIONS

Pyrolysis in conjunction with gas chromatography is one of the most efficient techniques to study the bulk composition of complex organic substances. Within Chinese herbal medicines, it is very useful for the discrimination of different chemical components at the genus, species and subspecies level. Compared with more conventional methods, Py-GC offers the advantages of speed and sensitivity. *Clematis chinensis Osbeck*, a Chinese herbal medicine, from different producing regions in China was studied by pyrolysis-gas chromatography and fingerprint techniques. Spectral comparison techniques, similarity index and spectral contrast angle, were used successfully for quantitative pyrogram analysis. Big eight peaks method, which would only consider those most dominant peaks, had been proven to be simple, reliable, and able to significantly reduce the workloads of spectral comparisons. Present study would provide a quantitative method for the quality control of Chinese herbal medicines.

ACKNOWLEDGEMENTS

Financial supports from Major Projects of Science and Technology of Guangdong Province (No. A301020101) and Natural Science Foundation of Guangdong Province (No. 011442) are gratefully acknowledged.

REFERENCES

[1] Bauer R. Drug Inform J, 1998; 32: 101-10
[2] World Health Organization. Guidelines for the Assessment of Herbal Medicines, Munich, 28.6.1991, WHO, Geneva, 1991
[3] Drug Administration Bureau of China. Requirements for Studying Fingerprint of Traditional Chinese Medicine Injections (Draft), 2000
[4] Hasler A, Sticher O and Meier BJ. J Chrom A 1992; 60: 41-8
[5] Zeng Z, Yang D, Song L, Yang T, Liu X, Yuan M and Zeng H. Chinese J Anal Chem 2002; 30: 849-52
[6] Yuan M, Zeng Z, Song L, Yang T, Liu X, Cao C and Zeng H. Chinese J Anal Chem 2003; 31: 455-8
[7] Zeng Z, Yang D, Song L, Yang T, Liu X, Yuan M and Zeng H. Chinese J Anal Chem 2003, 31: 1485-8
[8] Zeng Z, Yang D, Tao J, Yang T, Yuan M and Zeng H. Chinese J Anal Chem 2004; 32: 1035-8
[9] Reeves JB and Francis BA. J Anal Appl Pyrolysis, 1997, 40-41: 243-66
[10] Zhang J, Li K, Liu H and Chen Q. Chinese J Anal Chem, 2004; 32: 573-8

[11] Jiangsu New Medical College. The Dictionary of Traditional Chinese Medicines, 1977. pp: 3370
[12] Shao B, Qin G, Xu R, Wu H and Ma K. Phytochem, 1995; 38: 1473-9
[13] Wan KX, Vidavsky I and Gross ML. J Am Soc Mass Spectrom, 2002; 13: 85-8

The Bioactive Pigments from Marine Bacteria *Pseudomonas* sp

Hou-Jin Li[a]*, Wen-Jian Lan[b], Chuang-Hua Cai[c], Yi-Pin Zhou[c] and Yong-Cheng Lin[a]
[a]*School of Chemistry and Chemical Engineering, Sun Yat-sen University, Guangzhou 510275, China*
[b]*School of Pharmaceutical Sciences, Sun Yat-sen University, Guangzhou 510080, China*
**South China Sea Institute of Oceanology, The Chinese Academy of Sciences, Guangzhou 510301, China*
**Corresponding email: ceslhj@mail.sysu.edu.cn*

INTRODUCTION

Marine microorganisms are of considerable current interest as a new promising source of bioactive substances. Recently, marine microorganisms have been proven to produce a variety of chemically interesting and biologically significant secondary metabolites and some of them are expected to serve as lead compounds for drug development or pharmacological tools for basic life science studies. In our search for secondary metabolites of marine microorganisms collected from the South China Sea, many cytotoxic and/or novel compounds were isolated. These metabolites include different structural types: terpenoid, peptide and alkaloid, polyketide, polyester, and other types [1-4]. Marine microorganisms have provided seemingly endless novel structures.

A bacterial strain, *Pseudomonas* sp., collected from superficial seawater of the Daya Bay of the South China Sea, can produce red pigments. In a preliminary bioassay, the ethyl acetate extracts of the culture broth exhibited the cell growth inhibitory properties against the cancer cell lines Bel-7402, NCI-4460 at the dose of 10μg/ml, antimicrobial activities against the *Staphylococcus auereu*, *Pseudomonas pyocyanea*, *Escherichia coli*, *Bacillus subtilis* and *Saccharomyces cerevisae* and inhibitory activity of human DNA topoisomerase I *in vitro*. Here, we report the culture conditions, identification of the pigments and some other metabolites isolated from this bacterial strain.

RESULTS AND DISCUSSION

The bacteria, *Pseudomonas* sp., was cultivated in liquid medium containing peptone 5g/L, yeast extract 1g/L, 100% sea water, pH 7.5, at 28°C for 72h, then was harvested by centrifugation at 4000 rpm for 25min. The liquid was extracted with ethyl acetate, the precipitate was extracted with acidic methanol. The combined concentrates were subjected to flash column chromatography on silica gel followed by C-18 reversed-phase HPLC. The fraction afforded four pigments, the main bright red tripyrrole pigment prodigiosin, 4-methoxy-5- [(5-methyl-4- pentyl-2*H*-pyrrol-2-yliden) methyl]-2.2'-bi-1*H*-pyrrole (**1**), a new purple prodigiosin analogue compound: dimethyl prodigiosin (**2**), cycloprodigiosin (**3**) and a minor blue pigment (**4**). Additionally, two new lactones (**5-6**), which were derived from 3-hydroxy-decanoic acid (**7**) were also obtained. The structures of these compounds were elucidated by extensive analysis of IR, FAB-MS, LC-MS^n, 1D NMR and 2D NMR data and comparison with the literature data [5, 6].

Prodigiosin is a naturally occurring polypyrrole red pigment produced by a restricted group of microorganisms, including some *Streptomyces* and *Serratia* strains, characterised by a common pyrrolyldipyrrolylmethene skeleton. Prodigiosin, metacycloprodigiosin, nonylprodigiosin and undecylprodigiosin are all members of this family. Prodigiosin was first isolated from *S. marcescens*. As typical secondary metabolites, prodigiosin and related materials have no clearly defined physiological functions in the producing organisms. However, members of the Prodigiosin family have potent antimicrobial, antimalarial, immunosuppressive and cytotoxic activity. It induces apoptosis in several human cancer cell lines, hepatocellular carcinoma xenografts and human primary cancer cells, with no marked toxicity in non-malignant cell lines [7, 8].

We also identified 3-hydroxy-decanoic acid (**7**) and two new lactones (**5-6**) which derivated from it. Verma, et al. [9] investigated the correlation between antibiotic production, undecylprodigiosin and poly-3-hydroxybutyric acid (PHB) utilization in *S. coelicolor* A3 (2) M145. The results indicated a possible role of PHB as a carbon reserve material used for antibiotic production. PHB production increased rapidly during the exponential growth phase of the organism. Then, the cells utilized the polymer during the stationary phase of growth. The polymer may play a role in antibiotic biosynthesis, possibly by providing precursor molecules in the form of acetyl-CoA [9]. Whether there exists a correlation between the 3-hydroxy-decanoic acid and the production of prodigiosin and their analogues, it is still unknown.

The conditions in the culture medium were also preliminarily investigated. The production of prodigiosin, that causes a bright red pigmentation, is easily seen with the naked eyed. *Pseudomonas* sp. was cultivated in the liquid medium containing peptone 5g/L, yeast extract 1g/L, 100% sea water, pH 7.5, 28°C on a rotary shaker (rpm=140), the culture broth turned bright red after 24h. As a comparison, it took 72h in 50% seawater and 120h in 20% seawater. However, the culture did not turn red in distilled water. It was tentatively concluded that high salt concentration was much better for its growth and the production of red pigments. The quantitative correlation between them is still under investigation. Sevcikova and Kormanec [10] once reported that high salt concentration differentially affected production of two pigmented antibiotics, actinorhodin and undecylprodigiosin in *Streptomyces coelicolor* A3 (2), with actinorhodin being inhibited and undecylprodigiosin activated. The effect of high salt concentration on actinorhodin and undecylprodigiosin production is mediated at the transcriptional level by the differential expression of genes encoding corresponding pathway-specific transcriptional regulators. Proline is a precursor of undecylprodigiosin. Intracellular concentration of proline in salt-stressed medium increased more than 50-fold. This dramatic increase in the intracellular proline concentration under salt stress conditions may account for the observed induction of undecylprodigiosin. Thus, in salt-stressed cells, excess of intracellular proline is diverted into the production of undecylprodigiosin [10].

The structure of the minor blue pigment was tentatively determined by the ESI MS^n. ESI-MS: *m/z* 335 $[M+H]^+$. Previously, we cultivated *Pseudomonas* sp. at 18 °C, a lower temperature than the regular, and found that the culture turned blue, instead of turning red. A known blue pigment, violacein (**8**) was obtained from the culture extracts. Violacein showed significative cytotoxic activities against the cancer cell lines MCG803 and BEL-7402 with the IC_{50} values 4.6 and 6.8 μg/mL respectively [11]. There are some reports about the effects of temperature on bacterial prodigiosin production. *Sarratia marcescens* is a pigmented bacteria. Once the growth temperature of *Sarratia marcescens* is raised to 40 °C, the pigment stops being produced. To the best of our knowledge, it is the first account of blue pigment production under lower temperature. It is believed that the enzyme used in the production of prodigiosin is affected by the temperature.

EXPERIMENTAL SECTION

Isolation and characterisation of bacterial strain

Pigment producing strain of *Pseudomonas* sp. was isolated from

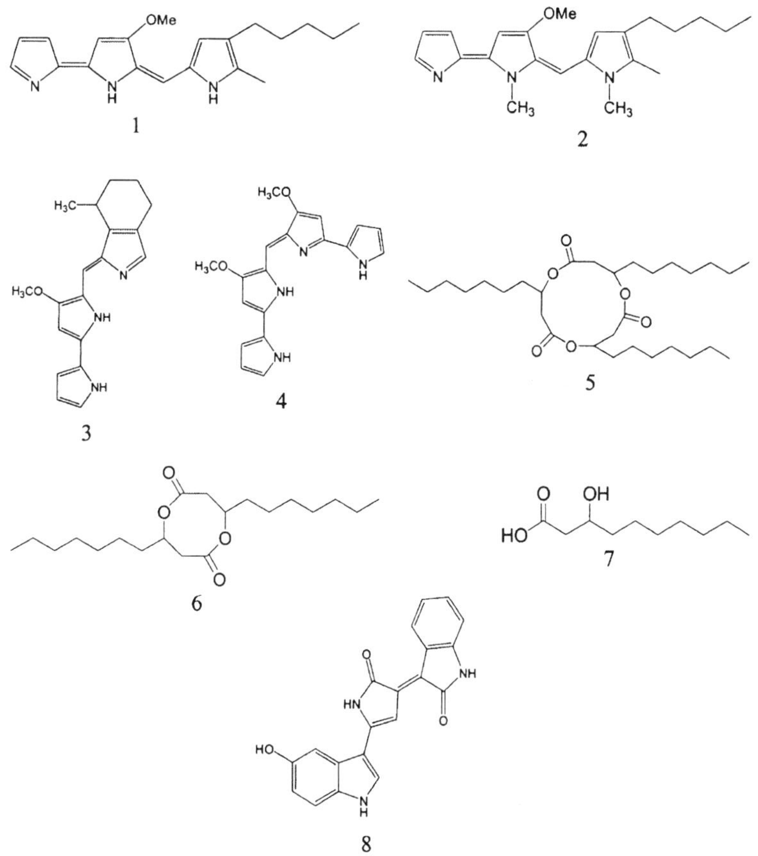

Figure 1. The structures of compounds isolated from marine bacteria *Pseudomonas* sp.

the superficial seawater of the Daya Bay of the South China Sea. Specimens were deposited at the School of Chemistry and Chemical Engineering, Sun Yat-Sen University, Guangzhou, China. The bacterial isolate was subjected to the various regular biochemical tests, lipase assay, and lactose fermentation to charaterise the genus. The isolate was further designed for the identification of *Pseudomonas* sp. based on the 16s rRNA sequence alignment.

Culture conditions

Pseudomonas sp. was cultivated in 250mL liquid medium containing peptone 5g/L, yeast extract 1g/L, 100% sea water, pH 7.5 in an Erlenmeyer flask. The flasks were incubated at 28°C on a rotary shaker (rpm=140) for 72h. 100 liters of a liquid medium was incubated. The bacteria were then harvested by centrifugation at 4000rpm for 15min. The liquid was extracted with ethyl acetate, the precipitate was extracted with acidic methanol (methanol-1N HCl (24:1)). Both extracts were mixed and concentrated by rotary evaporation.

Purification of metabolites

Atmospheric pressure liquid chromatography of the extract was performed on silica gel with a petroleum ether-chloroform-methanol gradient as the eluent. and was followed by preparative LC separations performed on Agilent 1100 HPLC. A C18 reversed phase column (300×7.8mm), MeOH/H_2O (70/30 v/v) was used as mobile phase at a flow rate of 2mL/min. The elution was monitored both using diode-array UV detector and by ESI-MS using a Therm Finnigan LCQ™ DECA XP LC-MS. The pure fractions were vacuum evaporated, the structures of these compounds were identified by spectroscopic analyses.

Compound characterization

Prodigiosin, ESI-MS: *m/z* 324 $[M+H]^+$. NMR ($CDCl_3$, 500MHz, ppm): 2-CH, δC 126.0, δH 7.17; 3-CH,110.7,6.29; 4-CH, 116.0,6.85; 5-C, 121.2; 6-C, 146.7; 7-CH, 91.6, 6.02; 8-C, 164.8; 9-C, 119.7; 10-CH, 115.0, 6.89; 11-C, 124.1; 12-CH, 127.5, 6.62; 13-C, 127.4; 14-C, 146.0; 15-CH_3, 11.5, 2.47; 16-CH_2, 24.3, 2.33; 17-CH_2, 28.8, 1.54; 18-CH_2, 30.9, 1.32; 19-CH_2, 21.5, 1.26; 20-CH_3, 13.0, 0.88; 21-OCH_3, 57.7, 3.94; 22-NH, 12.76; 23-NH, 12.58.

Dimethyl prodigiosin (**2**), ^1H NMR ($CDCl_3$, 500MHz, ppm): 7.24

(2-CH), 6.36 (3-CH), 6.96 (4-CH), 6.08 (7-CH), 6.92 (10-CH), 6.68 (12-CH), 2.55 (15-CH_3), 2.40 (16-CH_2), 1.56 (17-CH_2), 1.33 (18-CH_2), 1.26 (19-CH_2), 0.88 (20-CH_3), 4.01 (21-CH_3), 3.49(22-CH_3),1.43(23- CH_3).

Cycloprodigiosin (**3**) decomposes faster in room light. So, the red solid was obtained as cycloprodigiosin hydrochloride. UV (in $CHCl_3$), EI MS and the NMR data were identical to the literature data. ^{13}C NMR ($CDCl_3$, 125 MHz) δC 165.3, 147.4, 146.6, 146.0, 126.4, 124.2, 123.3, 122.9, 119.4, 116.2, 113.3, 111.7, 92.9, 58.9, 30.8, 26.6, 24.2, 21.2, 18.7.

Compound **5**, $C_{30}H_{54}O_6$, EA analysis found: C, 69.88; H, 11.06. FABMS: *m/z* 511 $[M+H]^+$. NMR ($CDCl_3$, 500MHz, ppm): C=O, δC 177.4; CH, δC 68.1, δH 4.03, m; CH_2, δC41.1, δH 2.56, dd, J=16.5, 3.0Hz, 2.46, dd, J=16.5, 9.0Hz; CH_2, δC 36.4, δH 1.44, m, 1.55, m; CH_2, δC31.8, δH 1.29, brs; CH_2, δC 29.4, δH 1.29, brs; CH_2, δC 29.2, δH 1.29, brs; CH_2, δC 25.4, δH 1.30, m, 1.56, m; CH_2, δC 22.6, δH 1.29; CH_3, δC 14.0, δH 0.88, t, J=6.5Hz; OH, δH 5.30, 2H, brs. Compound **6**, the NMR data were identical with Compound **5**. FABMS: *m/z* 341 $[M+H]^+$.

3-hydroxy-decanoic acid (**7**), $C_{10}H_{20}O_3$. EA analysis found: C, 63.88; H, 10.91. IR ν/cm^{-1} (KBr): 3426, 2957, 2926, 2855, 1720, 1571, 1405, 1303, 1128, 1084, 722. FABMS: *m/z* 189 $[M+H]^+$, 171, 153. ^1H NMR ($CDCl_3$, 500MHz, ppm): 6.20, brs, 2H; 4.02, m, 1H; 2.57, dd, J=16.5, 3.3, 1H, 2.46, dd, J=16.5, 9.0Hz, 1H; 1.56~1.44, m, 2H; 1.28, brs, 10H.

ACKNOWLEDGMENT

This work was supported by the National Natural Science Foundation of China (20502036), the Natural Science Foundation of Guangdong Province, China (05300667) and the Research Award Program for Young Teachers of Sun Yat-sen University (2004-36000-1131072).

REFERENCES

[1] Li HJ, Lin YC, Yao JH, Vrijmoed LLP, Jones EBG. Two new metabolites from

the mangrove endophytic fungus no. 2524. Journal of Asian Natural Products Research 2004; 6: 185-91

[2] Li HJ, Lin YC, Vrijmoed LLP, Jones EBG. A new cytotoxic sterol produced by an endophytic fungus from *Castaniopsis fissa* at the South China Sea Coast. Chinese Chemical Letters 2004; 15 : 419-22

[3] Lin YC, Wu XY, Feng SA, Jiang GC, Luo JH, Zhou SN, Vrijmoed LLP, Jones EBG, Krohn K, Steingrover K, Zsila F. Five unique compounds: Xyloketals from mangrove fungus *Xylaria* sp

[4] from the South China Sea coast. Journal of Organic Chemistry 2001; 66: 6252-6

[5] Lin YC, Wu XY, Feng S, Jiang GG, Zhou SN, Vrijmoed LLP, Jones EBG. A novel N-cinnamoylcyclopeptide containing an allenic ether from the fungus *Xylaria* sp (strain # 2508) from the South China Sea. Tetrahedron Letters 2001; 42: 449-51

[6] Boger DL, Patel M. Total synthesis of prodigiosin. Tetrahedron Letters 1987; 28: 2499-502.6. Gerber NN. Cycloprodigiosin from *Beneckea gazogenes*. Tetrahedron Letters 1983; 24: 2797-8

[7] Manderville RA. Synthesis, Proton-Affinity and Anti-Cancer Properties of the Prodigiosin- Group Natural Products. Current Medicinal Chemistry–Anti-Cancer Agents 2001; 1: 195-218

[8] Fürstner A. Chemistry and biology of roseophilin and the prodigiosin alkaloids: A survey of the last 2500 years. Angewandte Chemie International Edition 2003; 42: 3582-603

[9] Verma S, Bhatia Y, Valappil S P, Roy I. A possible role of poly-3-hydroxybutyric acid in antibiotic production in *Streptomyces*. Archives of Microbiology 2002; 179: 66-9

[10] Sevcikova B, Kormanec J. Differential production of two antibiotics of *Streptomyces coelicolor* A3(2), actinorhodin and undecylprodigiosin, upon salt stress conditions. Archives of Microbiology 2004; 181: 384-9

[11] Wen L, Yuan BH, Li HJ, Lin YC, Chan WL, Zhou SN. A blue pigment isolated from marine *Pseudomonas* sp.. Acta Scientiarum Naturalium Universitatis Sunyatseni 2005; 44: 63-5

Bioactive Natural Products from Marine Sponges

Shixiang Bao*, Biwen Wu and Huiqin Huang
Institute of Tropical Biosciences and Biotechnology / State Key Laboratory of Tropical Crop Biotechnology, Chinese Academy of Tropical Agricultural Sciences, Haikou 571101, P.R.China
**Corresponding email: bsxhhg@yahoo.com.cn*

INTRODUCTION

The ocean covers 70.8% of the earth and has abundant organisms. These marine organisms likely have unique metabolic pathway and genetic background because of their varied living environment and long evolutionary history. Thus the potential of marine organisms to provide new bioactive products is enormous. These active products have been detected mostly in marine algae, starfish, echinus and especially frequently in sponges. Sponges are the most primitive of the multi-celled animals that have existed for 700–800 million years and there are approximately 15,000 sponge species. Sponges often have associated symbiosis with other organisms and have not been attacked by them. In the 1950's, Ara-U was first isolated from marine sponge as a lead antitumor compound and has led to the synthesis of an antitumor compound, Ara-C. During the last 20 years, numerous novel products have been isolated from marine sponges and many of these have shown pronounced biological activity. This review summarizes the natural active products derived from marine sponges with antitumor, antiviral and antibacterial activities and also the strategies for producing sponge biomass.

SCREENING NATURAL PRODUCTS FROM MARINE SPONGES FOR BIOACTIVITY

Antitumor products

With the development of marine drug research, the role of antitumour compounds in marine sponges is being increasingly studied. Two novel polyketides, spiculoic acids A and B [1], have been isolated from extracts of the Caribbean marine sponge, *Plakortis angulospiculatus*. Their structures were elucidated by detailed analysis of spectroscopic data. Spiculoic acid A showed *in vitro* cytotoxicity against human breast cancer MCF-7 cells. Three new kapakahines E-G [2] have been isolated from the marine sponge, *Cribrochalina olemda*. Limited quantities of these compounds are required not only for NMR but also for FAB-MS/MS analysis for structure elucidation. Kapakahine E showed cytotoxicity against P388 marine leukemia cells.

A cyclodecapeptide, designated phakellistatin 12 [3], has been isolated from the Western Pacific Ocean sponge, *Phakellia* sp.. It is a new cancer cell growth inhibitory agent (P388 lymphocytic leukemia, ED_{50} 2.8μg/mL). Employing principally a combination of high resolution FAB with high field (500 MHz) ^{1}H, ^{13}C and 2-D NMR and chiral GC analyses, the structure, cyclo-Ile-Phe-Thr-Leu-Pro-Pro-Tyr-Ile-Pro-Pro, was obtained. Mycaperoxide H [4], a new cyclic norsesterterpene peroxide was isolated from a Thai marine sponge, *Mycale* sp.. The structure of mycaperoxide H was deduced by spectroscopic and chemical analysis. Mycaperoxide H was cytotoxic against HeLa cells with an IC_{50} value of 0.8ug/mL. A new cyclic heptapeptide phakellistatin 13 [5], isolated from the sponge, *Phakellia fusca* Thiele, collected at Yongxing Island of China, was elucidated as cyclo-(Pro1-Trp-Leu-Thr-Pro2-Gly-Phe) on the basis of MS, UV, IR, and high-field NMR (600 MHz) analysis. The compound was significantly cytotoxic against the human hepatoma BEL-7404 cell line with an $ED_{50} < 10^{-2}\mu$g/mL. Two new lysophosphatidylcholines and four new monoglycerides [6] were isolated from the marine sponge, *Stelletta* sp. by bioactivity-guided fractionation. The planar structures of the new compounds were established on the basis of NMR and MS analyses. The compounds were evaluated for cytotoxicity against a small panel of five human tumor cell lines. In addition, Renieramycin J [7], a new tetrahydro-isoquinoline alkaloid, has been isolated from a marine sponge, *Neopetrosia* sp., as a potent cytotoxin that induced morphological changes in 3Y1 cells. Such changes are characteristic of RNA and/or protein synthesis inhibitors.

There are other antitumor compounds from marine sponges that were reported recently. The Indo-Pacific marine sponge *Ircinia ramosa* has been found to contain two powerful (GI_{50} from 0.001 to < 0.0001 μg/mL) murine and human cancer cell growth inhibitors named irciniastatins

A and B [8]. A novel MMP inhibitor, ageladine A [9] with antiangiogenic activity, was isolated from a marine sponge, *Agelas nakamurai*. Discorhabdins S, T, and U [10], three new discorhabdin analogues, have been isolated from a deep-water marine sponge of the genus Batzella. These discorhabdin analogues showed *in vitro* cytotoxicity against PANC-1, P-388 and A-549 cell lines.

Antiviral products

Extensive research has turned to the antiviral effects of marine sponges and their active components. Various types of antiviral compounds include terpenoids, alkaloids, peptides and steroids. A marine sponge, *Petrosia similis*, afforded two bis-quinolizidine alkaloids, namely petrosin and petrosin-A [11]. Both showed anti-human immunodeficiency virus (HIV) inhibition with IC_{50} values of 41.3 and 52.9 μM, respectively. MAGI cell assays indicated that the compounds inhibited early steps of HIV replication. In extracellular HIV-1 Reverse Transcriptase inhibition assay, petrosin and petrosin-A inhibited HIV-1 RT at 10.6 and 14.8μM. This is the first report of petrosins with anti-HIV activity.

Two new polycyclic guanidine alkaloids, crambescidin 826 and dehydrocrambine A [12], were isolated from the marine sponge, *Monanchora* sp.. Their structures were elucidated by 2D NMR and mass spectrometry, and their relative stereochemistry established by analysis of coupling constants and ROESY spectra. The pentacyclic guanidine alkaloid, crambescidin 826, inhibited HIV-1 envelope-mediated fusion *in vitro* with IC_{50} of 1-3μM, while dehydrocrambine A, a tricyclic guanidine alkaloid, showed weaker inhibition with an IC_{50} of approximately 35 μM.

Two novel alkaloids, namely manadomanzamines A and B [13], were isolated from an Indonesian sponge, *Acanthostrongylophora* sp.. Their structures were elucidated and shown to contain a novel organic skeleton related to the manzamine-type alkaloids. The microbial community analysis for the sponge that produces these unprecedented alkaloids has also been completed. Manadomanzamines A and B exhibited strong activity against *Mycobacterium tuberculosis* with MIC values of 1.9 and 1.5 μg/mL, respectively. Manadomanzamines A and B also exhibit activities against HIV and AIDS opportunistic fungal infections.

Antibacterial products

The research has been focused on marine sponges for their

antibacterial compounds because of the fact that marine sponges are not easily decomposed. Nowadays, numerous active antimicrobial compounds have been isolated from marine sponges.

Novel antibiotics, YM-266183 and YM-266184 [14], were found in the culture broth of *Bacillus cereus* QN03323 which was isolated from the marine sponge, *Halichondria japonica*. The structures of both antibiotics were determined by several spectroscopic experiments as new thiopeptide compounds. They exhibited potent antibacterial activities against S*taphylococci* and *Enterococci* including multiple drug resistant strains, whereas they were inactive against Gram-negative bacteria.

Thirteen new metabolites, melophlins C-O [15], were identified from the marine sponge, *Melophlus sarassinorum*. Melophlin C displayed pronounced antibacterial activity against *Bacillus subtilis* and *Staphylococcus aureus*, together with antifungal activity against *Candida albicans*.

Structurally unique steroids, isocyclocitrinol A and 22-acetylisocyclocitrinol A [16], were isolated from the extract of a salt water culture of sponge-derived *Penicilliun citrinum*. Both of them showed weak antibacterial activity against *Staphylococcus epidermidis* and *Enterococcus durans*.

Known polyene amide, clathrynamide A, and three novel related metabolites, debromoclathrynamide A, (4E, 6E)-debromoclathrynamide A and (6E)-clathrynamide A [17], were isolated from an Okinawan marine sponge, *Psammoclemma* sp.. Their antifungal activities were also evaluated with a phytopathogenic fungus, indicating that both the bromine atom and double bond geometry affected the activity.

Enzyme inhibitors

Many enzyme inhibitors are also found in marine sponges. Three new alpha- glucosidase inhibitors, schulzeines A-C [18], were isolated from the marine sponge, *Penares schulzei*. Their structures were elucidated by spectral analysis and chemical degradations to be the isoquinoline alkaloids, schulzeines A-C. They inhibit alpha-glucosidases at IC_{50} values of 48-170 nM. Seven new polyhydroxylated sterols, as well as the known agosterols A, C, and D2 [19], were isolated from a marine sponge, *Acanthodendrilla* sp. as proteasome inhibitors.

APPROACHES TO LARGE SCALE PREPARATION OF MARINE SPONGE NATURAL PRODUCTS

Marine sponges contain various bioactive compounds at very low concentrations. One ton of wet marine sponge, for example, could only yield 310mg of the pure product, halichodrin B [20]. In addition, collecting wild marine sponges may destroy the marine ecology environment. So it is a challenge to produce natural active compounds from marine sponges on a large scale.

Sponge aquaculture

Cultivation of sponge in the sea from explants was first established for producing bath sponge over a century ago. Sponge manual cultivation is being developed to produce natural active compounds. Water, laden with nutrient particles, is drawn through aquiferous system. Sponges, being sessile filter feeders, feed non-selectively on all kinds of plankton, bacteria, decaying organic particles and dissolved organic matter as food. Because the uptake of particulate matter is non-selective, inert detritus such as clay is also taken up. Movement of non-food inerts into the aquiferous system can significantly reduce the nutrient intake and consequently reduce growth rate. arine sponge should be cultured at the salinity of seawater (35% wt/wt dissolved salts). A hypersaline environment tends to dehydrate the sponge cells whereas a lower than normal salinity could lead to dilution of the intracellular content. The normal cultivation pH for marine sponges is between 7.8 and 8.4, i.e. the pH of seawater. Sponges are sensitive to temperature, too high temperature normally stimulates sexual reproduction in sponges. The culture environment should maintain a temperature slightly lower than the summer temperature of the normal sponge habitat [21]. Although sponge cultivation does not require light for growth, many sponges harbor photosynthetic endosymbionts which require light for survival.

A great number of marine sponges require silica for building the needle-like spicules that constitute a part of the sponges' skeletal support. Lack of dissolved silica can easily limit sponge growth [21]. So a minimum concentration of dissolved silica must be maintained in the culture medium to prevent a limitation. Dissolved silica can be supplied as sodium metasilicate (Na_2SiO_3) and sodium fluorosilicate (Na_2SiF_6). Some of the silica needed by the sponge could be supplied via the diatoms, such as *Phaeodacylum trivornoyum,* which has siliceous exoskeleton relative to using other microalgal feeds. Explants of tropical sponge, *Pseudosuberites andrewsi*, were fed with the marine diatom, *P. trivornoyum* [22]. The food was supplied either as intact algae or as a filtered crude extract. Growth was found in both experiments. But it is unclear if this silica can be used to form spicules.

Although marine aquaculture is being developed to produce sponge biomass inexpensively, aquaculture has significant limitations. The culture conditions cannot be controlled for sustained rapid disease and infestations of parasites. In addition, it is not suitable for some sponges slowly growing. The present research focuses on optimizing the growth conditions for sustained rapid growth.

Sponge cell culture

Based on totipotency, individual cells can regenerate the whole sponge [21]. Sponges often associate with symbiotic populations, which may enter the culture system after sponges are dispersed. So it is not easy to obtain axenic sponge cells. Suppressing bacterial contamination becomes the first qualification to establish a culture of pure sponge cell strain. Antibiotics such as penicillin, streptomycin, rifamycin and amphotericin are often used to control microorganism contamination. The primary cell cultures of the marine demosponge, *Dysidea avara* and *Suberites domuncula,* were reportedly established [23], with microbial contamination controlled by the use of antibiotics. Cultures derived from sponge embryo, which has not been in contact with seawater, can also suppress microbial contamination [21].

Maintaining cell division and thus establishing continuous cell lines remains a major hurdle to overcome when using cultivated cells for metabolite production. The reason for this difficulty was reported by Custodio et al. (1998) [24]. They found that sponges maintain high telomerase activity in nature. The failure of disaggregated cells to thrive for long is probably associated with the apparent need for cell-cell contact for maintaining a proliferation capability. Telomerase activity was lost rapidly on disaggregating the cells and led rapidly to death. Based on this recognition, the primmorphs of the sponge, *Suberites donnmcula*, were established. Primmorphs are an organized clump of cells aggregated by primary cells. Depending on the sponges, primmorphs can range in size from 40μm to 3mm. Unlike dissociated sponge cells, primmorphs retain telomerase activity to maintain a continuous cell line culture. In addition, if formed in the presence of sponge symbionts, the symbiotic microorganisms will be included in the primmorphs. In contrast, cellular detritus and non-symbiotic microorganisms that may be present in suspension are excluded from primmorphs during formation by the aggregating sponge cells. This is good for sponge growth and marine sponge active compounds production. Primmorphs have been generated from many sponges [25] including *S. agminata, Ircinia muscarum, S. domuncula, G. cydonium, Axinalla polypoides, H. panicea, Stylissa massa, Halicolana oculata* and *P. andrewsi.* Alternatively, it may be possible to fuse normal sponge cells with immortal cells of other marine invertebrates to produce hybridomas capable of continuous growth.

Major questions remain concerning the primmorphs culture although sponge cell and primmorphs culture is being increasingly studied. Sponge nutrition can influence growth and metabolite production, chemical and biological mechanism of primmorphs formation and microbial contamination control. All these and other questions remain to be answered by further research.

Chemical synthesis of active compounds from sponge

Research on the synthesis of important bioactive marine natural compounds with complex structure is being extensively carried out. Chemical synthesis plays an important part in producing effective drug leads from structural modification of these active compounds. The synthesis of the cytotoxic sponge metabolite, haliclamine A, was achieved from a common thiophene intermediate [26]. A polybrominated diphenyl ether naturally occuring in a marine sponge was found to inhibit recombinant human ALR2 with an IC_{50} of 6.4 μM. A series of polyhalogenated analogues that were synthesized and tested *in vitro* to explore the structure-activity relationships displayed various degrees of inhibitory activity [27]. Anticancer agents such as 5-fluorouracil, methotrexate and more recently, the humanised anti-CD20 antibody rituximab and the tyrsine kinase inhibitor, imatinib, are examples of synthetic compounds [28]. There are other examples of successful synthetic compounds: asymmetric total synthesis of Callystatin A isolated from the sponge, *Callyspongia truncata* [29], total synthesis of Phorboxazole A isolated from marine sponge, *Phorbas* sp. [30] and from the synthesis of marine sponge, Alkaloid niphatoxin B, from Red sea sponge, *Niphates* sp. [31].

Active sponge products are often chiral compounds with complex structure, mostly cytol, and highly unsaturated; the process of chemical synthesis is also complex. Although many synthetic compounds have been obtained successfully, chemical synthesis in the laboratory is adequate for producing commercial quantities of metabolites of interest.

Fermentation of sponge-associated microorganisms

All sponge-associated symbiotic microbial populations are not only necessary for sponge growth but may also be the likely source of marine active products. These microorganisms are regenerated from biological sources, and the active metabolites can be obtained on a large scale from fermentation projects. Alteramide A [32], a new tetracyclic alkaloid, was isolated from a bacterium, *Alteromoas* sp., associated with the marine sponge, *Halichondria okadai*. Alteramide A was significantly cytotoxic against P-388, L1210 and KB cell lines. Two novel antimycin antibiotics, urauchimycins A and B, were also isolated from a fermentation broth of a *Streptomyces* sp., Ni-80 [33]. The bacterial strain has also been isolated from an unidentified sponge. These antibiotics are the first antimycin antibiotics with a branched side chain. The bacterial isolate, M22-1 [34], belonging to the genus *Vibrio,* was obtained from a homogenate of the sponge, *Hyatella* sp.. The same substance was found in the sponge extract, suggesting that the active component was synthesized by the associated microorganism. *Micrococcus*, isolated from marine sponge *Tedaniaigins* [35], contains the same substance which was also found in the sponge extract. Despite these data, it is still unclear if sponge-associated microorganisms are the exact sources of marine sponge active products.

REFERENCES

[1] Huang XH, van Soest R, Roberge M and Andersen RJ. Spiculoic acids A and B, new polyketides isolated from the Caribbean marine sponge *Plakortis angulospiculatus*. Org Lett 2004; 6:75-8

[2] Nakao Y, Kuo J, Yoshida WY, Kelly M and Scheuer PJ. More kapakahines from the marine sponge *Cribrochalina olemda*. Org Lett 2003; 5: 1387-90

[3] Pettit GR and Tan R. Antineoplastic agents 390. Isolation and structure of

phakellistatin 12 from a Chuuk archipelago marine sponge. Bioorg Med Chem Lett 2003; 13: 685-8
[4] Phuwapraisirisan P, Matsunaga S, Fusetani N, Chaitanawisuti N, Kritsanapuntu S and Menasvta P. Mycaperoxide H, a new cytotoxic norsesterterpene peroxide from a Thai marine sponge Mycale sp. J Nat Prod 2003; 66: 289-91
[5] Li WL, Yi YH, Wu HM, Xu QZ, Tang HF, Zhou DZ, et al.. Isolation and structure of the cytotoxic cycloheptapeptide phakellistatin 13. J Nat Prod 2003; 66: 146-8
[6] Zhao Q, Mansoor TA, Hong J, Lee CO, Im KS, Lee DS, et al.. New lysophosphatidylcholines and monoglycerides from the marine sponge *Stelletta* sp. J Nat Prod 2003; 66: 725-8
[7] Oku N, Matsunaga S, van Soest RW, Fusetani N. Renieramycin J, a highly cytotoxic tetrahydroisoquinoline alkaloid, from a marine sponge *Neopetrosia* sp.. J Nat Prod 2003; 66: 1136-9
[8] Pettit GR, Xu JP, Chapuis JC, Pettit RK, Tackett LP, Doubek DL, et al. Antineoplastic agents. 520. Isolation and structure of irciniastatins A and B from the Indo-Pacific marine sponge *Ircinia ramosa*. J Med Chem 2004; 47:1149-52
[9] Fujita M, Nakao Y, Matsunaga S, Seiki M, Itoh Y, Yamashita J, et al.. Ageladine A: an antiangiogenic matrixmetalloproteinase inhibitor from the marine sponge *Agelas nakamurai*. J Am Chem Soc 2003; 125: 15700-1
[10] Gunasekera SP, Zuleta IA, Longley RE, Wright AE, Pomponi SA. Discorhabdins S, T, and U, new cytotoxic pyrroloiminoquinones from a deep-water Caribbean sponge of the genus *Batzella*. J Nat Prod 2003; 66: 1615-7
[11] Venkateshwar Goud T, Srinivasa Reddy N, Raghavendra Swamy N, Siva Ram T, Venkateswarlu Y. Anti-HIV active petrosins from the marine sponge *Petrosia similis*. Biol Pharm Bull 2003; 26: 1498-501
[12] Chang L, Whittaker NF, Bewley CA. Crambescidin 826 and dehydrocrambine A: new polycyclic guanidine alkaloids from the marine sponge *Monanchora* sp. that inhibit HIV-1 fusion. J Nat Prod 2003; 66; 1490-4
[13] Peng J, Hu JF, Kazi AB, Li Z, Avery M, Peraud O, et al.. Manadomanzamines A and B: a novel alkaloid ring system with potent activity against mycobacteria and HIV-1. J Am Chem Soc 2003; 125: 13382-6
[14] Nagai K, Kamigiri K, Arao N, Suzumura K, Kawano Y, Yamaoka M, et al. YM-266183 and YM-266184, novel thiopeptide antibiotics produced by *Bacillus cereus* isolated from a marine sponge. I. Taxonomy, fermentation, isolation, physico-chemical properties and biological properties. J Antibiot (Tokyo) 2003; 56:123-8
[15] Wang CY, Wang BG, Wiryowidagdo S, Wray V, van Soest R, Steube KG, et al. Melophlins C-O, thirteen novel tetramic acids from the marine sponge *Melophlus sarassinorum*. J Nat Prod 2003; 66: 51-6
[16] Amagata T, Amagata A, Tenney K, Valeriote FA, Lobkovsky E, Clardy J, et

al.. Unusual C25 steroids produced by a sponge-derived *Penicillium citrinum*. Org Lett 2003; 5: 4393-6

[17] Ojika M, Itou Y, Sakagami Y. Structural studies and antifungal activity of unique polyene amides, clathrynamide A and three new derivatives, from a marine sponge, *Psammoclemma* sp. Biosci Biotechnol Biochem 2003; 67: 1568-73

[18] Takada K, Uehara T, Nakao Y, Matsunaga S, van Soest RW and Fusetani N. Schulzeines A-C, new alpha-glucosidase inhibitors from the marine sponge *Penares schulzei*. J Am Chem Soc 2004; 126: 187-93

[19] Tsukamoto S, Tatsuno M, van Soest RW, Yokosawa H and Ohta T. New polyhydroxy sterols: proteasome inhibitors from a marine sponge *Acanthodendrilla* sp. J Nat Prod 2003; 66: 1181-5

[20] Osinga R, Tramper J, Wijffels RH. Cultivation of Marine Sponges. Mar Biotechnol 1999; 1: 509-32

[21] Belarbi el H, Contreras Gomez A, Chisti Y, Garcia Camacho F, Molina Grima E. Producing drugs from marine sponges. Biotechnology Advances 2003; 21: 585-98

[22] Osinga R, Belarbi el H, Grima EM, Tramper J and Wijffels RH. Progress towards a controlled culture of the marine sponge *Pseudosuberites andrewsi* in a bioreactor. J Biotechnol 2003; 100: 141-6

[23] De Rosa S, De Caro S, Iodice C, Tommonaro G, Stefanov K and Popov S. Development in primary cell culture of demosponges. J Biotechnol 2003; 100: 119-25

[24] Custodio MR, Prokie I, Steffen R, Koziol C, Borojevic R, Brummer F, et al.. Primmorphs generated from dissociated cells of the sponge *Suberites domuncula*: a model system for studies of cell proliferation and cell death. Mech Ageing Dev 1998; 105: 45-59.

[25] Zhang W, Zhang X, Cao X, Xu J, Zhao Q, Yu X, et al.. Optimizing the formation of in vitro sponge primmorphs from the Chinese sponge *Stylotella agminata* (Ridley). J Biotechnol 2003; 100: 161-8.

[26] Michelliza S, Al-Mourabit A, Gateau-Olesker A, Marazano C. Synthesis of the cytotoxic sponge *metabolite haliclamine* A. J Org Chem 2002; 67: 6474-8

[27] de la Fuente JA, Manzanaro S, Martin MJ, de Quesada TG, Reymundo I, Luengo SM, et al.. Synthesis, activity, and molecular modeling studies of novel human aldose reductase inhibitors based on a marine natural product. J Med Chem 2003; 46: 5208-21

[28] Schwartsmann G, Da Rocha AB, Mattei J and Lopes R. Marine-derived anticancer agents in clinical trials. Expert Opin Investig Drugs 2003; 12: 1367-83

[29] Kobayashi M and Kitagawa I. Marine spongean cytotoxins. J Nat Toxins 1999; 8: 249-58

[30] Smith AB 3rd, Verhoest PR, Minbiole KP and Schelhaas M. Total synthesis of (+)-phorboxazole A. J Am Chem Soc 2001; 123: 4834-6

[31] Kaiser A, Marazano C and Maier M. First Synthesis of Marine Sponge *Alkaloid Niphatoxin* B. J Org Chem 1999; 64: 3778-82

[32] Shigemori H, M- A, Bae, K. Yazawa, T. Sasaki, and Y. Kobayashi. Alteramide A, a new tetracyclic alkaloid from a bacterium Alteromonas sp. associated with the marine sponge *Halichondria okadai.* J Org Chem 1992; 57: 4320-3

[33] Imamura N, Nishjima M, adachi K, Sano H. Novel antimycin antibiotics, urauchimycins A and B, produced by marine actinomycete. J antibiot 1993; 46:241-6

[34] Oclarit JM, Okada H, Ohta S, Kaminura K, Yamaoka Y, Iizuka T, et al. Anti-bacillus substance in the marine sponge, *Hyatella species*, produced by an associated *Vibrio* species bacterium. Microbios 1994; 78: 7-16

[35] Faulkner DJ. Marine natural products. Nat Prod Rep 1993; 10: 497-539

Anti-Inflammatory and Neurotrophic Alkaloids from Higher Plants and Fungi

Matthias Hamburger*
Institute of Pharmaceutical Biology, Department of Pharmaceutical Sciences, University of Basel, Klingelbergstrasse 50, CH-4056 Basel, Switzerland
**Corresponding email: matthias.hamburger@unibas.ch*

INTRODUCTION

Compounds of natural origin still play a major role as drugs and lead compounds for the development of new drugs. About 40 % of the world's best selling drugs derive directly or indirectly from molecules of biogenic origin [1]. In certain cases, the natural product may be directly developed to the new drug substance. Increasingly, however, the natural product is considered as a starting point for drug development. The compound may be systematically modified by medicinal chemists to improve its properties as a drug with regard to potency, selectivity, or pharmacokinetic properties. In that case, a derivative of the natural product will be ultimately developed. Recent examples include, among others, the camptothecin derivatives, irinotecan and topotecan. In other cases, the natural product serves as a starting point for the development of new synthetic drug substances which share with the original compound only very few features such as a common pharmacophore. The anti-Alzheimer drug, rivastigmin, which had the acetylcholinesterase inhibitor physostigmin as the natural product lead, falls into the latter category.

Over the past few years, we have been involved with the search for new compounds with anti-inflammatory properties, and with activity on targets implicated in age-related disorders of the central nervous system (CNS). Inflammatory diseases, in particular rheumatic disorders, afflict a large number of people worldwide [2]. Neurodegenerative disorders, such as Alzheimer's and Parkinson's disease are chronic and fatally progressive, and affect primarily the growing fraction of elderly persons in our societies [3]. In both cases, current therapies are not satisfactory: Although a wide range of anti-inflammatory drugs is available, most of them have considerable side effects with long term use. The interest in new drug substances with new modes of action and fewer side effects remains high [4]. Significant advances in the understanding of the molecular and cellular processes underlying Alzheimer's and Parkinson's diseases have been achieved [5]. Current therapy, in contrast, is still symptomatic and merely retards the progression of the illness to a certain extent. Hence, there is a dire need for treatments which have preventive and/or causal modes of action [6].

HPLC-BASED DISCOVERY OF NATURAL PRODUCTS

In the past, bioactive natural products were typically discovered, purified and characterized in a preparative, bioactivity-guided appropach involving substantial quantities of material, manpower, and time [7]. As a consequence of the paradigm shift in drug discovery towards screening of large compound libraries in a highly automated high-throughput process, natural product-based discovery programs have come under increasing pressure. Recognizing the need for rapid and miniaturized natural products discovery, various concepts and approaches have been proposed and, in part, implemented [8, 9]. Among these, HPLC-based activity profiling (Figure 1) has been successfully used for the identification of new lead structures [10, 11]. The approach combines analytical gradient HPLC with on-line detection devices such as diode array and mass spectrometry and with a simultaneous micro-fractionation for microtiter-based bioassays. This allows an overlay of bioactivity profile and HPLC chromatogram, and, hence, a rapid correlation of HPLC peaks with activity and spectroscopic data. Database searches may allow tentative structural proposals for known molecules and a critical appraisal of their potential. Only as a last step will a compound be isolated on a preparative scale.

We have implemented HPLC-based discovery as our current routine for compound discovery. Besides the examples discussed below, we successfully used this approach for the characterization of some tanshinone-type diterpenoids in *Salvia miltiorrhiza* Bunge (Lamiaceae) as inhibitors of monoamine oxidase A (MAO A) and of inducible NO synthase (iNOS) expression in RAW 267.4 cells [12, 13].

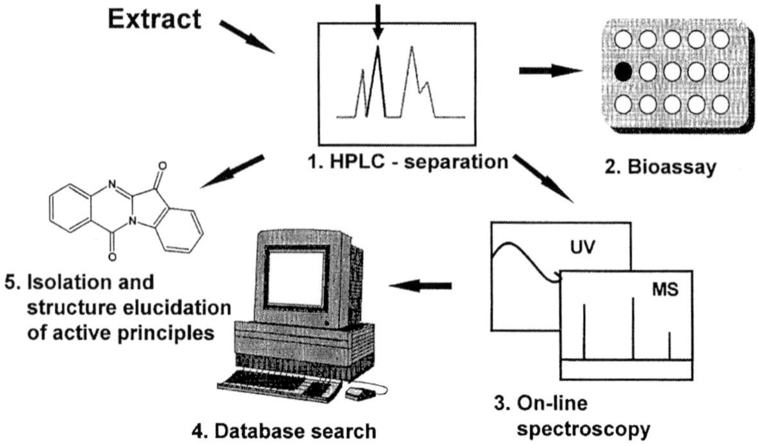

Figure 1. Principle of HPLC-based activity profiling of extracts

ANTI-INFLAMMATORY AND ANTI-ALLERGIC ALKALOIDS IN WOAD

We use ethnomedicinal information and historical documents in the selection of plants to be investigated. Convergent reports on the uses from independent sources are considered as an important criterion for prioritizing samples. The woad plant (*Isatis* sp., Brassicaceae) is such a case [14]. *Isatis tinctoria* L. has a long and well-documented history as dye plant and as a medicinal herb in Europe, where it was cultivated up to the 19th century. The leaves and roots of the related *Isatis indigotica* L. are used in Chinese Medicine. Two aspects aroused our interest in *Isatis*: the well-documented use over centuries as a medicinal plant, and a certain common thread in the historical uses, in particular the association of woad preparations with ailments involving inflammatory processes. We deemed the plant of sufficient promise for a broad-based evaluation of its potentialities as an anti-inflammatory. We assessed the bioactivity profile of *Isatis tinctoria* in a systematic screening of lipophilic and polar extracts of various plant parts, both fresh and dried. A broad-based pharmacological screening in a panel of cell-based and mechanism-based assays involving over 20 inflammation-related targets was employed [10, 14]. Of the six *Isatis* extracts tested, a dichloromethane extract from dry leaves significantly inhibited cyclooxygenase-2, 5-lipoxygenase, expression of inducible NO synthase, leucocytic elastase and histamine and serotonin release from mast cells.

Search for the COX-2 inhibitory principle in woad was carried out by HPLC-based activity profiling (Figure 2). In a first step, the COX-2 inhibitory activity was located in the time window 24-33 min which was then analyzed at a higher resolution. The inhibitory principle was finally localized in the minor peak at 25 min and readily identified by on-line spectra (PDA and ESI MS) as tryptanthrin (**1**). The identity was subsequently confirmed by preparative purification and by synthesis [10]. Although tryptanthrin has been known for a long time, its potential as an anti-inflammatory substance was new. Given the compound's rather remarkable potency in the cell-based assay for COX-2 catalyzed 6-keto-PGF_{1a} synthesis in stimulated Mono Mac 6 cells (IC_{50} 64 nM), we investigated its pharmacology in more detail [15]. An IC_{50} of 0.25 μM

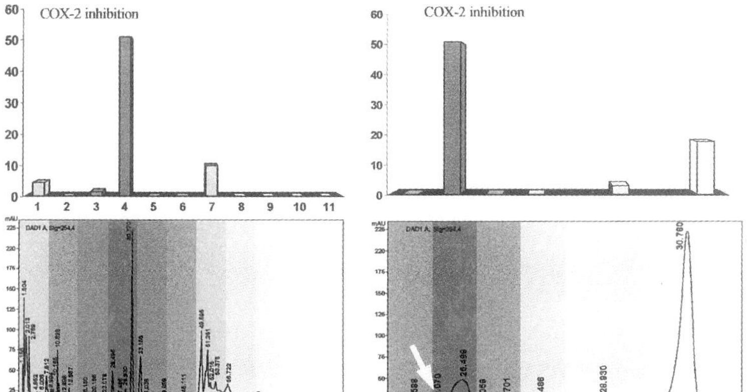

Figure 2. Activity profiling of a lipophilic extract from *Isatis tinctoria* leaves.

Left: HPLC profile (254 nm) of extract obtained by gradient elution, and (above) COX-2 inhibition (%) in Mono Mac 6 cells. The grey shades indicate time windows for microfractions.
Right: Expanded time window 24-32 min and (above) activity in bioassay.The white arrow indicates the tryptanthrin peak. Extract amount injected: 200 μg; one quarter of each fraction was used for bioassay (Reprinted with permission from [10]).

was found in RAW 264.7 cells (IC_{50} of nimesulide 0.21 μM). Assays with purified COX-2 and COX-1, in comparison to the non-selective inhibitors, diclofenac and indomethacin, confirmed that tryptanthrin was a selective inhibitor of the COX-2 isoform (Figure 3). COX-2 was inhibited with an IC_{50} of 0.83 μM, whereas COX-1 activity remained unaffected at concentrations up to 50 μM. In contrast, the non-selective reference compounds displayed comparable inhibitory potencies towards the two isoenzymes.

The pronounced activity of the *Isatis* extract on purified 5-LOX prompted us to test the extract and tryptanthrin in a cell-based assay with human neutrophils. The A23187-induced LTB_4 synthesis was taken as an indirect measure for 5-LOX activity. Tryptanthrin showed a potency comparable to that of the clinically used 5-LOX inhibitor, zileuton (IC_{50} values of 0.15 and 0.35 μM, respectively) [15]. The inhibitory properties of tryptanthrin on nitric oxide synthesis were reported by another group in research parallel to our investigations [16]. Taken together, tryptanthrin inhibits the formation of three major groups of pro-inflammatory substances in cell-based assays at nM to low-μM concentrations. The compound has no obvious structural resemblance to any of the selective COX-2 inhibitors of synthetic origin currently in use or in development (Figure 4) [17]. The closest structural similarity was with the alkaloid, rutaecarpine, a potent COX-2 inhibitor from *Evodia rutaecarpa*. It is interesting, but maybe not too surprising, that a pharmacophore-based mapping of natural products with reported COX-inhibitory properties did not pick up tryptanthrin as a hit [18].

Tryptanthrin fulfils most requirements for a promising lead compound. It has a fairly unique profile of activity and considerable potency *in vitro*, and exhibits physico-chemical properties that are favorable for drugs, such as its molecular mass, lipophilicity, number of proton donor and acceptor sites [19]. The tryptanthrin scaffold is easily amenable to synthesis and offers multiple sites for structural modification and, hence optimization.

In the search for woad constituents responsible for the inhibitory effect on histamine release, we identified benzylidenyl indolin-2-one derivative **2** (Figure 5) [20]. The compound had a marked effect on ionophore-mediated histamine release from rat peritoneal mast cells (IC_{50} 15 μM). The compound compared favorably with the mast cell stabilizing drug, disodium cromoglycate (IC_{50} 1.5 mM). Other *Isatis* alkaloids such

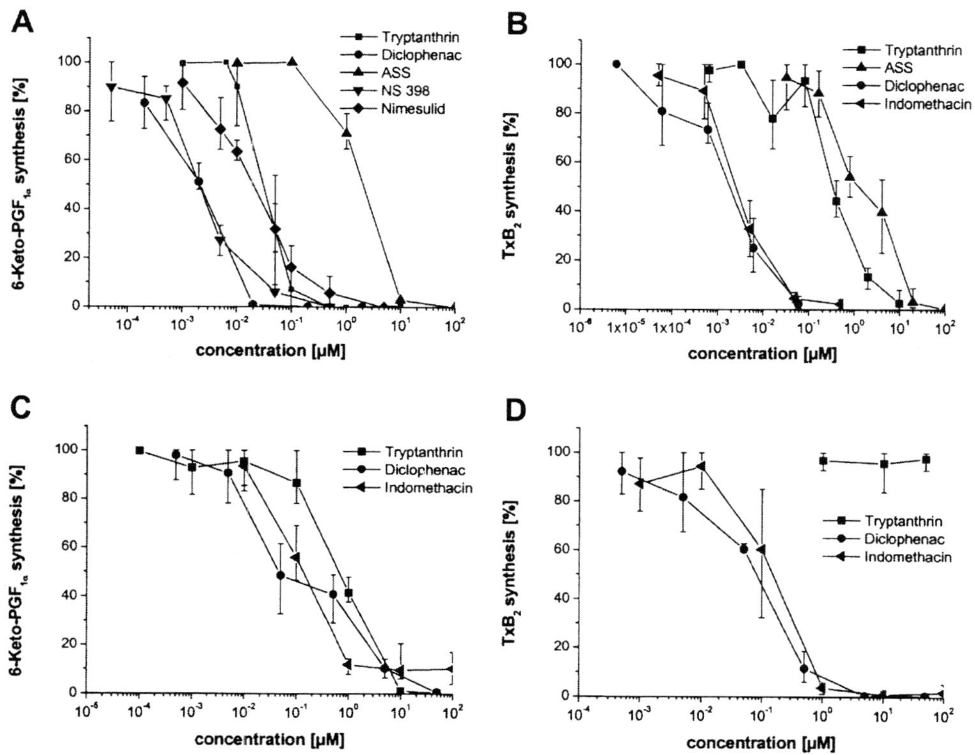

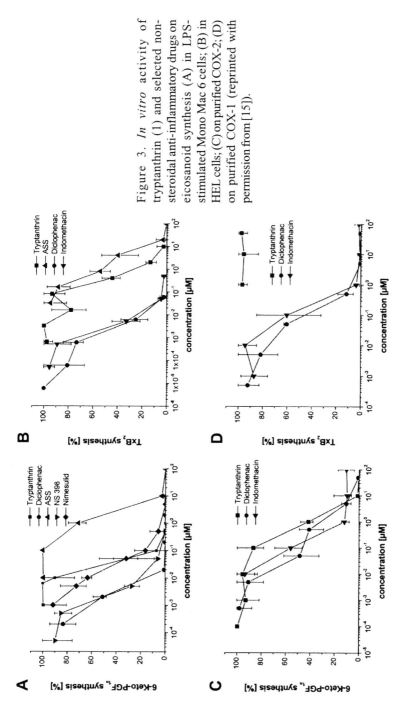

Figure 3. *In vitro* activity of tryptanthrin (1) and selected non-steroidal anti-inflammatory drugs on eicosanoid synthesis (A) in LPS-stimulated Mono Mac 6 cells; (B) in HEL cells; (C) on purified COX-2; (D) on purified COX-1 (reprinted with permission from [15]).

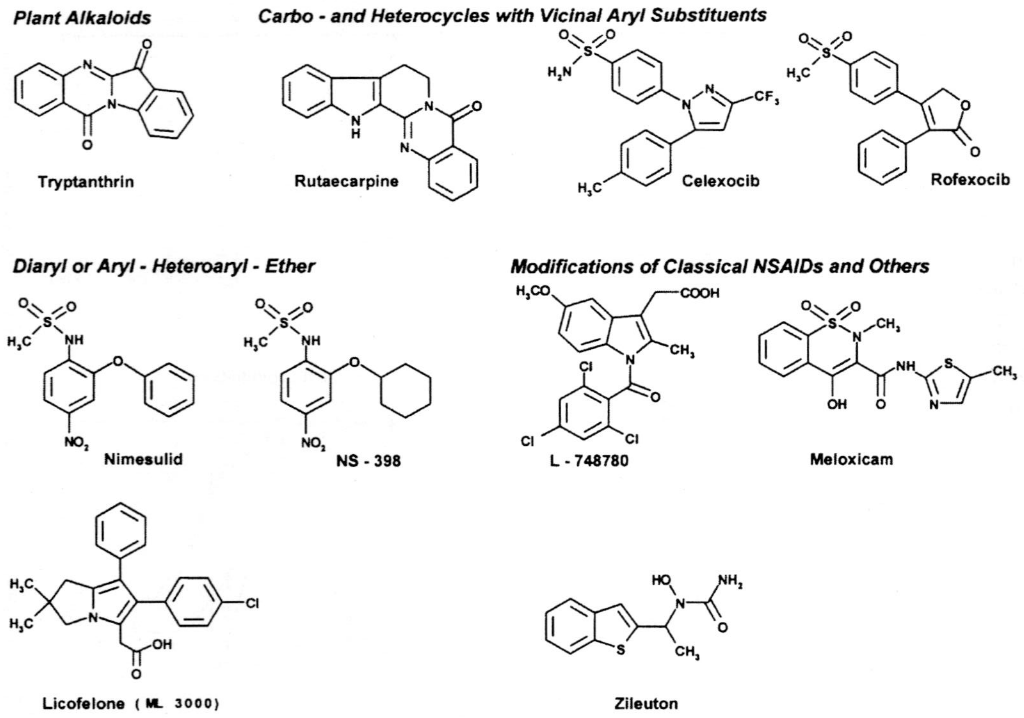

Figure 4. Structures of tryptanthrin, rutaecarpin, and synthetic compounds representing major classes of COX-2, 5-LOX, and dual COX-2/5-LOX inhibitors

as tryptanthrin (**1**), indirubin (**3**) and deoxyvasicinone (**4**) had no effect on histamine release when tested at 40 μM concentration. In the search for structure-activity relationships we synthesized related indolin-2-ones **5-9**. None of the compounds tested active. It is interesting to note that these compounds were reported by researchers at Sugen as inhibitors of various kinases [21]. Although compound **2** had not been tested in Sugen's kinase panel, the lack of activity of **5-9** in our assay indicated that kinase inhibition cannot be the mechanism underlying the inhibitory effect of **2** on histamine release from mast cells. Like tryptanthrin, indolin-2-one **2** possesses favorable physico-chemical properties for a drug lead.

NEURITOGENIC PYRIDONE ALKALOIDS FROM ENTOMOGENOUS FUNGI

Entomogenous fungi are a small group of highly specialized organisms. More than 700 species are known to date, including Deuteromycetes, Ascomycetes, Zygomycetes, Oomycetes, Chrytridiomycetes, Trichomycetes and Basidiomycetes [22]. Different relationships between insects and fungi have been reported, ranging from mycophagous alimentary diets of arthropods to facultative pathogenicity,

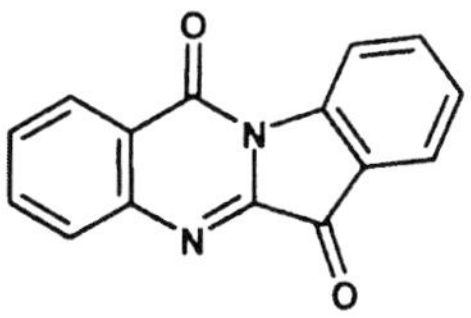

1

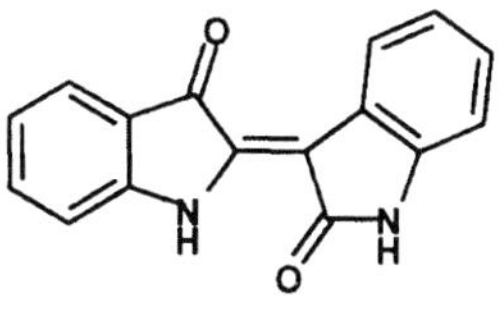

3

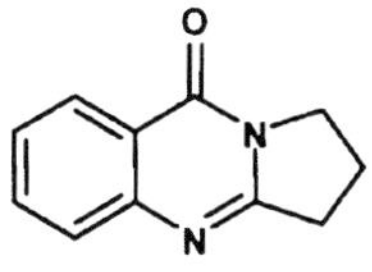

4

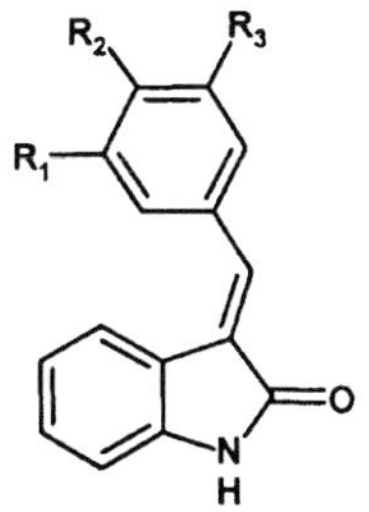

	R_1	R_2	R_3
2	- OCH_3	- OH	- OCH_3
5	- H	- OH	- H
6	- $C(CH_3)_3$	- OH	- $C(CH_3)_3$
7	- H	- $CH(CH_3)_2$	- H
8	- H	- H	- H
9	- $C(CH_3)_3$	- OCH_3	- H

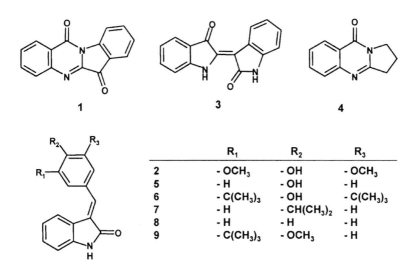

Figure 5. Structures of *Isatis* alkaloids **1-4**, and synthetic benzylidenyl indolin-2-ones **5-9**

parasitism, commensalism and symbiosis [23]. In some cases, the host-insect relationship has been found to be associated with bioactive fungal metabolites. In East Asia, entomopathogenic fungi have found medicinal uses. The crude drug, "Dong chong xia cao", which consists of fruiting bodies of the fungus *Cordyceps sinensis* (Berk.) Sacc. (Ascomycetes) together with its larval host, *Hepialus armoricanus* (Hepialidae), is highly valued in Chinese Medicine as a tonic [24].

Given the highly specific host-fungus interaction [25] and the medicinal importance of species such as *Cordyceps sinensis*, we hypothetized that entomopathogenic fungi might be a promising source for bioactive molecules of pharmaceutical interest. Current knowledge on secondary metabolites of these fungi is rather limited. In a collaboration with the Center for Entomogenous Fungi, Anhui Agricultural University, we embarked on a screening of Deuteromycetes for activities against pharmacological targets which are of clinical relevance in the context of degenerative disorders of the central nervous system (CNS). The assay panel included tests for radical scavenging, monoamine oxidase (MAO) inhibitory, NMDA (N-methyl-d-aspartate) antagonistic, and neuritogenic activities, together with assays for cytotoxic and antimicrobial activities [26]. I here describe in more detail our studies on a class of compounds which we found to possess promising neuritogenic properties.

Neurotrophins, such as NGF (nerve growth factor) and BDNF (brain-derived neurotrophic factor), and their receptors are important targets for the therapy of human diseases, with potential applications in chronic and acute neurodegeneration [27]. The therapeutic application of neurotrophins is severely hampered due to their proteinaceous nature. Small molecule modulators of neurotrophic function which are able to penetrate the blood-brain barrier could be a promising alternative for therapeutic intervention [28].

For the detection of neuritogenic properties, we used the PC12 cell line as a well-established model for neurochemical and neurobiological studies [29]. Upon exposure to NGF, PC12 cells stop dividing and acquire numerous properties of mature sympathetic neurons, including the extension of neuritis [30]. A mycelial extract of *Paecilomyces militaris*, RCEF 0095, induced pronounced neurite outgrowths and formation of networks and was selected for further investigation [26]. As a result, a novel pyridone alkaloid, militarinone A (**10**), was identified as the major neuritogenic metabolite in the extract [31]. A related pyridone alkaloid, **11,** and two acyl tetramic acids, **12-13**, were also obtained (Figure 6) [32]. In contrast to **10**, militarinones B-D showed no neuritogenic activity but rather some cytotoxic properties.

The isolation of militarinones A-D (**10-13**) was the first report of concomitant occurrence of pyridone alkaloids and acyl tetramic acids. As such, this finding resolved some of the controversy about biosynthesis of fungal pyridone alkaloids. It is generally accepted that the biosynthesis involves the aromatic amino acids, phenylalanine (or tyrosine), and an activated polyketo acid. However, the condensation of these precursors, and the subsequent steps leading to the pyridone moiety have remained uncertain. Early studies on tenellin biosynthesis [33] led to the hypothesis that phenylalanine would undergo a transformation to 3-amino-2-phenyl-propionic acid prior to its condensation with an activated polyketo acid to the pyridone alkaloid (Figure 7, pathway A). This pathway could not be confirmed in subsequent feeding experiments [34]. An alternative pathway was proposed in which the condensation of phenylalanine and the polyketo acid would provide an acyltetramic acid (Figure 7, pathway B). Upon *p*-hydroxylation of the aromatic ring, a reactive quinone methide intermediate would lead to ring expansion to the corresponding pyridone. However, feeding experiments with the hypothetical tetramic acid precursor for tenellin produced no incorporation into the target molecule [35].

The identification of militarinones A-D supports pyridone alkaloid biosynthesis via an acyl tetramic acid intermediate. The proposed biogenetic relationship between these compounds is depicted in Figure 8 [32]. Tyrosine and the polyketo acid would react to acyltetramic acid **12**, which in turn would be oxidized to the corresponding 6-hydroxy derivative **13**. The fact that tetramic acids **12** and **13** bear already a *p*-OH group at C-4' is in disaccord with the hypothesis of a *p*-hydroxylation being the initial step for the ring expansion of the pyrrolidin-2, 4-dione. Dehydration at C-5 and C-6 and tautomerisation could lead to a quinone methide intermediate which would react to the corresponding pyridine, **11**. This compound could undergo further reactions such as N-oxidation, and several oxidative and reductive steps in the aromatic ring would lead to the *cis*-(1,4-dihydroxy-cyclohexyl) moiety of **10**. Militarinone A (**10**) is the major pigment in *P. militaris*, and this fact supports its putative role as the end product of the pyridone alkaloid pathway in *P. militaris*.

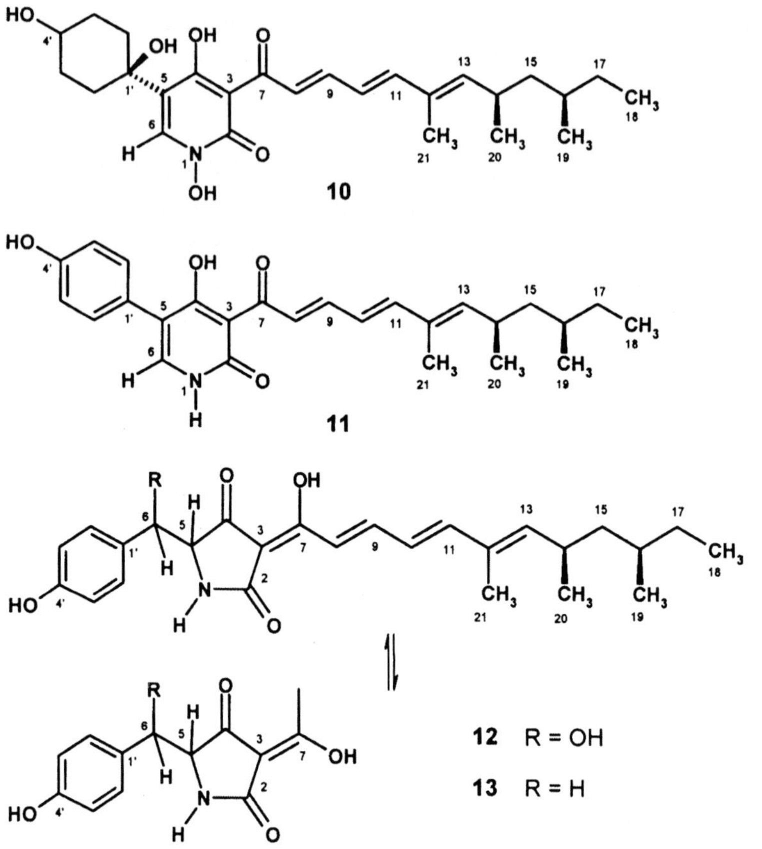

Figure 6. Structures of militarinones A-D (**10-13**) from *Paecilomyces militaris* RCEF 0095

Given that pyridone alkaloids represented a new structural class of neuritogenic small molecules, we decided to analyze in more detail the signal transduction pathways underlying the effect of militarinone A in PC12 cells. NGF-triggered signalling in the PC12 cell line has been extensively studied. The binding of NGF to its high-affinity tyrosine kinase A (Trk A) receptor activates mainly two cascades of cellular signaling responses mediating neurotrophic effects, namely MAPK and phosphoinositol 3 (PI_3) kinase pathways [36, 37]. Ultimately, effects on gene transcription and regulation of the cytoskeletal machinery are triggered.

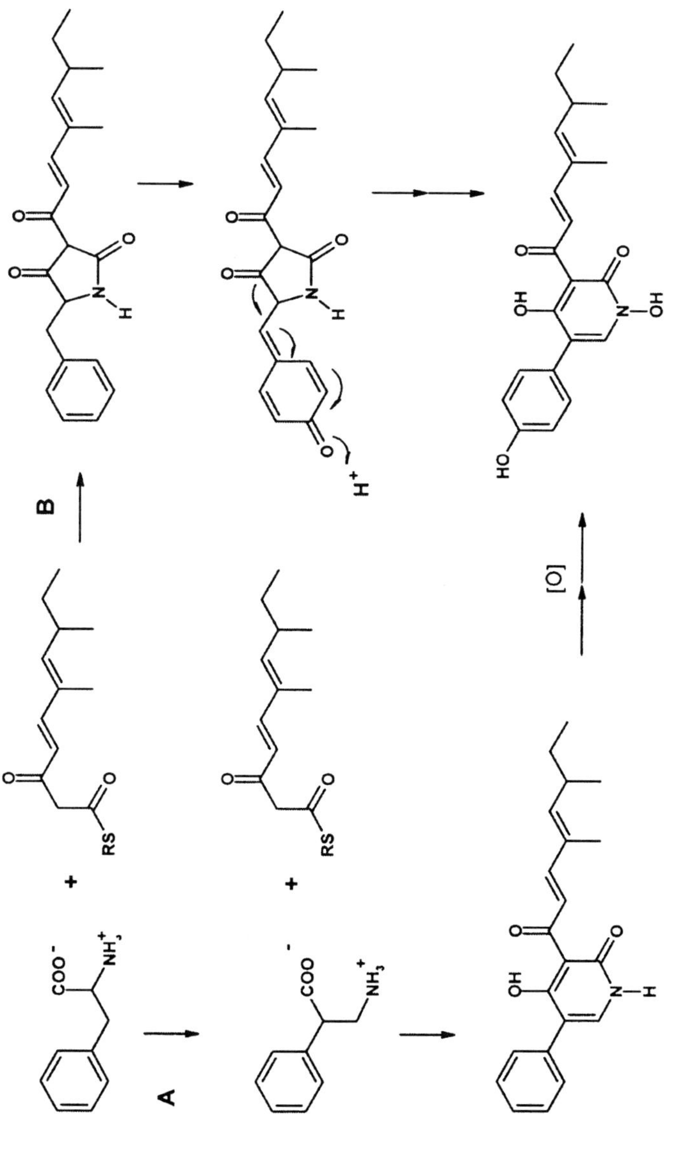

Figure 7. Biosynthetic pathways for fungal pyridone alkaloids – earlier proposals

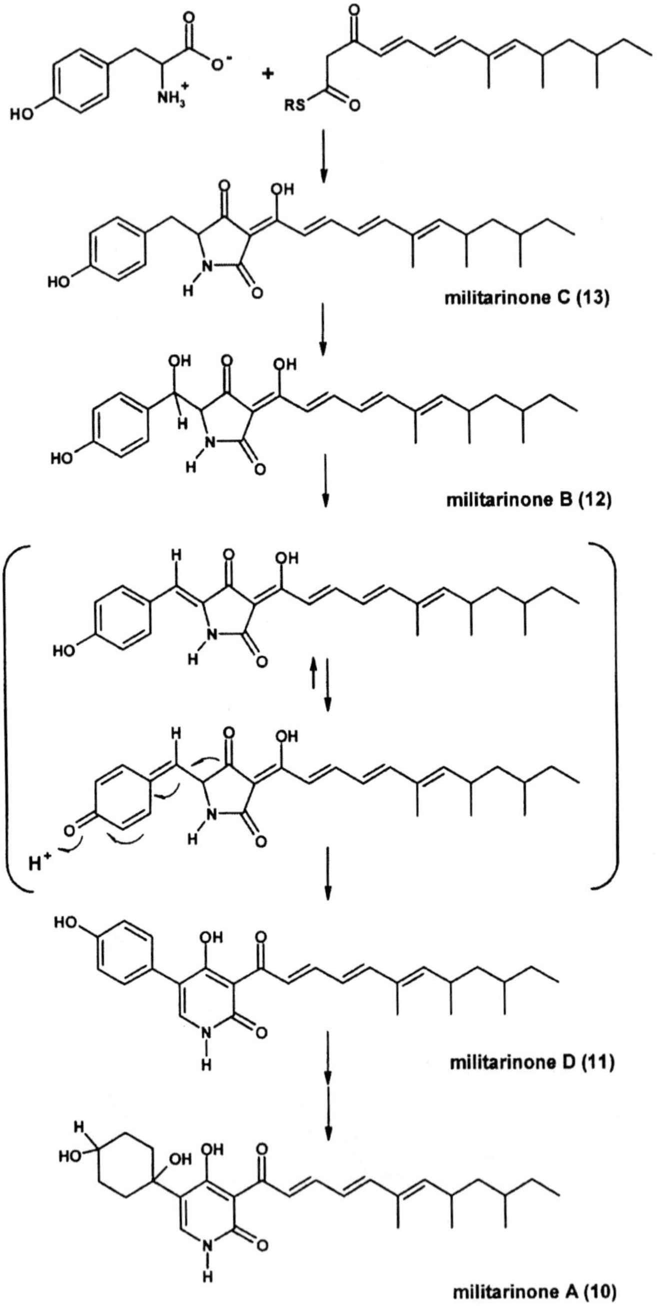

Figure 8. Proposed biogenetic relationship of militarinones A-D

When PC12 cells were treated with militarinone A, the compound rapidly accumulated in the cell membrane and was slowly released into the cytoplasma. The high affinity of militarinone A for membrane bilayers was due to the distinct amphiphilic properties of the molecule. In primed PC12 cells, an early activation of protein kinase B (Akt), representing a downstream target of (PI_3) kinase, and a delayed phosphorylation of extracellular signal-regulated kinases 1 and 2 (ERK1/2), and of transcription factor, cAMP-responsive element binding protein (CREB), was found. The NGF-dependent activation of c-Jun amino terminal kinase (SAPK/JNK1) was potentiated [38]. Morphological differentiation of cells and the phosphorylation of specific signal molecules were blocked by the MAP kinase (MEK1) inhibitor, PD098059, the PI_3-kinase inhibitor, wortmannin, and the adenylyl cyclase inhibitor, 9-cyclopentyladenin. Thus, militarinone A activated distinct downstream proteins on the NGF signaling pathways and potentiated the effects of NGF. We currently study possible target proteins upstream of MAPK and PI_3 kinase to identify the molecular target of militarinone A, and with the aid of related pyridone alkaloids, we try to delineate the pharmacophoric structure in this compound class.

CONCLUSIONS AND PERSPECTIVES

A number of natural products with promising pharmacological profiles and drug-like physico-chemical properties have been identified from plant and fungal sources. Even though the woad alkaloids tryptanthrin (**1**) and benzylidenyl indolin-2-one **2** are known compounds, their potency *in vitro* and their amenability to structural variation renders these molecules attractive starting points for medicinal chemistry. Militarinone A (**10**) represents a new structural class of neuritogenic compounds. The complex functional response appears to be caused by a fairly specific activation of signaling pathways of NGF. Here, more work is needed to clarify the exact potential of these pyridone alkaloids as drug leads or as tools for basic research in cell biology.

ACKNOWLEDGEMENTS

The author wishes to express his gratitude to the graduate students, postdocs and technicians who were involved in the work, namely Drs. H.

Danz, K. Dittmann, U. Riese, K. Schmidt, B. Schubert, S. Stoyanova, and S. Adler, H. Graf, B. Hoffmann, G.-U. Rüster, and E. Ziegler. The collaborating partners are kindly acknowledged for their contributions to these projects, namely Prof. Z. Li, Center for Entomogenous Fungi, Anhui Agricultural University, Anhui; Prof. G. Dannhardt, Institute of Pharmacy, University of Mainz, Prof. H.U. Simon, Institute of Pharmacology, University of Bern; Dr. W. Günther, Institute of Organic and Macromolecular Chemistry, University of Jena.

REFERENCES

[1] Newman DJ, Cragg GM, Snader KM. The influence of natural products upon drug discovery. Nat Prod Rep 2000; 17: 215-34

[2] Mutschler E, Geisslinger G, Kroemer HK, Schäfer-Korting M. Arzneimittelwirkungen - Lehrbuch der Pharmakologie und Toxikologie, 8th ed.; Wissenschaftliche Verlagsgesellschaft: Stuttgart 2001; p. 224-57

[3] Cummings JL. Alzheimer disease. J Amer Med Assoc 2002; 287: 2335-8

[4] Kontogiorgis CA, Hadjipavlou-Litina DJ. Non steroidal anti-inflammatory and anti-allergy agents. Curr Med Chem 2002; 9: 89-98

[5] Selkoe DJ. Alzheimer's disease: genes, proteins, and therapy. Physiol Rev 2001; 81: 741-66

[6] Lahiri DK, Farlow MR, Sambamurti K, Greig NH, Giacobini E, Schneider LS. A critical analysis of new molecular targets and strategies for drug developments in Alzheimer's disease. Curr Drug Targets 2003; 4: 97-112

[7] Hamburger M, Hostettmann K. Bioactivity in plants: the link between phytochemistry and medicine. Phytochemistry 1991; 30: 3864-74

[8] Hamburger M. Tracking bioactivity in plant extracts – new concepts and approaches. In Society for Medicinal Plant Research – 50 Years 1953-2003; Sprecher, E.; Caesar, W., Eds.; Wissenschaftliche Verlagsgesellschaft: Stuttgart 2003; 109-22

[9] Van Elswijk DA, Irth H. Analytical tools for the detection and characterization of biologically active compounds from nature. Phytochem Rev 2002; 1: 427-39

[10] Danz H, Stoyanova S, Wippich P, Brattström A, Hamburger M. Identification and isolation of the cyclooxygenase-2 inhibitory principle in *Isatis tinctoria*. Planta Med 2001; 67: 411-6

[11] Fitch RW, Garaffo HM, Spande TS, Yeh HJC, Daly JW. Bioassay-guided isolation of epiquinamide, a novel quinolizidine alkaloid andnicotinic agonist from an Ecuadoran poison frog *Epidobates tricolor*. J Nat Prod 2003; 66: 1345-50

[12] Dittmann K, Riese U, Hamburger, M. HPLC-based bioactivity profiling of plant extracts: a kinetic assay for the identification of monoamine oxidase-

A inhibitors using human recombinant monoamine oxidase-A. Phytochemistry 2004; 65: 3261-8

[13] Dittmann K, Gerhäuser C, Klimo K, Hamburger M. HPLC-based activity profiling of *Salvia miltiorrhiza* for MAO A and iNOS inhibitory activities. Planta Med 2004; 70: 909-13

[14] Hamburger M. *Isatis tinctoria* – From the rediscovery of an ancient medicinal plant towards a novel anti-inflammatory phytopharmaceutical. Phytochem Rev 2002; 1: 333-44

[15] Danz H, Stoyanova S, Thomet OAR, Simon HU, Dannhardt G, Ulbrich H, Hamburger M. Inhibitory activity of tryptanthrin on prostaglandin and leukotriene synthesis. Planta Med 2002; 68: 875-80

[16] Ishihara T, Kohno K, Ushio S, Iwaki K, Ikeda M, Kurimoto M. Tryptanthrin inhibits nitric oxide and prostaglandin E(2) synthesis by murine macrophages. Eur J Pharmac 2000; 407: 197-204

[17] Dannhardt G, Kiefer W. Cyclooxygenase inhibitors – current status and future prospects. Eur J Med Chem 2001; 36: 109-26

[18] Rollinger JM, Haupt S, Stuppner H, Langer T. Combining ethnopharmacology and virtual screening for lead structure discovery: COX-inhibitors as application example. J Chem Inf Comp Sci 2004; 44: 480-8

[19] Lipinski CA, Lombardo F, Dominy BW, Feeney PJ. Experimental and computational approaches to estimate solubility and permeability in drug discovery and development settings. Adv Drug Deliv Rev 1997; 23: 3-25

[20] Rüster GU, Hoffmann B, Hamburger M. Inhibitory activity of indolin-2-one derivatives on compound 48/80-induced histamine release from mast cells. Pharmazie 2004; 59: 236-7

[21] Sun L, Tran N, Tang F, App H, Hirth P, McMahon G, Tang C. Synthesis and biological evaluation of 3-substituted indolin-2-ones: a novel class of tyrosine kinase inhibitors that exhibit selectivity towards particular receptor tyrosine kinases. J Med Chem 1998; 41: 2588-603

[22] Hajek AE, St. Leger RJ. Interactions between fungal pathogens and insect hosts. Annu Rev Entomol 1994; 39: 293-322

[23] Ferron P. In Comprehensive Insect Physiology, Biochemistry and Pharmacology; Kerkut GA, Gilbert LI. Eds.; Pergamon Press: Oxford, 1985; 12: 314-46

[24] Zhu J-S, Halpern GM, Jones K. The scientific rediscovery of an ancient chinese herbal medicine: *Cordyceps sinensis*. Part I. J Altern Complem Med 1998; 4: 289-303

[25] Khachatourians GG. Biochemistry and molecular biology of entopathogenic fungi. In *The Mycota VI: Human and Animal Relationships*; Howard, DE, Ed.; Springer-Verlag: Berlin, 1996; p. 331-63

[26] Schmidt K, Li ZZ, Schubert B, Huang B, Stoyanova S, Hamburger M. Screening of entomopathogenic deuteromycetes for activities on targets

involved in degenerative diseases of the central nervous system. J Ethnopharmac 2003; 89: 251-60

[27] Saragovi HU, Gehring K. Development of pharmacological agents for targeting neurotrophins and their receptors. Trends Pharmac Sci 2000; 21: 93-8

[28] Pollack SJ, Harper SJ. Small molecule Trk receptor agonists and other neurotrophic factor mimetics. Curr Drug Targets CNS Neurol. Disord. 2002; 1: 59-80

[29] Greene LA, Rukenstein A. The quantitative bioassay of nerve growth factor with PC-12 cells. In Nerve Growth Factors; Rush, R. A., Ed.; John Wiley & Sons: Chichester, 1989; 139-47

[30] Gollapudi L, Neet KE. Different mechanisms for inhibition of cell proliferation via cell cycle proteins in PC12 cells by nerve growth factor and staurosporine. J Neurosci Res 1997; 49: 461-74

[31] Schmidt K, Günther W, Stoyanova S, Schubert B, Li Z, Hamburger M. Militarinone A, neurotrophic pyridone alkaloid from *Paecymonyces militaris*. Org Lett 2002; 4: 197-9

[32] Schmidt K, Riese U, Li Z, Hamburger M. Novel tetramic acids and pyridone alkaloids, militarinones B, C and D, from the insect pathogenic fungus *Paecylomyces militaris*. J Nat Prod 2003; 66: 378-83

[33] Leete E, Kowanko N, Newmark RA, Vining L, McInnes AG, Wright JLC. Use of C-13 nuclear magnetic resonance to establish that biosynthesis of tenellin involves an intramolecular rearrangement of phenylalanine. Tetrahedron Lett 1975; 47: 4103-6

[34] Cox RJ, O'Hagan D. Synthesis of isotopically labeled 3-amino-2-phenylpropionic acid and its role as a precursor in the biosynthesis of tenellin and tropic acid. J Chem Soc Perkin Trans I, 1991; 2537-40

[35] Moore MC, Cox RJ, Duffin GR, O'Hagan D. Synthesis and evaluation of a putative acyl tetramic acid intermediate in tenellin biosynthesis in *Beauveria bassiana*. A new role for tyrosine. Tetrahedron 1998; 54: 9195-206

[36] Friedman WJ, Greene LA. Neurotrophin signaling via Trks and p75. Exp Cell Res 1999; 253: 131-42

[37] Greene LA, Kaplan DR. Early events in neurotrophin signalling via Trk and p75 receptors. Curr Opin Neurobiol 1995; 5: 579-87

[38] Riese U, Ziegler E, Hamburger M. Militarinone A induces differentiation in PC12 cells via MAP and Akt kinase signal transduction pathways. FEBS Lett 2004; 577: 455-9

Effects of Bovine Kidney Heparan Sulphate and Shark Cartilage Chondroitin-6-Sulphate on Palatal Fibroblast Activities

George W Yip[a]*, Xiao-Hui Zou[b], Weng-Chiong Foong[b], Tong Cao[b], Boon-Huat Bay[a] and Hong-Wei Ouyang[c]
[a]Department of Anatomy, Yong Loo Lin School of Medicine, [b]Faculty of Dentistry, and [c]Division of Bioengineering, National University of Singapore, 4 Medical Drive, Block MD 10, Singapore 117597
**Corresponding email: georgeyip@nus.edu.sg*

INTRODUCTION

Glycosaminoglycans are large, unbranched, polyanionic molecules that are made up of repeating disaccharide subunits consisting of a hexosamine and a hexuronic acid [1]. Heparan sulphate and chondroitin sulphate are sulphated glycosaminoglycans that are present in the soft palate. The subunits of heparan sulphate comprises N-acetylglucosamine and either iduronate or glucuronate [2]. On the other hand, chondroitin sulphate subunits contains N-acetylgalactosamine and glucuronate [1]. Heparan sulphate and chondroitin sulphate are capable of binding to over one hundred molecules [2-4]. These include the fibroblast growth factor family and members of the transforming growth factor family, which have been shown to be involved in the wound healing process [5, 6].
With an annual incidence of 1.5 to 2 per 1,000 births, congenital cleft palate is one of the commonest birth defects [7]. Surgical correction of the palatal cleft is currently the treatment of choice. Optimal wound healing is an important requirement for successful surgical treatment of cleft palate, and this depends on cell proliferation, cell adhesion and cell migration. However, the molecular regulation of palatal wound healing is not fully understood.

The aim of this study was to determine the effects of bovine kidney heparan sulphate and shark cartilage chondroitin-6-sulphate in palatal wound healing. Using the rabbit as a model organism, we examined the effects of these two molecules on palatal fibroblast proliferation, adhesion and migration.

EXPERIMENTAL SECTION

Primary Culture of Rabbit Palatal Fibroblasts. All animal experiments were approved by the Ethics Committee, Faculty of Dentistry, National University of Singapore. Soft palatal mucosa was removed from adult New Zealand white rabbits using Number 18 blade under sterile conditions. The mucosa was rinsed three times using 0.9% sodium chloride solution, 200 units/mL penicillin and 200 μg/mL streptomycin. The sub-epithelial tissues were then collected, cut into small pieces and incubated for 150 min at 37°C in 0.9% sodium chloride solution, 0.25% collagenase type I and 0.2% albumin. Palatal fibroblasts were collected by centrifugation for 10 min at 250 *g*, washed and suspended in culture medium made up of Dulbecco's modified Eagle's medium, 10% fetal bovine serum, 100 units/mL penicillin and 100 μg/mL streptomycin. The cells were cultured in a humidified atmosphere with 5% CO_2 at 37°C. Chlorate, bovine kidney heparan sulphate, or shark cartilage chondroitin-6-sulphate (purchased from Sigma, St. Louis, MO, USA) were added to the culture medium. Culture medium was changed once in three days. *In vitro* assays were performed on Passage 2 cells derived from at least three independent animals.

Cell Proliferation Assay. Palatal fibroblasts were seeded at a density of 30,000 cells per well in 24-well plates and cultured for three days. Fibroblast numbers were then measured at Day 3 using the Cell Titer 96® AQ_{ueous} Non-Radioactive Cell Proliferation Assay (Promega, Madison, WI, USA) according to the manufacturer's protocol.

Cell Adhesion Assay. Cells were plated at 60,000 cells per well and cultured for 8 h. The number of adherent fibroblasts was determined using the Cell Titer 96® AQ_{ueous} Non-Radioactive Cell Proliferation Assay.

***In vitro* Wound Model.** Three horizontal lines were scraped across the bottom of each well of 6-well plates containing 90% confluent palatal fibroblasts using a sterile 100-μL plastic pipette tip ('wounding'). The distance between the wound edges was measured at 0 and 12 h after wounding.

Scanning Electron Microscopy. Palatal fibroblasts were cultured on glass slides and wounded as described above for scanning electron microscopy. The cells were fixed in 3% glutaraldehyde and 2% paraformaldehyde in 100 mmol/L cacodylate for 30 min, followed by 2%

osmium tetroxide. The cells were then dehydrated in a graded series of methanol, transferred to acetone, and dried in a Balzers critical point dryer. Cells were coated with 20 nm gold particles with a Balzers sputter coater, and examined using a Philips XL-30 field emission gun scanning electron microscope.

Statistical Analysis. Results were compared among treatment groups by Student's *t*-test or one-way analysis of variance (ANOVA) with Tukey's post test using GraphPad Prism v4.02 for Windows (GraphPad Software, San Diego, CA, USA).

RESULTS AND DISCUSSION

To determine if sulphated glycosaminoglycans play a role in regulating palatal fibroblast proliferation, fibroblasts were cultured in the presence of chlorate, an inhibitor of glycosaminoglycan sulfation. Chlorate has been extensively used in cell, organ and whole embryo cultures to inhibit sulfation without any significant effect on glycosaminoglycan or protein synthesis or on cell viability [2, 8, 9]. Chlorate competes with sulphate in the synthesis of 3'-phosphoadenosine 5'-phosphosulphate, the sulphate donor in glycosaminoglycan sulfation.

Palatal fibroblasts cultured in the presence of chlorate appeared healthy, with the same spindle-shaped appearance as those in the control groups that were cultured in chlorate-free medium. However, there was a significant reduction in proliferation of cells treated with chlorate (Figure 1). This decrease in cell proliferation could be prevented by supplementing the chlorate-containing culture medium with either exogenous bovine kidney heparan sulphate or shark cartilage chondroitin-6-sulphate over the three-day culture period (Figure 1).

Chlorate was also used to determine if sulphated glycosaminoglycans are necessary for palatal fibroblast adhesion. Fibroblasts treated with chlorate showed a significant reduction in cell adhesion (Figure 2). Supplementation of chlorate-containing culture medium with either bovine kidney heparan sulphate or shark cartilage chondroitin-6-sulphate resulted in an increase in cell adhesion (Figure 2).

To examine the effects of sulphated glycosaminoglycans on cell migration and wound closure, we made use of an *in vitro* wound closure model [10]. The distance between the wound edges measured 12 h after

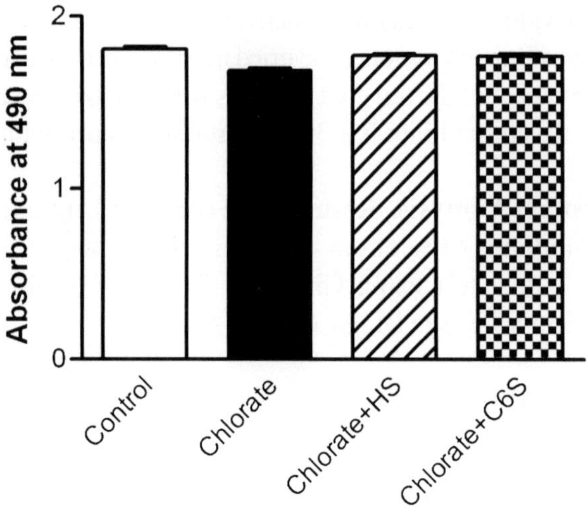

Figure 1. Heparan sulfate and chondroitin sulfate regulate palatal fibroblast proliferation. Cells were cultured for 3 d in normal culture medium, medium supplemented with 30 mmol/L chlorate, or medium containing 30 mmol/L chlorate plus 100 ng/mL bovine kidney heparan sulfate (HS) or 100 ng/mL shark cartilage chondroitin-6-sulfate (C6S). The number of cells present varies among treatment groups (one-way ANOVA; $p = 0.0016$). Cell numbers are significantly reduced in the chlorate-treated group compared with the untreated group (Tukey's test; $p < 0.01$). In contrast, there is no significant difference between the control group and cells treated with chlorate plus heparan sulfate or chondroitin-6-sulfate ($p > 0.05$ in each case). Values represent mean ± standard error (SEM) of 3 experiments.

wounding is used as an indicator of the rate of wound closure. The average wound gap obtained by scraping across the bottom of 6-well plates was 489 μm, and this decreased by 44.9% after a 12-h culture period for the untreated group (Figure 3). In contrast, the rate of wound closure and cell migration was significantly reduced in the chlorate-treated group, with only 5.3% decrease in wound gap distance. The chlorate-induced slow down in wound closure and cell migration was partially abolished by adding either bovine kidney heparan sulphate or shark cartilage chondroitin-6-sulphate to the culture medium (Figure 3).

Taken together, the results of this study show that heparan sulphate and chondroitin-6-sulphate play important roles in palatal wound healing. They regulate fundamental cellular processes of proliferation, adhesion

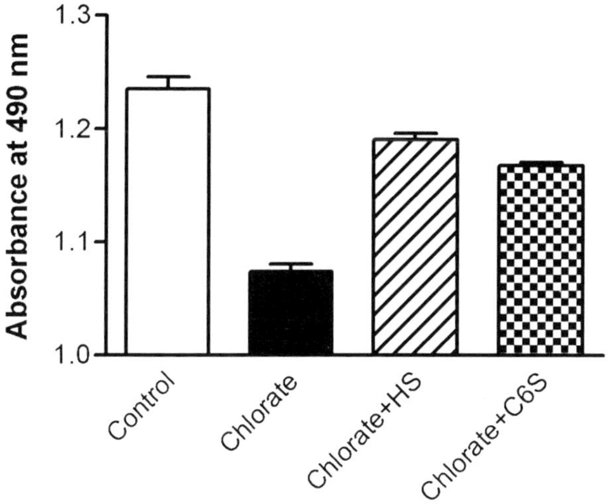

Figure 2. Sulfate group on glycosaminoglycans regulates cell adhesion. Palatal fibroblasts were cultured for 8 h. Statistical comparison using one-way ANOVA shows a significant difference in adhesion among cells cultured in normal medium, medium supplemented with 30 mmol/L chlorate, or medium with 30 mmol/L chlorate plus either 100 ng/mL bovine kidney heparan sulfate or 100 ng/mL shark cartilage chondroitin-6-sulfate ($p < 0.0001$). The number of adherent cells in the chlorate-treated group is significantly smaller than in the control group (Tukey's test; $p < 0.001$). Addition of either heparan sulfate or chondroitin sulfate to the chlorate-containing medium significantly improves cell adhesion ($p < 0.001$ in both cases). Values represent mean ± SEM of 3 experiments.

and migration, activities that are critically dependent on the presence of the sulphate group on these molecules.

Basic fibroblast growth factor is a stimulatory factor for gingival fibroblast proliferation [11]. Receptors for fibroblast growth factors have been reported in myofibroblasts in full-thickness palatal mucoperiosteal wounds [6]. Heparan sulphate has been shown to potentiate fibroblast growth factor signaling [3, 12, 13]. Recent studies have shown that chondroitin sulphate may also help to mediate the cell proliferative effect of basic fibroblast growth factor [14].

Integrins are transmembrane heterodimeric proteins that regulate cell-cell and cell-extracellular matrix adhesion [15]. They act by localizing focal adhesion kinase (FAK) at focal adhesion points and tyrosine phosphorylation of FAK [16, 17]. In addition, FAK facilitates cell

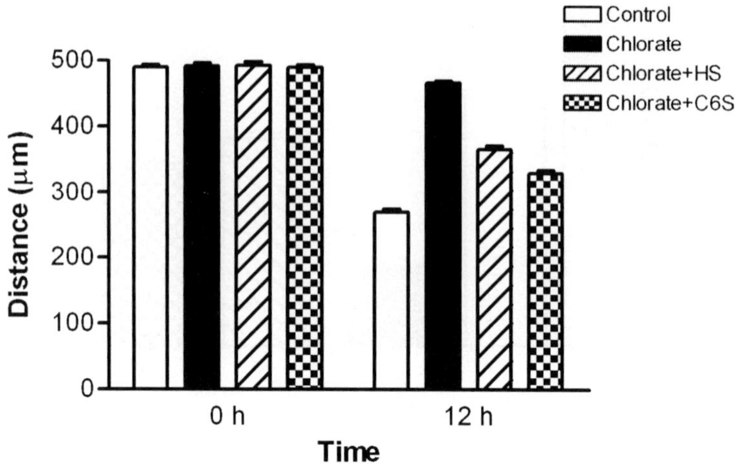

Figure 3. Heparan sulfate and chondroitin sulfate influences cell migration and wound closure *in vitro*. Palatal fibroblasts cultured in the normal medium, medium containing 30 mmol/L chlorate, and medium with 30 mmol/L chlorate plus 100 ng/mL bovine kidney heparan sulfate or 100 ng/mL shark cartilage chondroitin-6-sulfate show a significant difference in wound edge distance 12 h after 'wounding' (one-way ANOVA; $p < 0.0001$). The wound gap in the chlorate-treated group is significantly wider than in the control group (Tukey's test; $p < 0.001$). This chlorate-induced reduction in wound closure speed was partially abolished by adding either heparan sulfate or chondroitin-6-sulfate to the chlorate-supplemented medium ($p < 0.001$ in each case). Values represent mean ± SEM of 9 wounds per group.

migration by down-regulating the activity of RhoA, a member of the Rho family of GTPases [18-20]. A common feature is that chondroitin sulphate has been show to bind to $\alpha4\beta1$ integrin, enhance FAK phosphorylation and inhibit RhoA activity for cell adhesion and migration [21, 22].

Although supplementation of the culture medium with either heparan sulphate or chondroitin sulphate could significantly abolish the effects of chlorate on cell proliferation, adhesion and migration, each species acting alone does not fully restore these cellular processes to normal. A combination of heparan sulphate and chondroitin sulphate might lead to better improvement in wound healing compared to the individual molecular species. We are currently investigating this, and the possible roles played by other sulphated glycosaminoglycans in palatal wound healing.

ACKNOWLEDGMENT

Financial support was obtained from the Academic Research Fund, National University of Singapore (Grant R-222-000-005-112). We thank Y. G. Chan for her expert technical assistance.

REFERENCES

[1] Varki A, Cummings R, Esko J, Freeze H, Hart G, Marth J. Essentials of Glycobiology. New York: Cold Spring Harbor Laboratory Press; 1999

[2] Conrad HE. Heparin-binding proteins. San Diego: Academic Press; 1998

[3] Bernfield M, Gotte M, Park PW, et al. Functions of cell surface heparan sulphate proteoglycans. Annu Rev Biochem 1999;68: 729-77

[4] Oohira A, Matsui F, Tokita Y, Yamauchi S, Aono S. Molecular interactions of neural chondroitin sulphate proteoglycans in the brain development. Arch Biochem Biophys 2000;374: 24-34

[5] Yokozeki M, Moriyama K, Shimokawa H, Kuroda T. Transforming growth factor-beta 1 modulates myofibroblastic phenotype of rat palatal fibroblasts in vitro. Exp Cell Res 1997;231: 328-36

[6] Kanda T, Funato N, Baba Y, Kuroda T. Evidence for fibroblast growth factor receptors in myofibroblasts during palatal mucoperiosteal repair. Arch Oral Biol 2003;48: 213-21

[7] Mitchell JC, Wood RJ. Management of cleft lip and palate in primary care. J Pediatr Health Care 2000;14: 13-9

[8] Yip GW, Ferretti P, Copp AJ. Heparan sulphate proteoglycans and spinal neurulation in the mouse embryo. Development 2002;129: 2109-19

[9] Zou XH, Foong WC, Cao T, Bay BH, Ouyang HW, Yip GW. Chondroitin sulphate in palatal wound healing. J Dent Res 2004;83: 880-5

[10] Guo F, Gao Y, Wang L, Zheng Y. p19Arf-p53 tumor suppressor pathway regulates cell motility by suppression of phosphoinositide 3-kinase and Rac1 GTPase activities. J Biol Chem 2003;278: 14414-9

[11] Fujisawa K, Miyamoto Y, Nagayama M. Basic fibroblast growth factor and epidermal growth factor reverse impaired ulcer healing of the rabbit oral mucosa. J Oral Pathol Med 2003;32: 358-66

[12] Schlessinger J, Lax I, Lemmon M. Regulation of growth factor activation by proteoglycans: what is the role of the low affinity receptors? Cell 1995;83: 357-60

[13] Spivak-Kroizman T, Lemmon MA, Dikic I, et al. Heparin-induced oligomerization of FGF molecules is responsible for FGF receptor dimerization, activation, and cell proliferation. Cell 1994;79: 1015-24

[14] Milev P, Monnerie H, Popp S, Margolis RK, Margolis RU. The core protein of the chondroitin sulphate proteoglycan phosphacan is a high-affinity

ligand of fibroblast growth factor-2 and potentiates its mitogenic activity. J Biol Chem 1998;273: 21439-42

[15] Hynes RO. Integrins: versatility, modulation, and signaling in cell adhesion. Cell 1992;69: 11-25

[16] LaFlamme SE, Auer KL. Integrin signaling. Semin Cancer Biol 1996;7: 111-8

[17] Schaller MD. Biochemical signals and biological responses elicited by the focal adhesion kinase. Biochim Biophys Acta 2001;1540: 1-21

[18] Ren XD, Kiosses WB, Sieg DJ, Otey CA, Schlaepfer DD, Schwartz MA. Focal adhesion kinase suppresses Rho activity to promote focal adhesion turnover. J Cell Sci 2000;113 (Pt 20): 3673-8

[19] Arthur WT, Burridge K. RhoA inactivation by p190RhoGAP regulates cell spreading and migration by promoting membrane protrusion and polarity. Mol Biol Cell 2001;12: 2711-20

[20] Wakatsuki T, Wysolmerski RB, Elson EL. Mechanics of cell spreading: role of myosin II. J Cell Sci 2003;116: 1617-25

[21] Iida J, Meijne AM, Oegema TR, Jr., et al. A role of chondroitin sulphate glycosaminoglycan binding site in alpha4beta1 integrin-mediated melanoma cell adhesion. J Biol Chem 1998;273: 5955-62

[22] Yang J, Price MA, Neudauer CL, et al. Melanoma chondroitin sulphate proteoglycan enhances FAK and ERK activation by distinct mechanisms. J Cell Biol 2004;165: 881-91

Inhibition of Aflatoxin Biosynthesis in *Aspergillus Flavus* by Phenolic Natural Products

Russell J Molyneux*, Noreen Mahoney, Jong H Kim and Bruce C Campbell
Plant Mycotoxins Research Unit, Western Regional Research Center, Agricultural Research Service, U.S. Department of Agriculture, 800 Buchanan Street, Albany, California 94710
**Corresponding email: molyneux@pw.usda.gov*

INTRODUCTION

Natural products are extremely important factors for crop production around the world. The compounds biosynthesized by plants may provide valuable nutritional and health benefits, protecting consumers from age-related diseases such as cancer, diabetes and arteriosclerosis. However, these benefits can be offset by contamination with synthetic chemicals such as herbicides and pesticides, and by natural toxins present in the plant itself or produced by infection with micro-organisms. Microbial metabolites are a particular concern because they commonly occur and can be formed both in the field and in storage. Treatment with antibiotics to prevent growth of microorganisms, or with other chemicals to destroy the toxins, are unlikely to be acceptable to consumers because of the added costs and risk of introducing other noxious compounds or changing the quality characteristics of the crop. The efficient and economical production of foodstuffs therefore requires that natural methods be used wherever possible to control contamination.

Among the mycotoxin contamination problems, probably the most important worldwide are due to aflatoxins (Figure 1), produced by various strains of the fungi, *Aspergillus flavus* and *A. parasiticus* which can infect major agricultural crops such as corn (maize), peanuts, and tree nuts. In subsistence-farming countries, *Aspergillus* infection is common and exacerbated by poor storage conditions. In a recent episode involving moldy maize in eastern Kenya, at least 112 persons, predominantly children, were reported to have died from acute toxicity since May 2004 with many more hospitalized, and the death toll is expected to increase

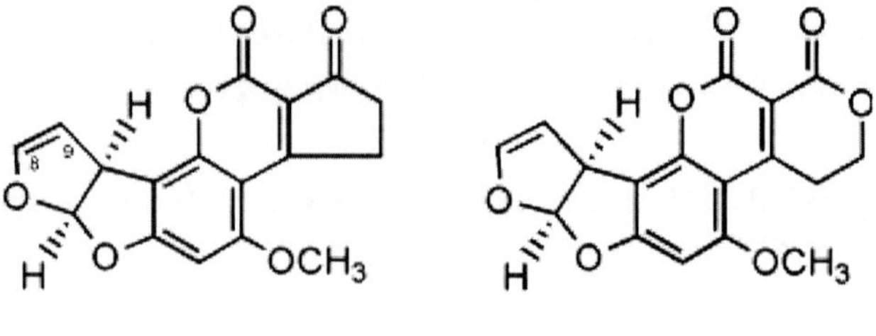

1 2

Figure 1. Chemical structures of aflatoxin B_1, **1**, and aflatoxin G_1, **2**; aflatoxins B_2 and G_2 are the 8, 9-dihydro derivatives of **1** and **2**, respectively.

[1]. The situation reached crisis proportions because much of the available maize stocks have had to be destroyed to avoid further poisoning, resulting in widespread hunger. In developed countries, contamination is much less widespread but because aflatoxins, in conjunction with exposure to hepatitis B or C, are known risk factors for liver cancer in humans, they are highly regulated both within the U.S. and in many other countries even though epidemiological studies have cast doubt on whether lowering of aflatoxin levels is effective in reducing the rate of this disease [2].

In the United States, almonds, pistachios and walnuts are one of the most important agricultural crops affected by aflatoxin contamination. A large proportion of their value comes from the sales overseas, with edible tree nut exports totaling $1.7 billion for the period from August 2003 to June 2004 [3]. In California, where most of the production is located, tree nuts are the single most valuable agricultural crop. However, potential food safety and quality concerns jeopardize this market because aflatoxins are highly regulated, especially in the European Community (EC), which has imposed exceptionally low tolerance levels of 2 ng/g for aflatoxin B_1 and 4 ng/g total aflatoxins [4]. Rejection, or even destruction, of shipments entering the European Community can therefore result in large economic losses to producers and exporters if aflatoxins exceed regulatory amounts. In contrast to the EC regulations, the Food and Drug Administration has set a maximum guidance level limit of 20 ppb (20 ng/g) for tree nuts, including shells, intended for human consumption within the United States [5].

In order to ensure that aflatoxin levels in tree nuts are safe for consumption and in conformance to regulations, we have investigated endogenous natural products that suppress aflatoxin biosynthesis and potentially inhibit *Aspergillus* colonization and growth. Previously, we have shown that naphthoquinones present in walnut hulls can either suppress or stimulate levels of aflatoxin production and also delay germination of *A. flavus* [6]. The biotransformation of aflatoxin B_1 to mutagenic compounds was also found to be inhibited by natural products from other plants and spices [7]. These results supported the concept that manipulation of natural product levels can be used as a method to control aflatoxin contamination. Subsequent experiments have shown that walnuts are much less susceptible to aflatoxigenesis than almonds and pistachios [8]. Since all tree nuts are well-protected against fungal infection by the hull, shell, and the seed coat surrounding the kernel, the source of this difference is unlikely to be due merely to such physical barriers but rather to the presence of phytochemical constituents. The identification of such compounds would enable their levels to be enhanced either through selection of cultivars or by directed crop breeding programs, without adversely affecting quality characteristics. However, it is important to note that the compounds must be phytoanticipins, inherently present in the plant tissues, rather than phytoalexins produced in response to fungal attack [9]. The time-lag involved in phytoalexin production is unlikely to enable significant resistance to aflatoxin formation to develop and furthermore many phytoalexins are themselves inherently toxic and therefore undesirable in foods. Walnut production in California consists of about 11 major cultivars of *Juglans regia* (family Juglandaceae) and these cultivars display a differential susceptibility to aflatoxin formation. This observation provides a means by which specific phytochemical constituents may be identified through comparison of their respective amounts with levels of aflatoxin produced. In particular, the cultivar 'Tulare' has been shown to exhibit complete suppression of aflatoxigenesis even when the kernel is directly inoculated *in vitro* with *A. flavus* [8]. This resistance was shown to be located entirely within the seed coat and not in the kernel itself. In addition, the aflatoxin suppression increased as the walnut matured. These observations provided a basis for bioactivity-directed fractionation of phytochemicals within the seed coat tissue.

The objectives of our studies were therefore: to identify the seed coat constituents which confer resistance to aflatoxin contamination; to correlate the levels of these components in 'Tulare' walnut throughout the growing season, in comparison with the more susceptible cultivars; and, to use the compounds identified to investigate possible mechanisms whereby aflatoxigenic *Aspergillus* species are rendered atoxigenic. Ultimately, this information could lead to an understanding of the role that aflatoxins play in the ecology of *Aspergillus* species, a question which has yet to be resolved.

IDENTIFICATION OF AFLATOXIN RESISTANCE FACTORS IN WALNUT PELLICLE

Since earlier experiments had shown that all of the anti-aflatoxigenic activity was in the seed-coat, extraction and fractionation was focused upon this material [10]. Ground seed-coat was sequentially extracted with hexane, acetone, methanol and water and the latter three extracts tested at an incorporation level of 0.25% (w/v) in the defined medium, Vogel's medium 'N' (VMN). The hexane extract was not tested because of solubility problems in VMN but testing of the seed-coat material after initial extraction showed that its activity was the same as an unextracted control sample, indicating that compounds of interest were not soluble in this solvent. In contrast, the acetone, methanol and water extracts reduced the aflatoxin content by 100%, relative to control, indicating that the compounds responsible had a range of polar characteristics, and the majority of the soluble material was extracted by acetone and methanol, comprising 56% of the original weight. The unextractable residual seed-coat amounted to 23.5% by weight and aflatoxin inhibitory activity was 77% of control, indicating that even very polar solvents did not extract all of the inhibitory compounds. The fact that a large proportion of the activity was extractable by a range of solvents, but that some of the inhibitory activity was not extractable, suggested that a complex of relatively high molecular weight substances was responsible. Low molecular weight compounds did not appear to be involved and this was confirmed by GC-MS analysis of the extracts as their trimethylsilyl derivatives. Only methyl gallate was detected and the trace amounts present were insufficient to explain the potent activity of the extracts.

The properties of the bioactive extracts were consistent with them being mixtures of hydrolysable tannins, which are well established as being major constituents of walnut seed-coat. For example, juglanin was isolated from seed-coat removed in walnut processing [11] and shown to consist of a glucose core esterified with one molecule of gallic acid and one molecule of hexahydroxydiphenic acid (Figure 2) [12]. Structurally, it was isomeric with a constituent of divi-divi tannin, corilagin, the 1-*O*-galloyl derivative of 3, 6-hexahydroxydiphenoylglucose (**3**) [13], but differed in melting point and optical rotation. The structure of juglanin, (α_D^{20} +52) has never been established with certainty but comparison of its properties with the known mono-galloyl derivatives of 4,6-hexahydroxydiphenoylglucose suggests that it may be identical with 1-*O*-

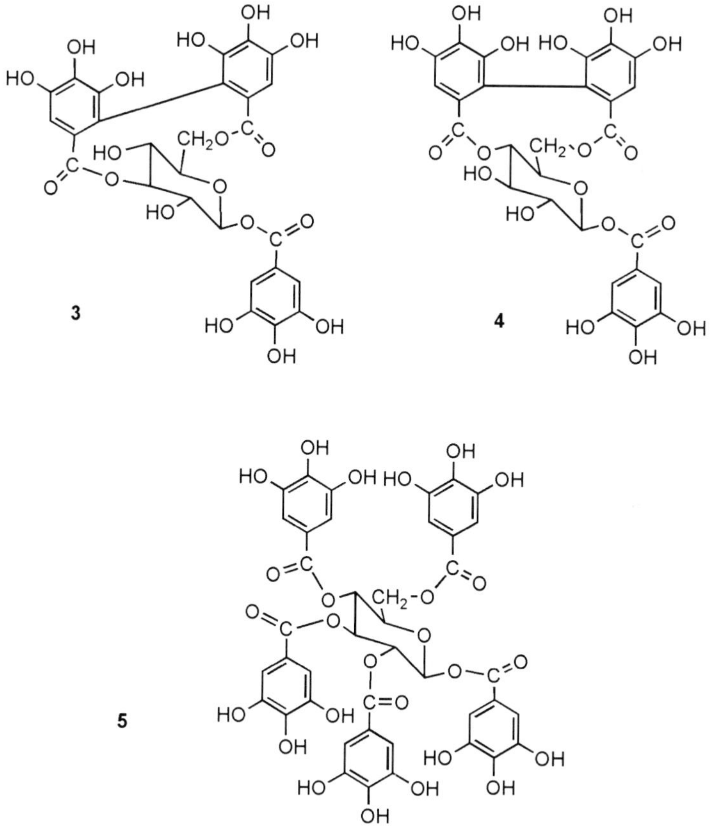

Figure 2. Structure of corilagin **3**, and presumed structure of juglanin, **4**, a constituent of walnut seed coat. Hydrolyzable tannins are biosynthesized from the precursor pentagalloylglucose, **5**, through oxidative dimerization of adjacent galloyl ester moieties to hexahydroxydiphenoyl ester substituents and partial hydrolysis of some of the undimerized galloyl ester substituents.

galloyl-4,6-(*S*)-hexahydroxydiphenoyl-α-D-glucopyranose (**4**), (α_D^{20} +54.9), previously isolated from leaves of *Cornus officinalis*. It is noteworthy that strictinin, (α_D^{20} -3.1), the β-anomer of the latter compound, has recently been isolated from walnuts, together with 11 other known hydrolysable tannins and three new compounds named glansrins A-C [14]. The co-occurrence of such a complex mixture of tannins is consistent with our findings that the anti-aflatoxigenic activity is extractable with a range of solvents. The variety and complexity of structures present is explained by the known biosynthesis of hydrolysable tannins, in which the precursor, pentagalloylglucose, **5** (Figure 2), subsequently undergoes oxidative dimerization of adjacent galloyl groups to hexahydroxydiphenic acid moieties, with hydrolytic cleavage of other galloyl groups [15, 16].

A potential role for hydrolyzable tannins in plants is to protect them against microbial attack and inhibit decay. However, *Aspergillus flavus* and *A. parasiticus* are commensals and not pathogens, so it would only be coincidental if the tannins acted in such a way against these fungi. Furthermore, bioassays for such activity have demonstrated that *in vitro* growth inhibition of filamentous fungi requires levels of 0.1-0.5 g/L [17, 18]. The effect of walnut tannin on growth of *A. flavus* was evaluated by extracting the seed coat with methanol, passing the extract through Sephadex LH-20 to remove low molecular weight compounds [19], and adding the tannin fraction to walnut kernel medium. Radial growth was initially delayed, with less dense mycelium relative to control, but this effect was completely surmounted over the course of the experiment. In contrast, the tannin incorporation markedly suppressed aflatoxin formation, with 0.05% (w/v) tannin incorporation reducing the level to 50% of control and only 0.5% (w/v) being required to eliminate production of the toxin (Figure 3). This established that the hydrolyzable tannin was either directly or indirectly capable of inhibiting aflatoxin biosynthesis without substantially affecting fungal growth.

It is known that a number of *Aspergillus* and *Penicillium* species are able to utilize tannins as their sole carbon source and may even thrive in the presence of very high tannin levels such as those present in tanning vats [17]. This capability ensues from the presence of an extracellular fungal tannase, capable of hydrolyzing the ester linkages in the tannin to liberate the glucose moiety as a nutrient for fungal growth [20]. The ester moieties are released as gallic acid, **7** and ellagic acid, **8**; the initial hydrolysis

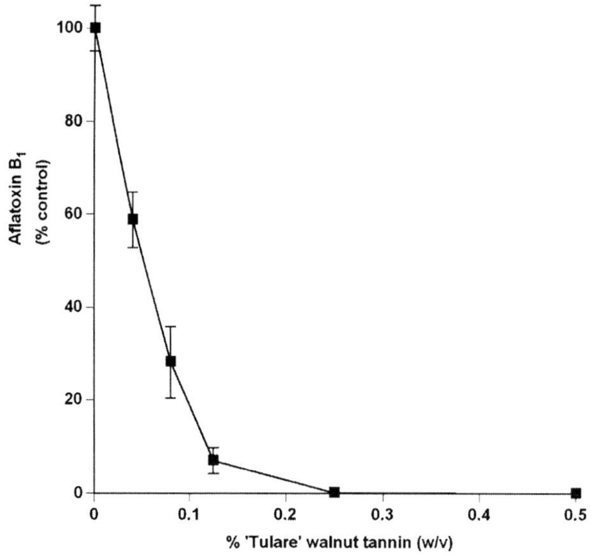

Figure 3. Suppression of aflatoxin formation by *Aspergillus flavus* grown on walnut endosperm media containing various levels of walnut hydrolyzable tannin extracted from seed coat of the 'Tulare' cultivar with methanol and purified by Sephadex LH-20 chromatography.

product, hexahydroxydiphenic acid, **9**, cannot be isolated due to spontaneous lactonization (Figure 4). The presence of a tannase in the strain of *A. flavus* isolated from pistachio used in our experiments (NRRL 25347) was not established and it was therefore tested in a simple plate assay. When the fungus was grown on tannic acid media, a broad zone of clearing was observed [21] (Figure 5) proving the presence of the enzyme in this particular strain. Because the fungus has the capability to hydrolyze the tannin, the suppression of aflatoxin production may be due to the tannin itself, to one or both of the hydrolysis products, gallic and ellagic acids, or even to a combination of these. Pure gallic and ellagic acids were therefore tested as aflatoxigenesis inhibitors.

Inhibition by gallic acid and ellagic acid, added at 0.2% to VMN, over an 11-day time-course showed that aflatoxin production peaked at days 5-7. The effect of gallic acid was potent, with aflatoxin reduced to 4% of control on day 6. In contrast, ellagic acid had little effect, reducing aflatoxin to only 84% on the same day (Figure 6), although it did delay

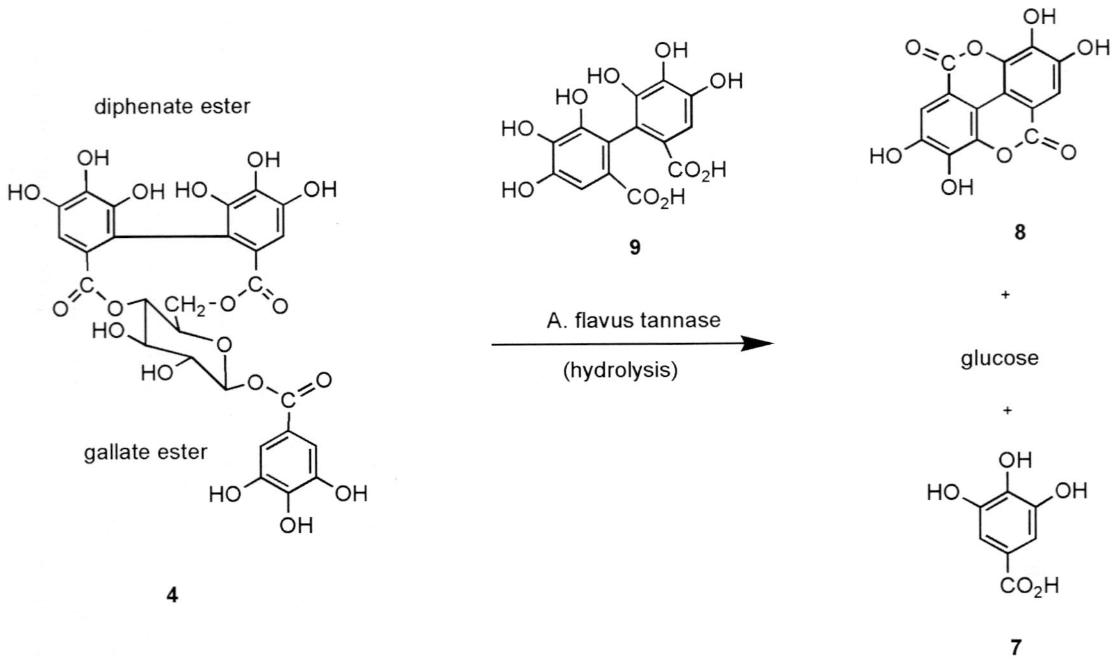

Figure 4. Depiction of hydrolysis by a fungal tannase of juglanin, **4**, to yield gallic acid, **7**, and ellagic acid, **8**; the initial hydrolysis product, hexahydroxydiphenic acid, **9,** cannot be isolated due to spontaneous lactonization.

Figure 5. Zone of clearing produced *by A. flavus* NRRL 25347 grown on tannic acid media, establishing the presence of an extracellular tannase.

fungal growth and produced less mycelium whereas gallic acid had no effect on radial growth of the fungus. As with walnut tannin itself, this growth effect was not permanent and was eventually overcome, appearing to be normal at later time stages. However, at an incorporation level of 0.05%, ellagic acid actually stimulated aflatoxin to 148% of control. It was not possible to evaluate the effect of hexahydroxydiphenic acid, **8**, the precursor of ellagic acid, because this compound cannot be isolated in unlactonized form. Nevertheless, the results suggest that ellagic acid is not a contributor to aflatoxin inhibitory activity whereas incorporation of gallic acid into media consisting of walnut kernel reduced aflatoxin to 50% of control at only 0.02% (w/v) incorporation and to 13.8% at 0.2% (w/v) incorporation. This was supported by testing commercial tannic acid, a hydrolysable tannin possessing only gallate ester moieties, which completely inhibited aflatoxin production at 0.4% incorporation without any effect on fungal growth.

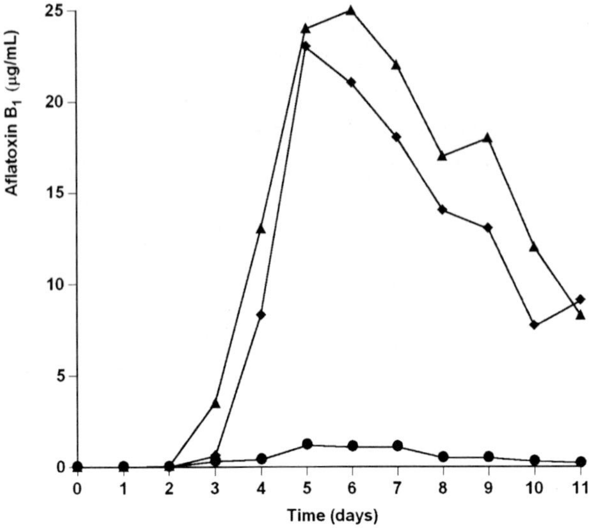

Figure 6. Effect of gallic acid (●) and ellagic acid (◆) on aflatoxin production by *Aspergillus flavus* in Vogel's medium N (VMN) (▲, control), over an 11-day time-course.

CORRELATION OF GALLIC ACID LEVEL IN WALNUT SEED COAT WITH AFLATOXIN RESISTANCE AND COMPARISON WITH CULTIVARS

In order to measure the gallic acid content, ground walnut seed coat was treated with anhydrous methanolic HCl, hydrolyzing the tannin and forming methyl gallate and ellagic acid, which can be analyzed by HPLC with UV monitoring at 280 and 252 nm, respectively [22]. Linear calibration curves were generated for each compound using standards and the identity of the compounds confirmed by retention time and by their UV spectra obtained with a diode-array detector. Samples of 'Tulare', an aflatoxin-resistant walnut variety, and 'Chico', a more susceptible variety [8], were then collected and analyzed on a biweekly basis throughout the growing season for the years 2002 and 2003.

The gallic and ellagic acid analytical results for 2003 were typical and are illustrated in Figure 7. After an initial rapid increase for both cultivars, which reached a maximum three weeks after first formation of the walnuts, there was a noticeable decline in gallic acid content. This was consistent with the known biosynthetic pathway for all hydrolyzable tannins [15, 16], which proceeds from the common precursor pentagalloyl glucose, **5**, with subsequent hydrolysis of some galloyl moieties and oxidative dimerization of others to form hexahydroxydiphenate ester moieties (Figure 2). A combination of these reactions would give a simple hydrolyzable tannin such as juglanin, **4**, with an overall reduction in gallic acid content relative to the maximum level. Subsequent reactions can increase the structural complexity of the tannins as the plant matures due to addition of gallic acid residues through oxidative couplings or depside linkages to the core structure. Following this initial rise, gallic acid level in 'Tulare' was generally twice as great as that in 'Chico' (Figure 7A). In contrast to the gallic acid level, the ellagic acid content for both cultivars increased fairly steadily, leveling off or declining slightly as the walnut matured (Figure 7B). This was also consistent with biosynthetic theory since gallic ester moieties are precursors of the hexahydroxydiphenate moieties. Although ellagic acid levels were much higher than those of gallic acid, the content in the two cultivars was relatively similar, in contrast to the marked difference in gallic acid content. The lack of potent anti-aflatoxigenic activity for

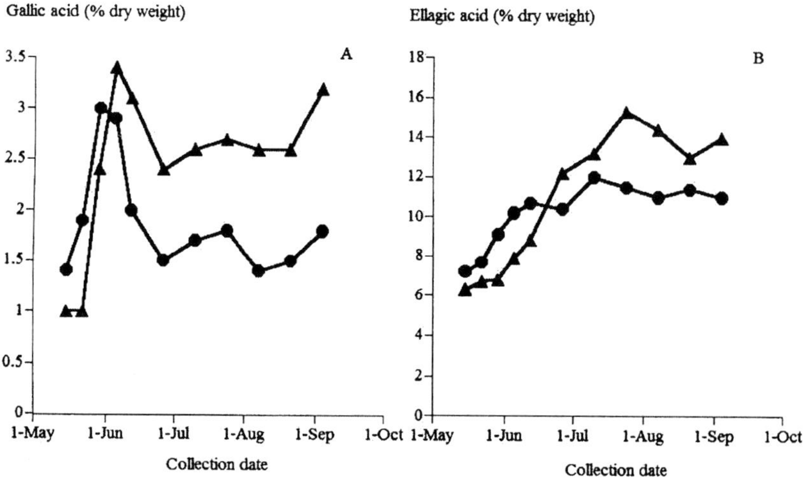

Figure 7. Variation in gallic acid and ellagic acid content of hydrolysable tannins in the walnut cultivars 'Tulare' (▲) and 'Chico' (●) during the 2003 growth season.

ellagic acid and the similar content in aflatoxin resistant and susceptible varieties indicates that it is not responsible for the inhibitory effect on *A. flavus* and that gallic acid is the primary atoxigenic component.

The gallic and ellagic acid contents of a number of other English walnut (*Juglans regia*) cultivars from the 2003 season were also analyzed solely at the mature stage. The gallic acid levels ranged from 1.4 to 3.4% with only the cultivars 'Red Zinger' (3.4%) and 'Tulare' (3.2%) exceeding 3% dry weight. The seed coat of both of these varieties had been shown to reduce aflatoxin B_1 levels comparably *in vitro* [8]. Only the cultivar 'Chandler' (1.4%) had a gallic acid level less than that of 'Chico' (1.8%) and four cultivars had contents of 2.0-2.6%. These values correlated well with aflatoxin inhibitory activity of seed coat from the various cultivars [8] and indicated that gallic acid levels below 2.2% do not prevent aflatoxin formation. The black walnut species, *J. hindsii* cv. 'Rawlins' (1.0%) and *J. nigra* cv. 'Thomas' (1.1%), had much lower gallic acid levels than all English walnut cultivars and these values were consistent with the high aflatoxin levels produced *in vitro* with black walnut seed coat. For all varieties, the increase in ellagic acid content generally paralleled that of gallic acid, with values of 10.0-15.9% for English walnuts and much lower levels of 2.6-3.1% for black walnuts. The ratio of ellagic acid to gallic acid was consistent with a biosynthetic pathway in which oxidative dimerization of gallate ester moieties are less prone to occur in the high gallic acid cultivars with consequently lower ellagic acid levels.

MECHANISM OF ANTI-AFLATOXIGENIC ACTION

It is significant that hydrolysable tannins can completely eliminate aflatoxin production by *A. flavus* but the molecular mode of action responsible for this effect is not established. Elucidation of the precise reasons for biosynthesis of such a complex molecule as aflatoxin should enable strategies to be developed to control its formation and thereby ensure human health and safety. Although the specific steps and corresponding genes for aflatoxin biosynthesis have been extensively studied [23, 24], the role played by aflatoxins in the biology of the fungus is not understood and the selective advantage for fungi to synthesize aflatoxins is unknown. There are many naturally atoxigenic strains of *A. flavus* and these can co-exist with toxigenic strains and under certain circumstances out-compete the latter [25]. Other strains of *A. flavus* have incomplete biosynthetic pathways to aflatoxins, only synthesizing precursors, and these strains also appear to thrive well under natural conditions. These precursors are phenolic compounds and potentially have inherent antioxidant activity, thus contributing to alleviation of oxidative stress. Our finding of the existence in plants of discrete water-soluble compounds that suppress aflatoxin formation provides the tools to investigate in detail the potential advantages to the fungus of aflatoxin production.

An abbreviated representation of the biosynthesis of aflatoxins is shown in Figure 8. The earliest aromatic precursor, norsolorinic acid, **9**, is bright yellow-orange in color and many of the subsequent metabolites are fluorescent. However, no visible evidence of such compounds was observed in the extracts of the media treated with walnut seed coat, hydrolysable tannin or gallic acid during aflatoxin analysis, showing that these expected intermediates did not accumulate due to disruption of the biosynthetic pathway prior to aflatoxin formation. This indicates that the genes affected are most likely to be controlling the fatty acid synthase or polyketide synthase enzymes involved in the synthesis of the polyketide precursors of the aromatic metabolites. The effect of gallic acid on specific genes involved in aflatoxin biosynthesis is currently under active investigation.

Because gallic acid and hydrolyzable tannins are potent antioxidants and free radical scavengers [14], we hypothesized that they act by mitigating oxidative stress on the fungus. Infection of plant cells by *A. flavus* can induce rapid generation of reactive oxygen species (ROS) during plant-defense responses [26], producing hydrogen peroxide [27]. Other sources of oxidative stress include reactive radicals and organic peroxides generated from intracellular respiration and metabolism, as well as environmental stresses. Aflatoxin production increases under high temperature or drought, environmental conditions that elicit oxidative stress signaling pathways in microbial and plant cells [28, 29]. There appears to be an intimate association between oxidative stress and induction of aflatoxin biosynthesis and if this is so, then the rationale for aflatoxigenesis is that it is a consequence of increased oxidative stress resulting in formation of peroxides and epoxides through enhanced lipid peroxidation [30]. Jayashree and Subramanyam (1999) [31] had shown that 2-methoxy-4-allylphenol (eugenol) reduces aflatoxin production to 30% of control at

0.45 mM, without any decrease in fungal growth. However, at lower concentrations, both fungal growth and aflatoxin production were increased. We have observed a similar effect in our previous investigation of the antiaflatoxigenic activity of naphthoquinones from walnut hulls [6]. The hydrolysable tannins and gallic acid are unique in not exhibiting this

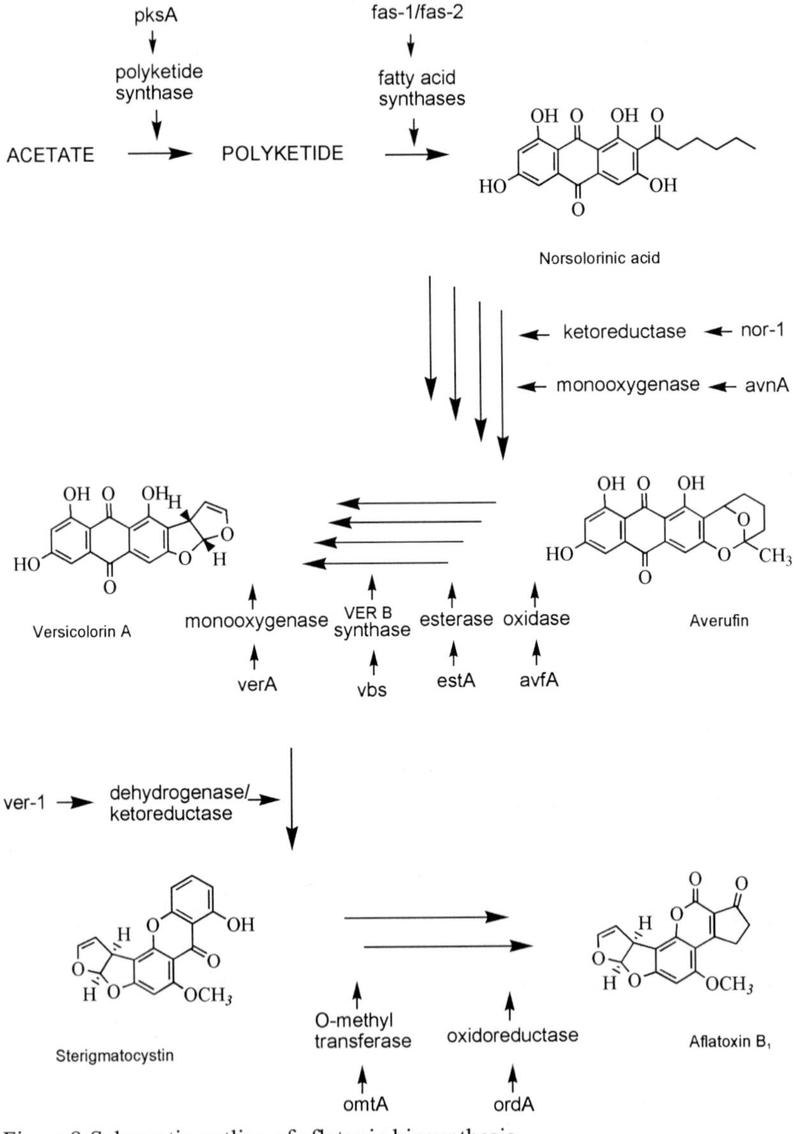

Figure 8. Schematic outline of aflatoxin biosynthesis.

reversibility of action at lower concentrations. We have therefore used these and related compounds to examine functional genomics of oxidative stress responses in aflatoxigenesis and reduction of aflatoxin biosynthesis by antioxidants.

Functional elucidation in *A. flavus* is impractical due to the lack of an effective and efficient gene delivery/transformation system. However, a number of stress response pathways, including oxidative stress, have been well characterized in the yeast, *Saccharomyces cerevisiae* [32]. We have therefore used this yeast as a model system to examine the phenotypic response to oxidative stress, with 22 strains, including a broad spectrum of single gene deletion mutants, being exposed to hydrogen peroxide (H_2O_2) [33]. Responses of particular mutants treated during H_2O_2 exposure with either tannic acid, as a representative hydrolysable tannin, or gallic acid, indicated that these compounds alleviated oxidative stress, as shown in Figure 9. Under control conditions in the absence of H_2O_2, all colonies from serially diluted cells were visible when grown on tannic acid or gallic acid up to 0.4% (w/v), showing that these compounds were not toxic to the yeast. On exposure to 3.3 mM hydrogen peroxide, growth of wild-type *S. cerevisiae* cells was completely inhibited, except for the sample with undiluted (~10^6) cells. The toxicity of H_2O_2 was overcome by treatment with the test compounds and the extent of this effect correlated with the antioxidant concentration increasing from 0.05% to 0.4%. Gallic acid was more potent than tannic acid, with almost complete regression of toxicity at a level of 0.4%.

To identify the functional role of particular *S. cerevisiae* genes in antioxidative stress responses, cell-growth of 22 different deletion mutants under oxidative stress and potential alleviation by tannic and gallic acids were examined, with glutathione also included as an antioxidant control. The deletion mutants examined included strains defective in gene regulation (*yap1Δ*) and antioxidative stress response (*ctt1Δ, cta1Δ, glr1Δ, osr1Δ, trr1Δ, trr2Δ, tsa1Δ, grx1Δ, grx2Δ, trx1Δ, trx2Δ, gsh1Δ, gsh2Δ, sod1Δ, sod2Δ*). Examples of yeast dilution bioassays used to visualize responses of the deletion mutants exposed to H_2O_2 and antioxidant compounds are shown in Figure 10. Hydrogen peroxide was moderately to severely lethal to all yeast strains but whereas the antioxidants fully alleviated oxidative stress in the wild-type, the degree of alleviation varied among the deletion mutants, depending upon the deleted gene and

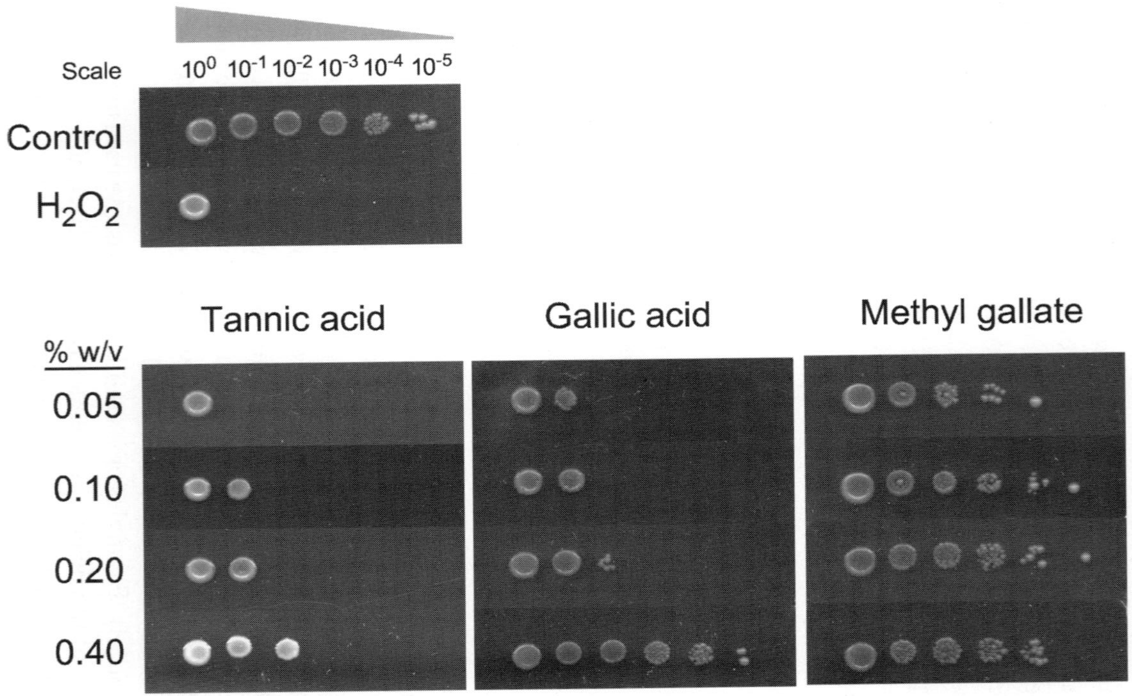

Figure 9. Alleviation of H_2O_2-induced oxidative stress by tannic acid, gallic acid and methyl gallate in *Saccharomyces cerevisiae*.

compound tested. Two strains, *yap1Δ* and *sod1Δ*, did not show appreciable recovery when treated with glutathione or the antioxidants, suggesting that these strains are highly sensitive to oxidative stress. Yap1 induces downstream expression of at least four antioxidative stress genes such as *GLR1*, *TRR1*, *TRX* and *GSH1*, the individual deletion mutants of which recovered when treated with antioxidants. Tannic acid alleviated toxicity in all strains except *yap1Δ* and *sod1Δ*, while gallic acid alleviated toxicity in all strains except *glr1Δ*, *tsa1Δ*, and *sod2Δ*. Tannic and gallic acids therefore clearly possess activity with respect to oxidative stress genes.

To gain further insights into a potential link between oxidative stress and aflatoxin biosynthesis, 43 orthologs of *S. cerevisiae* genes involved in gene regulation, signal transduction and antioxidation were identified in an *A. flavus* Expressed Sequence Tag (EST) library. A functional complementation of an antioxidative stress gene from *A. flavus*, mitochondrial superoxide dismutase (*sodA*), was performed in a *sod2Δ* yeast mutant missing the ortholog. The fact that *sodA* functionally complements the yeast mitochondrial superoxide dismutase mutant demonstrates the potential for *S. cerevisiae* to serve as a model system to perform this type of high throughput analysis. This complementation indicates that *S. cerevisiae* deletion mutants may serve as a model system to study *A. flavus* and has identified a number of candidate genes in such signaling pathways from an *A. flavus* EST library that could serve as a focus for future efforts. Using functional complementation analysis and knockout mutants, understanding of the relationship between aflatoxin biosynthesis and oxidative stress responses will be greatly facilitated.

IMPLICATIONS AND CONCLUSION

The results of our investigation demonstrate the ability of hydrolyzable tannins in tissues surrounding the edible portion of the walnut to eliminate aflatoxin formation. A high gallate ester level in the tannins of seed coat tissue is required to suppress or eliminate aflatoxin biosynthesis. An increase in the gallate ester content could be achieved by conventional breeding or genetic manipulation. Analysis of seed coat tissue of walnut germplasm may provide a useful indicator of selections with high potential for inheritability of this trait. Alternatively, a directed genetic approach could

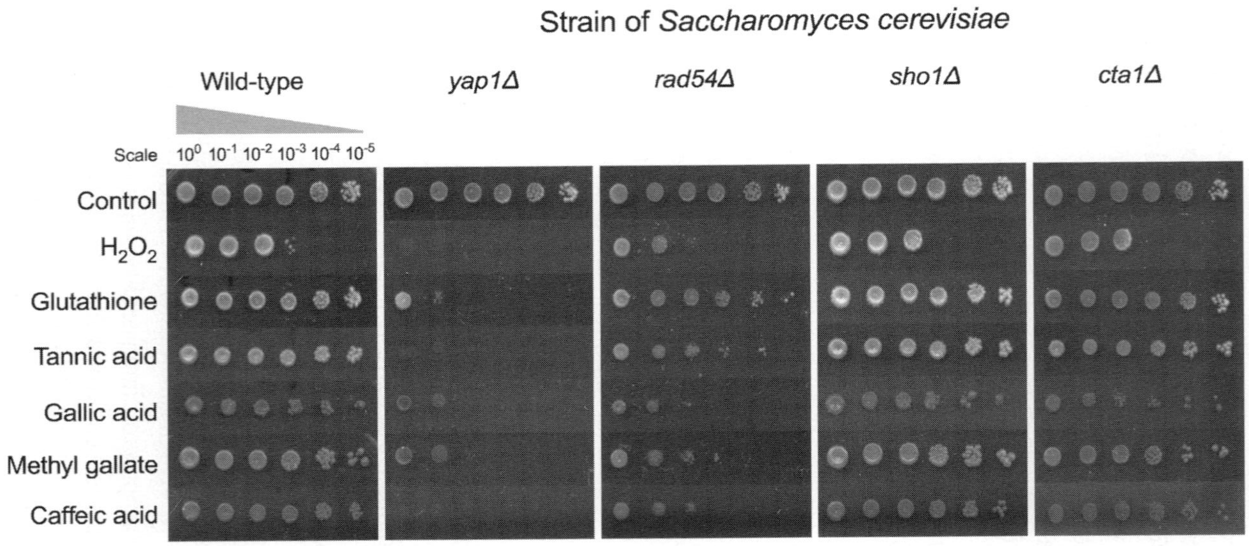

Figure 10. Alleviation of H_2O_2-induced oxidative stress by tannic acid, gallic acid and methyl gallate in *Saccharomyces cerevisiae* gene deletion strains; glutathione included as a positive control.

be adopted, focused on the biosynthetic pathway to gallic acid and the hydrolyzable tannins, through the shikimate pathway [34]. Stimulation of galloyl ester biosynthesis with accompanying down-regulation of subsequent transformations such as oxidative dimerization and gallate hydrolysis should lead to pellicle tannin constituents with the optimum gallic acid content to prevent aflatoxin biosynthesis. We are currently investigating the question of whether or not repeated exposure of the fungus to hydrolyzable tannins results in permanent atoxigenicity.

The antioxidant activity of hydrolyzable tannins may also have important implications with respect to consumer acceptance of walnuts since sufficiently high levels should prevent or delay the development of rancidity through suppression of lipid peroxidation [35]. It therefore appears that there should be no major contraindications to attempts to increase gallic acid levels in walnut pellicle in order to suppress aflatoxin formation. A primary objective of our future research will be to extend these findings to other tree nut species such as almonds and pistachios, and eventually to other crops that are also adversely affected in domestic and export markets by aflatoxin contamination. Furthermore, the demonstrated feasability of the *S. cerevisiae* high throughput model system to study the role of specific compounds in regulating genes responsible for aflatoxins in *A. flavus* may finally provide a rationale for their biosynthesis by the fungus. Our current hypothesis is that aflatoxins and their precursors are produced by the fungus in response to oxidative stress but in the presence of gallic acid and hydrolysable tannins the biosynthesis of aflatoxins is down-regulated because the exogenous antioxidants relieve the oxidative stress.

REFERENCES

[1] ProMED mail. Aflatoxin poisoning Kenya (Makueni) (05). ProMED mail 2004; 10 Jun: 220040610.1557. http://www.promedmail.org (accessed 1 September 2004)

[2] Henry SH, Bosch FX, Troxell TC, Bolger PM. Reducing liver cancer: Global control of aflatoxin. Science 1999; 286: 2453-4

[3] Foreign Agricultural Service. 2004. FAS Online. Horticultural and Tropical Products Division; USDA: Washington, DC. http://www.fas.usda.gov/htp/horticulture/nuts.html (accessed 1 September 2004)

[4] Commission of the European Community. Off J Eur Commun Legislation 1998; L201: 93-101

[5] Food and Drug Administration. Compliance Policy Guides Manual. 1996; Sec. 555.400: 268; Sec. 570.500: 299

[6] Mahoney N, Molyneux RJ, Campbell BC. Regulation of aflatoxin production by naphthoquinones of walnut (*Juglans regia*). J Agric Food Chem 2000; 9: 4418-21

[7] Lee SE, Campbell BC, Molyneux RJ, Hasegawa S, Lee HS. Inhibitory effects of naturally occurring compounds on aflatoxin B_1 biotransformation. J Agric Food Chem 2001; 49: 5171–7

[8] Mahoney N, Molyneux RJ, McKenna J, Leslie CA, McGranahan G. J. Resistance of 'Tulare' walnut (*Juglans regia* cv. Tulare) to aflatoxigenesis. J Food Sci 2003; 68: 619 22

[9] Mansfield JW. Antimicrobial compounds and resistance: The role of phytoalexins and phytoanticipins. In: Mechanisms of Resistance to Plant Diseases (Slusarenko, A.J., Fraser, R.S.S., Van Loon, L.C., eds), 2000. Kluwer Academic Publishers, Dordrecht, The Netherlands, pp. 325-70

[10] Mahoney N, Molyneux RJ. Phytochemical inhibition of aflatoxigenicity in *Aspergillus flavus* by constituents of walnut (*Juglans regia*). J Agric Food Chem 2004; 52: 1882-9

[11] Jurd L. Plant polyphenols. I. The polyphenolic constituents of the pellicle of the walnut (*Juglans regia*). J Am Chem Soc 1956; 78: 3445-8

[12] Jurd L. Plant polyphenols. III. The isolation of a new ellagitannin from the pellicle of the walnut. J Am Chem Soc 1958; 80: 2249-52

[13] Schmidt OTh, Schmidt DM, Herok J. Natural tannins. XIX. Constitution and configuration of corilagin. Justus Liebigs Ann Chem 1954; 587: 67-74

[14] Fukuda T, Ito H., Yoshida T. Antioxidative polyphenols from walnuts (*Juglans regia* L.). Phytochemistry 2003; 63: 795-801

[15] Haslam E. The metabolism of gallic acid and hexahydroxydiphenic acid in higher plants. Prog Chem Org Nat Prod 1982; 41: 1-46

[16] Haslam E, Cai Y. Plant polyphenols (vegetable tannins): Gallic acid metabolism. Nat Prod Rep 1994; 11: 41-66

[17] Scalbert A. Antimicrobial properties of tannins. Phytochemistry 1991; 30: 3875-83

[18] Latte KP, Kolodziej H. Antifungal effects of hydrolysable tannins and related compounds on dermatophytes, mold fungi and yeasts. Zeitschr Naturforsch C: J Biosci 2000; 55: 467-72

[19] Hagerman AE. Tannin Handbook. (accessed 1 September 2003)

[20] Yamada H, Adachi O, Watanabe M, Sato N. Part 1. Formation, purification and catalytic properties of tannase of *Aspergillus flavus*. Agric Biol Chem 1968; 32: 1070-8

[21] Bradoo S, Gupta R, Saxena RK. Screening of extracellular tannase-producing fungi: Development of a rapid and simple plate assay. J Gen Appl Microbiol 1996; 42: 325-9

[22] Lei Z, Jervis J, Helm RF. Use of methanolysis for the determination of total ellagic and gallic acid contents of wood and food products. J Agric Food Chem 2001; 49: 1165-8

[23] Minto RE, Townsend CA. Enzymology and molecular biology of aflatoxin biosynthesis. Chem Rev 1997; 97: 2537-55

[24] Payne GA, Brown MP. Genetics and physiology of aflatoxin biosynthesis. Annu Rev Phytopathol 1998; 36: 329-62

[25] Cotty PJ, Bhatnagar D. Variability among atoxigenic *Aspergillus flavus* strains in ability to prevent aflatoxin contamination and production of aflatoxin biosynthetic pathway enzymes. Appl Environ Microbiol 1994; 60: 2248-51

[26] Levine A, Tenhaken R, Dixon R, Lamb C. H_2O_2 from the oxidative burst orchestrates the plant hypersensitive disease resistance response. Cell 1994; 79: 583-93

[27] Bolwell GP. Role of active oxygen species and NO in plant defence responses. Curr Opin Plant Biol 1999; 2: 287-94

[28] Estruch F. Stress-controlled transcription factors, stress-induced genes and stress tolerance in budding yeast. FEMS Microbiol Rev 2000; 24: 469-86

[29] Wang W, Vinocur B, Altman A. Plant responses to drought, salinity and extreme temperatures: Towards genetic engineering for stress tolerance. Planta 2003; 218: 1-14

[30] Jayashree T, Subramanyam C. Oxidative stress as a prerequisite for aflatoxin production by *Aspergillus parasiticus*. Free Radic Biol Med 2000; 29: 981-5

[31] Jayashree T, Subramanyam C. Antiaflatoxigenic activity of eugenol is due to inhibition of lipid peroxidation. Lett Appl Microbiol 1999; 28: 179-83

[32] Toone WM, Jones N. Stress-activated signalling pathways in yeast. Genes Cells 1998; 3: 485-98

[33] Kim JH, Campbell BC, Yu J, Mahoney N, Chan KL, Molyneux RJ, Bhatnagar D, Cleveland TE. Examination of fungal stress response genes using *Saccharomyces cervisiae* as a model system: Targeting genes affecting aflatoxin biosynthesis by *Aspergillus flavus* Link. Appl Microbiol Biotechnol 2004; 67: 807-15

[34] Ossipov V, Salminen J.-P, Ossipova S, Haukioja E, Pihlaja K. Gallic acid and hydrolysable tannins are formed in birch leaves from an intermediate compound of the shikimate pathway. Biochem Syst Ecol 2003; 31: 3-16

[35] Dufour C, da Silva E, Potier P, Queneau Y, Dangles O. Gallic esters of sucrose as efficient radical scavengers in lipid peroxidation. J Agric Food Chem 2002; 50: 3425-30

Effects of the Cultured *Cordyceps* Exopolysaccharide Fraction (Epsf) on Some Parameters of Mouse Immune Function *In Vivo* and *In Vitro*

Wei-yun Zhang*, Jin-yu Yang, Jia-ping Chen, Pei-hua Shi and Li-jun Ling
Medical School, Nanjing University, Nanjing 210093, China
**Corresponding email:* zhangwy@nju.edu.cn

INTRODUCTION

Cordyceps sinensis (Cs), called Dong-cong-xia-cao (winter-worm, summer-grass) in China, is a complex of an insect larva parasitized by a fungus and has been used as a precious tonic medicine since ancient time because people has known it may strengthen and improve lung and kidney functions, restore health after prolonged sickness and maintain overall body health. Recently, researches reported that Cs possess many important pharmacological functions including anti-oxidation, antisenescence, antiatherosclerosis and immunomodulation [1, 2]. However, wild Cs herb is a rare natural species and cannot meet the increasing demand for tonic and therapeutic uses. Some researchers have found that the function of the worm in *Cordyceps* is just to provide a growth medium for the fruiting body [3]. Moreover, some investigators have demonstrated cultured *Cordyceps* exert similar pharmacological efficacy as the natural herb in many aspects [4, 5, 6, 7]. Therefore, the cultivated mycelium of Cs fungal species by fermentation has been widely used as a substitute of the wild herb.

Furthermore, it is confirmed that many biological activities of *Cordyceps* are associated with its polysaccharides [8, 9, 10]. Researchers thought some of fungus polysaccharides could scavenge free radical[11], induce differentiation of cancer cells [12], enhance animal or human's antitumor ability via activating different immune responses in the host [13, 14, 15]. In the present study, we have prepared the exopolysaccharide from the supernatant of cultured *Cordyceps* fungi CSG1, and preliminary explore its immunomodulatory functions.

MATERIALS AND METHODS

Preparation of the exopolysaccharide extract

The *Cordyceps* fungus (CSG1) is preserved in the microbiology lab of medical school of Nanjing University. It was cultured 3~7 days with a shaker at 25 ~28 °C, 120 rpm in the medium including mycological peptone, yeast extract, glucose, KH_2PO_4, and $MgSO_4$. 400 ml of culture supernatant was used to obtain the exopolysaccharide. The supernatant was first concentrated *in vacuo* to about 100 ml. It was then subjected to alcohol precipitation by slowly adding to them 3 times volume of 95 % alcohol and kept standing overnight at room temperature. Sediments were collected, dried and dissolved in distilled water again. Finally, centrifugation was carried out to collect the supernatant, which was lyophilized to get about 270 mg of EPSF.

Animal and drug administration

Twenty-four of ICR mice (18~22 g) were purchased from the Animal Center of Nanjing Medical University. The experimental animals were randomly divided into four groups: control group (C), low-dose group (15 mg/kg) (L), mid-dose (30 mg/kg) (M), and high-dose group (60 mg/kg) (H). Three male and three female mice were for each group and allowed one week to adapt to their environment before the experiments. All the animals were maintained on a 12 h light/dark cycle in an air-conditioned room and commercial pellet diet and water were given *ad libitum* during the study. Animals were peritoneally administered everyday by EPSF at three different doses for seven days. Injection volume was 0.1 ml for each mouse. Control animals received the same volume of normal saline.

Phagocytosis of mouse peritoneal macrophages

In vivo study, peritoneal macrophages were aseptically collected from each mouse with RPMI-1640 medium after 7 days treatment. The phagocytosis of mouse peritoneal macrophages was tested by neutral red uptake. The collected macrophages were centrifuged and washed once with culture medium, and resuspended at 2×10^6 cells/ml in culture medium, then 100 μl aliquots of macrophages were added into each well of 96 well microwell plate (Nunc). After 4 h incubation at 37 °C in a humidified 5 % CO_2-air mixture to allow the cells to attach, the supernatant was discarded and 0.075 % of neutral red dye was added to each well (100 μl/well). The plate was incubated for another 1 h. Then the plate was washed three times with PBS (pH 7.2) and let it drain. Finally, 100 μl of lysis solution was pipetted into each well. After the plate was placed at room temperature overnight, the microplate was read on a Bio-Rad microplate reader (Model 550) using test wavelength of 540 nm.

In vitro study, peritoneal macrophages were aseptically collected from six mice and was adjusted to a concentration of 2×10^6 cells/ml. To each well of a 96-well microplate, 50 μl of the macrophage suspension and 50 μl of different concentrations of EPSF were added and the EPSF ultimate concentrations was 6.25, 12.5, 25, 50, 100 mg/L, respectively. After incubation at 37 °C in a 5% CO_2 atmosphere for 24h, the supernatant was discarded and 0.075 % of neutral red dye was added to each well (100 μl/well). Then, phagocytosis of peritoneal macrophages was tested by neutral red uptake.

Proliferation of spleen and thymus lymphocytes

In *vivo study,* the spleens and thymuses were removed immediately after animals were sacrificed. B- and T-lymphocyte single cell suspension was prepared respectively and adjusted to 2×10^6 cells/ml. The 100 μl aliquots of lymphocytes were seeded into each well of a 96 well flat bottom microwell plate (Nunc) in the presence of 10 mg/L mitogen Concanavalin A (Con A, Sigma). After incubating the lymphocytes for 48 h at 37°C in a humidified 5% CO_2 incubator, their number was assayed by the MTT. Briefly, the plate was incubated for additional 4 h after addition of 10 μl MTT per well, and then 100 μl of 20 % sodium dodesyl sulfate (Sigma, SDS) were pipetted into each well. After overnight incubation at 37°C in 5% CO_2 atmosphere, the microplate was read on a Bio-Rad microplate reader (Model 550) using test wavelength of 570 nm.

In vitro study, six ICR mice were used for lymphocyte preparation. Mouse spleens and thymuses were aseptically removed after the mice were euthanized. Lymphocytes were prepared and adjusted to a concentration of 2×10^6 cells/ml. And mitogen Con A was added to reach the concentration of 20 mg/L. To each well of a 96-well microplate, 50 μl of the cell suspension and 50 μl of different concentrations of EPSF were added and the ultimate concentrations was 6.25, 12.5, 25, 50, 100 mg/L, respectively. After incubation at 37°C in a 5% CO_2 atmosphere for 48h,

MTT method is used to assay.

RESULTS AND DISCUSSION

Effect of EPSF on the phagocytosis of mouse macrophages

In the present study, the neutral red uptake capacity of macrophages *in vivo* study was increased significantly by EPSF at all the three different doses when compared with untreated mice ($p<0.001$) (Figure 1A). *In vitro* study, EPSF may significantly elevate the phagocytosis of peritoneal macrophage in the concentration of 50 and 100 mg/L ($p<0.05$) (Figure 1B). These results indicated that EPSF could stimulate murine macrophage function both *in vivo* and *in vitro*.

It is well recognized that immune responses, especially cellular immunity, play an important role in the elimination of locally growing and circulating bacterium or tumor cells. Phagocytosis of macrophages represents the indispensable step of the immunological defense system[16], and enhancement of phagocyte function is applicable for therapy of infectious disease and cancer. Thus we deduce that EPSF might enhance animal's or human's defense ability against infectious disease and cancer.

Effect of EPSF on spleen lymphocyte proliferation

To determine the effects of EPSF on lymphocyte proliferation, mitogen-induced proliferation of mouse lymphocytes was measured by MTT method. As shown in Figure 2A, EPSF significantly stimulated the proliferation of spleen lymphocytes at the dose of 60 mg/kg ($p<0.01$). *In vitro* experiment, spleens lymphocytes were treated directly with five

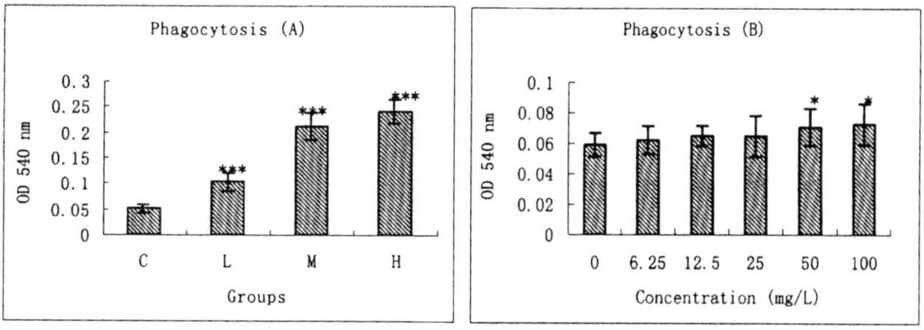

Figure 1. Effects of EPSF on the phagocytosis of mouse macrophages
A. *In vivo* study. B. *In vitro* study. $p<0.05$, *** $p<0.001$, compared with control

different concentrations of EPSF and their proliferation ability was also assayed by MTT method. The results showed that EPSF significantly stimulated the proliferation of spleen lymphocytes in the concentration of 100 mg/L ($p<0.001$) (Figure 2B).

Lymphocytes help protect the body against diseases and fight infections as well. They assist providing specific responses to attack the invading organisms when dangerous invading microorganisms or tumors have penetrated the general defense systems of the body. Spleen is mainly composed of B-lymphocytes and thus the polysaccharide EPSF might stimulate the function of B-lymphocytes.

Effect of EPSF on thymus lymphocyte proliferation

Thymus gland is composed of T-lymphocytes, which differ from B-lymphocytes in function and the surface molecules. T-lymphocytes may attack virus-infected cells, foreign tissue, and cancer cells.

In vivo study, thymus proliferation of mice did not show significant variation between treated mice and control mice (Figure 3A). However, when T-lymphocytes were directly treated with EPSF *in vitro* experiment, its proliferation ability was significantly promoted in the concentration of 100 mg/L ($p<0.05$) (Figure 2B). The results suggested that the proliferation ability of thymus lymphocytes was accelerated only when cells were treated directly by EPSF.

REFERENCES

[1] Zhu JS, Halpern GM, Jones, K. The scientific rediscovery of an ancient Chinese herbal medicine: *Cordyceps sinensis*: Part I. J Altern Complem

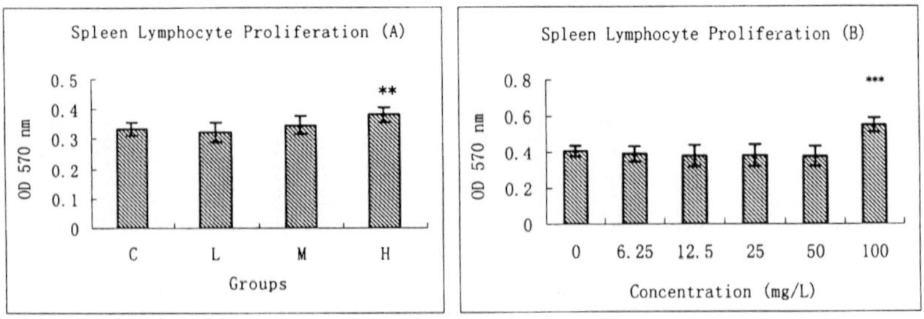

Figure 2. Effects of EPSF on the proliferation of mouse spleen lymphocytes
A. *In vivo* study. B. *In vitro* study. **$p<0.01$, ***$p<0.001$, compared with control

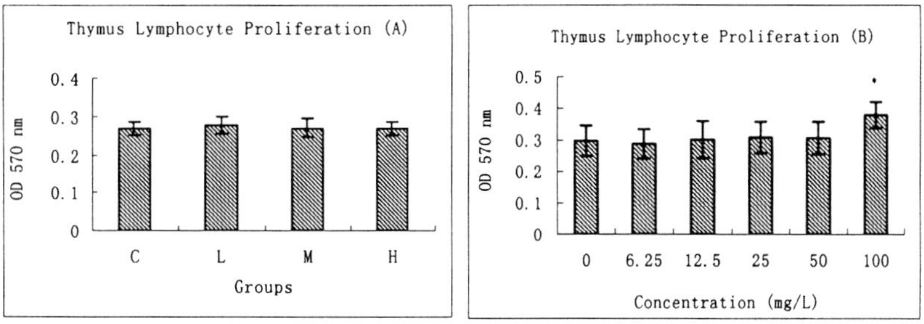

Figure 3. Effects of EPSF on the proliferation of mouse thymus lymphocytes
A. *In vivo* study. B. *In vitro* study. *$p<0.05$, compared with control

Med 1998; 4: 289-303

[2] Zhu JS, Halpern GM, Jones, K. The scientific rediscovery of a precious ancient Chinese herbal regimen: *Cordyceps sinensis*: Part II. Zhu JS, Halpern GM, Jones, K. J Altern Complem Med 1998; 4: 429-57

[3] Li SP, Su ZR, Dong TTX, Tsim KWK. The fruiting body and its caterpillar host of *Cordyceps sinensis* show close resemblance in main constituents and anti-oxidation activity. Phytomedicine 2002; 9:319-24

[4] Li SP, Su ZR, Dong TTX, Tsim KWK. Anti-oxidation activity of different types of natural *Cordyceps sinensis* and cultured *Cordyceps* mycelia. Phytomedicine 2001; 8: 207-12

[5] Yamaguchi Y, Kagota S, Nakamura K, Shinozuka K, Kunitomo, M. Antioxidant activity of the extract from fruiting bodies of cultured *Cordyceps sinensis*. Phytother Res 2002; 14: 647-9

[6] Yamaguchi Y, Kagota S, Nakamura K, Shinozuka K, Kunitomo, M. Inhibitory effects of water extracts from fruiting bodies of cultured *Cordyceps sinensis* on raised serum lipid peroxide levels and aortic cholesterol deposition in atherosclerotic mice. Phytother Res 2002; 14: 650-2

[7] Liang YJ, Liu Y, Yang JW, Liu XC. Studies on pharmacological activities of cultivated *Cordyceps sinensis*. Phytother Res 1997; 11: 237-9

[8] Kiho T, Yamane A, Hui J, Usui S, Ukai S. Polysaccharides in fungi. XXXVI. Hypoglycemic activity of a polysaccharide (CS-F30) from the cultural mycelium of *Cordyceps sinensis* and its effect on glucose metabolism in mouse liver. Biol Pharm Bull 1996; 19: 294-6

[9] Huang BM, Ju SY, Wu CS, Chuang WJ, Sheu CC, Leu SF. *Cordyceps sinensis* and its fractions stimulate MA-10 mouse Leydig tumor cell steroidogenesis. J Androl 2001; 22: 831-7

[10] Li SP, Zhao KJ, Ji ZN, Song ZH, Dong TTX, Lo CK, Cheung JKH, Zhu SQ, Tsim KWK. A polysaccharide isolated from *Cordyceps sinensis*, a traditional Chinese medicine, protect PC12 cells against hydrogen peroxide-induced injury. Life Sci 2003; 73: 2503-13

[11] Sun C, Wang JW, Fang L, Gao XD, Tan RX. Free radical scavenging and antioxidant activities of EPS2, an exopolysaccharide produced by a marine filamentous fungus *Keissleriella sp.* YS 4108. Life Sci 2004; 9: 1063-73

[12] Chen YY, Chang HM. Antiproliferative and differentiating effects of polysaccharide fraction from fu-ling (*Poria cocos*) on human leukemic U937 and HL-60 cells. Food Chem Toxicol 2004; 5: 759-69

[13] Wang YY, Khoo KH, Chen ST, Lin CC, Wong CH, Lin CH. Studies on the immuno-modulating and antitumor activities of *Ganoderma lucidum* (Reishi) polysaccharides: functional and proteomic analyses of a fucose-containing glycoprotein fraction responsible for the activities. Bioorgan Med Chem 2002; 10: 1057-62

[14] Fujimiya Y, Suzuki Y, Oshiman K, Kobori H, Moriguchi K, Nakashima H, Matumoto Y, Takahara S, Ebina T, Katakura R. Selective tumoricidal effect of soluble proteoglucan extracted from the basidiomycete, *Agaricus blazei* Murill, mediated via natural killer cell activation and apoptosis. Cancer Immunol Immunother 1998; 46: 147-59

[15] Kodama N, Komuta K, Sakai N, Nanba H. Effects of D-Fractionÿa polysaccharide from *Grifola frondosa* on tumor growth involve activation of NK cells. Biol Pharm Bull 2002; 25: 1647-50

[16] Stuelp-Campelo PM, de Oliveira MB, Leao AM, Carbonero ER, Gorin PA, Iacomini M. Effect of a soluble alpha-D-glucan from the lichenized fungus Ramalina celastri on macrophage activity. Int J Immunopharmacol 2002; 2: 691-8

Lactobacillus rhamnosus Induces Differential Anti-proliferative Responses and Interleukin-6 Expression Levels in SV-40 and Malignant Uroepithelial Cells

Ying-Jing Yong[a], Ratha Mahendran[b], Yuan-Kun Lee[a] and Boon-Huat Bay[c]*

Departments of [a]Microbiology, [b]Surgery and [c]Anatomy, Yong Loo Lin School of Medicine, National University of Singapore, S 117 597, Singapore

**Corresponding email: antbaybh@nus.edu.sg*

INTRODUCTION

Lactic acid bacteria has been shown to inhibit bladder cancer growth *in vivo* [1,2] and prevent recurrence of human superficial bladder cancer [3]. We have also recently reported that *Lactobacillus rhamnosus* strain GG (LGG) was as efficacious as BCG (a known modality of treatment for bladder cancer) in inhibiting the proliferation of MGH bladder cancer cells *in vitro* [4]. Lactobacillus is also known to induce Interleukin-6 (IL-6) production [5]. IL-6, a 26 kDa inducible protein, is a pleiotropic cytokine which exerts a wide range of biological activities including the regulation of immune response, hematopoiesis and acute phase reactions [6,7].

Since the discovery of IL-6 as a growth factor for multiple myeloma a decade ago, this cytokine has also been reported to act as an autocrine/paracrine growth factor for various human tumors, including genitourinary tumors [8]. In human bladder cell lines, Okamoto and colleagues [9] demonstrated that IL-6 functions as an autocrine growth factor in human bladder cancer cells (253J, RT4 and T24) *in vitro* but not in primary cultured uroepithelial cells. Previous to that, there were reports of IL-6 mRNA expression in *de novo* bladder cancers [[0] and IL-6 production and release by bladder carcinoma cells *in vitro* [11].

To examine the potential role of LGG as an alternative approach to bladder cancer therapy, we studied the anti-proliferative effects of LGG and the expression of IL-6 in the simian virus 40 (SV40) immortalized non-tumorigenic human uroepithelial cells (HUC-1) and its neoplastic counterpart, HUC-T2 cells. The HUC-1 cells have been repeatedly found to be non-tumorigenic when inoculated into athymic nude mice [12] and the HUC-T2 cells were derived by 3-methylcholanthrene chemical transformation of the parental HUC-1 cells.

MATERIALS AND METHODS

Lactobacillus strain

Lactobacillus rhamnosus strain GG (ATCC 53103) (LGG) was obtained from The National Collections of Industrial and Marine Bacteria LTD (Scotland, UK). The bacterium was propagated in MRS broth (Oxoid Ltd; Hampshire, England) at 37°C with 5% CO_2. The bacterial cells were enumerated using an experimentally derived conversion factor [13]. The conversion factor of LGG used in the experiment was 2.27 x 10^8 CFU/ml/OD. Bacterial cells were harvested when lactobacillus culture reached an optical density (OD_{550nm}) of 7.00 Absorbance.

Cell Culture

Normal uroepithelial cells immortalized by SV40 transformation, HUC-1 (ATCC CRL-9520) and tumorigenic HUC-T2 cell line derived by chemical transformation of the HUC-1 cell line were obtained from the American Type Culture Collection (ATCC). Monolayer cultures were grown in tissue culture flasks at 37°C in Ham's medium supplemented with 10% fetal bovine serum and 50U/ml Penicillin G

Cell proliferation assay

Monolayers of HUC-1 and HUC-T2 cells were grown in 6-well tissue culture plates (Falcon) with 1x10^5 cells per well in 4 ml of culture medium with 50U/ml PenG. Cells were plated for 15 hours before the addition of 1x 10^7 or 1x10^8 bacteria. In control flasks, no bacteria was added. The cells were allowed to grow for 24 hours, 48 hours and 72 hours before harvesting. The cell monolayers were washed 3 times with phosphate buffered saline (PBS), harvested by scraping and suspended in phosphate-buffered saline before counting with a haemocytometer under a 100X light microscope.

Cytokine Assay

Culture supernatants from the 2 cell lines which were grown in the presence of 1 x 10^7 and 1 x 10^8 LGG were collected after 72 hours of incubation, centrifuged at 12000rpm for 10 minutes to remove cells and bacteria and stored at –30°C for cytokine analysis. The quantitative determination of TNF-α was performed using EIA kits (Assay Designs, Inc).

Statistical analysis

The GraphPad Prism and SPSS statistical software packages were used. One-way ANOVA with post hoc Scheffe's test and Student's t-test were used to examine the level of significance. *P* value of < 0.05 was taken to indicate statistical significance.

RESULTS

Differential anti-proliferative effects of LGG

The number of HUC-T2 cells was higher than that of HUC-1 cells at all time points although not statistically significant (Figures 1 and 2). The mean cell numbers for HUC-1 and HUC-T2 cells respectively were 63020 ± 7236 cells versus 139800 ± 41590 cells (24 hours), 82710 ± 14120 cells versus 196000 ± 51320 cells (48 hours) and 120600 ± 23380 cells versus 277500 ± 84470 cells (72 hours). When 1 x 10^7 LGG was added to the cultures, a significant decrease in cell numbers after 72 hours of treatment, was observed only in HUC-T2 cells compared to the respective control (Figure 2, P= 0.017). At a higher dosage of LGG (1 x 10^8 bacteria), a significant decrease was observed after 72 hours of incubation with HUC-1 cells (P= 0.039) as compared to the respective control (Figure 1) whereas the number of HUC-T2 cells in the culture wells was significantly lower than the respective control by 48 hours (P= 0.032) (Figure 2).

IL-6 secretion induced by LGG

Basal production of IL-6 was significantly higher in HUC-T2 cells compared with HUC-1 cells (P =0.002; Figure 3). There was no significant increase in IL-6 production in the presence of both 1 x 10^7 and 1 x 10^8

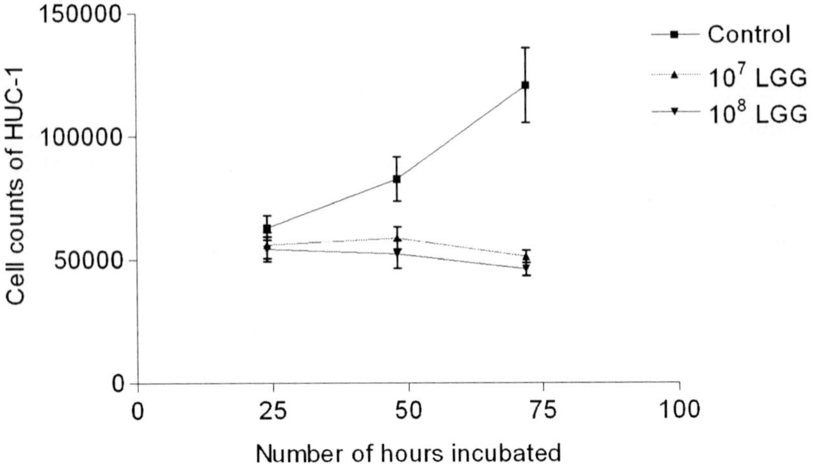

Figure 1. Cell proliferation assay of HUC-1 cells in the presence of 10^7 and 10^8 viable LGG in Pen G medium. Results are shown as mean ± SEM from triplicate results.

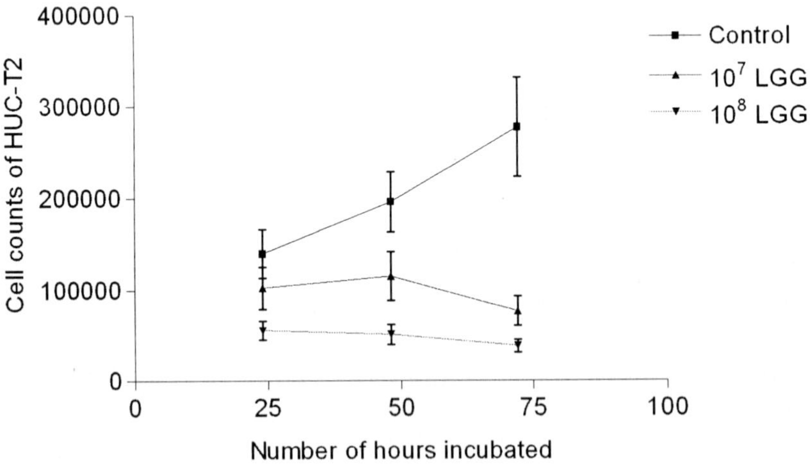

Figure 2. Cell proliferation assay of HUC-T2 cells in the presence of 10^7 and 10^8 viable LGG in Pen G medium. Results are shown as mean ± SEM from triplicate results.

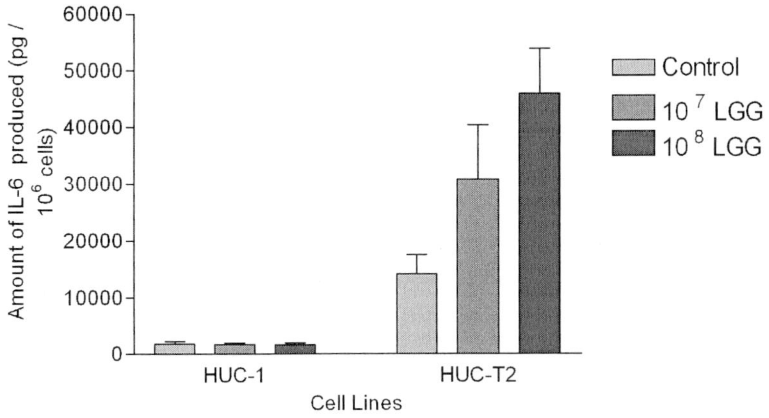

Figure 3. IL-6 production by HUC-1 and HUC-T2 cell lines after 72 hours exposure to LGG. Results are shown as mean ± SEM from triplicate results.

bacteria in HUC-1 cells. However, a significant increase in IL-6 in HUC-T2 cells in the presence of 1 x 10^8 bacteria was observed. (P=0.038).

DISCUSSION

Malignant HUC-T2 cells have a higher proliferation rate than the parental HUC-1 non-tumorigenic cells as shown by the cell proliferation study. The HUC-T2 cells also produce significantly higher IL-6 than its benign counterpart. This data would appear to support the notion that IL-6 promotes growth in bladder cancer cells. However, when LGG was shown to induce IL-6 production by HUC-T2 cells, there was no corresponding increase in cell proliferation but rather, an inhibitory effect on cell growth was observed. In the HUC-1 cells, inhibition of cell proliferation by LGG was not accompanied by any alteration of IL-6 levels. Okamoto et al. [9] had previously reported that IL-6 functions as an autocrine growth factor in 253J, RT4 and T24 bladder cancer cells but not in normal urothelial cells. The role of IL-6 in cancer progression is more complicated than it seems and the simultaneous triggering of multiple pathways by this cytokine may tip the balance towards a biological response such as proliferation or growth arrest [14].

IL-6 has been reported to be a proliferation factor in a variety of other genitourinary cancers. IL-6 has been identified as an autocrine growth factor in renal cell carcinomas [15] and as proliferation factor in human prostate cancer *in vivo* and *in vitro* [16-18]. The fact that IL-6 stimulates growth in prostate cancer cells but not in epithelial cells derived from benign prostatic hypertrophy [16], has lead researchers to the perception that IL-6 could contribute to prostate cancer progression [19]. IL-6 induced cell proliferation is believed to be mediated by signal transducers and activators of transcription (STAT) 3 molecules [17, 20]. STAT proteins are latent transcription factors which when activated by phosphorylation in response to extracellular ligands, will be translocated to the nucleus and regulate cell cycle progression [21]. On the other hand, IL-6 has been shown to inhibit proliferation in T47 D breast cancer cells [14] and lung cancer cell lines [22].

In conclusion, LGG has a greater anti-proliferative and cytotoxic effect on malignant uroepithelial HUC-T2 cells as compared to the non-tumorigenic parental cell line. Moreover, up-regulation of IL-6 production in HUC-T2 cells but not HUC-1 cells by LGG implies that inflammatory responses evoked by LGG would be specifically localized to tumor cells and not generalized to the non-tumorigenic parts of the bladder.

Hence the potential of LGG in intravesical bladder cancer therapy holds great promise.

ACKNOWLEDGEMENT

This work was supported by a grant from the Singapore National Medical Research Council.

REFERENCES

[1] Asano M, Karasawa E and Takayama T. Antitumor activity of Lactobacillus casei (LC 9018) against experimental mouse bladder tumor (MBT-2). J Urol 1986; 136: 719-21

[2] Lim BK, Mahendran R, Lee YK and Bay BH. Chemopreventive effect of Lactobacillus rhamnosus on growth of a subcutaneously implanted bladder cancer cell line in the mouse. Jpn J Cancer Res 2002; 93: 36-51

[3] Aso Y, Akaza H, Kotake T, Tsukamoto T, Imai K and Naito S. Preventive effect of a Lactobacillus casei preparation on the recurrence of superficial bladder cancer in a double blind clinical trial. Eur Urol 1995; 27: 104-9

[4] Seow SW, Rahmat JN, Ismail KM, Mahendran R, Lee YK and Bay BH. Commonly consumed Lactobacillus spp. Are more cytotoxic to bladder cancer cells than Mycobacterium bovis, Bacillus Calmette Guerin. J Urol. 2002; 168: 2236-9

[5] Miettinen M, Vuopio-Varkila J and Varkila K. Production of human tumor necrosis factor alpha, interleukin-6, and interleukin-10 is induced by lactic acid bacteria. Infect Immun. 1996; 64: 5403-5

[6] Tilg H, Dinarello CA and Mier JW. Immunol Today 1997; 18: 428-32

[7] Hirano T. Interleukin-6 and its receptor: ten years later. Int Rev Immunol 1998; 16: 249-84

[8] Akira S and Kishimoto T. The evidence for interleukin-6 as an autocrine growth factor in malignancy. Semin Cancer Biol 1992; 3: 17-26

[9] Okamoto M, Hattori K and Oyasu R. Interleukin-6 functions as an autocrine growth factor in human bladder carcinoma cell lines in vitro. Int J Cancer 1997; 72: 149-54

[10] Meyers FJ, Gumerlock PH, Kawasaki ES, Wang AM., deVere White RW and Erlich HA. Bladder cancer. Human leukocyte antigen II, interleukin-6, and interleukin-6 receptor expression determined by the polymerase chain reaction. Cancer 1991; 67: 2087-95

[11] De Reijke TM, Vos PCN, De Voer EC, Bevers RFM, De Muinck Keizer WH, Kurth KH and Schamhart DHJ. Cytokine production by the human bladder caricinoma cell line T24 in the presence of bacillus Calmette Guerin (BCG). Urol Res 1993; 21: 349-52

[12] Meisner LF, Wu SQ, Christian BJ and Reznikoff CA. Cytogenetic instability with balanced chromosome changes in an SV40 transformed human uroepithelial cell line. Cancer Res 1988, 48:3215-20

[13] Greene J D and Klaenhammer T R, Factors involved in adherence of lactobacillus to human Caco-2 cells. Appl Environ Microbiol 1994; 60: 4487-94

[14] Badache A and Hynes NE. Interleukin 6 inhibits proliferation and, in cooperation with an epidermal growth factor receptor autocrine loop, increases migration of T47D breast cancer cells. Cancer Res 2001; 61: 383-91

[15] Miki S, Iwano M, Miki Y, Yamamoto M, Tang B, Yokokawa K, Sonada T and Hirano T. Interleukin-6 (IL-6) functions as an in vitro autocrine growth factor in renal cell carcinomas. FEBS Lett 1989; 250: 607-10

[16] Okamoto M, Lee C and Oyasu R. Interleukin-6 as a paracrine and autocrine growth factor in human prostatic carcinoma cells in vitro. Cancer Res 1997; 57: 141-6

[17] Lou W, Ni Z, Dyer K, Tweardy FJ and Gao AC. Interleukin-6 induces prostate cancer cell growth accompanied by activation of stat3 signaling pathway. Prostate 2000; 42: 239-42

[18] Giri D, Ozen M and Ittmann M. Interleukin-6 is an autocrine growth factor in human prostate cancer. Am J Pathol 2001; 159:v2159-65

[19] Smith PC, Hobisch A, Lin DL, Culig Z and Keller ET. Interleukin-6 and prostate cancer progression. Cytokine Growth Factor Rev 2001; 12: 33-40. Review

[20] Horiguchi A, Oya M, Marumo K and Murai M. STAT3, but not ERKs, mediates the IL-6-induced proliferation of renal cancer cells, ACHN and 769P. Kidney Int 2002; 61: 926-38

[21] Bromberg JF. Activation of STAT proteins and growth control. BioEssays 2002; 23: 161-9

[22] Takizawa H, Ohtoshi T, Ohta K, Yamashita N, Hirohata S, Hirai K, Hiramatsu K and Ito K. Growth inhibition of human lung cancer cell lines by interleukin 6 in vitro: a possible role in tumor growth via an autocrine mechanism. Cancer Res 1993; 53: 4175-81

Cruciferous Vegetables and Chemoprotection – A Role of ITC-Mediated Apoptosis

Peter Rose*

Department of Biochemistry, National University of Singapore, Singapore 119260

**Corresponding email: phytochemistry@hotmail.co.uk*

INTRODUCTION

Epidemiological evidence suggests that a diet high in fruits and vegetables can prevent the development of a wide range of human pathologies including cancer. A significant finding from many of these studies was the strong correlations between a reduction in stomach, oesophageal, prostate, lung, breast and colon cancers, with the high consumption of cruciferous vegetables, particularly members of the agronomically important genus, the Brassicaceae [1-4]. The chemoprotective effects are believed to be due in part to chemicals derived from a group of sulphur-containing precursor molecules known as glucosinolates (GSL). Ordinarily, GSLs are stored within plant tissues and it is only upon tissue damage that bioactive principles are generated [4]. Interestingly, these chemicals are produced through the enzymatic hydrolysis of the parental GSL storage compounds by plant myrosinases [thioglucoside glucohydrolase; EC 3.2.3.1], leading to the conversion of the precursor GSLs to several chemical agents, including the bioactive isothiocyanates (ITCs). To date, both cruciferous vegetable extracts and their representative ITCs can prevent chemically induced tumors in rodent models while also reducing the risk of cancers in humans, as revealed in several recent epidemiological studies.

The cruciferous vegetables and ITCs have a wide range of biological functions, including antibacterial and antifungal properties [5-8]. Moreover, early records show that members of the Brassicaceae, particularly cabbage, were used in the treatment of cancers as described by Cato the Censor (234 – 149 B.C) [9]. Research conducted in the 1960s demonstrated that synthetic ITCs and their structural analogues were potent antineoplastic and cytotoxic agents against cancer cells [10-12]. In addition, watercress has been reported to have a potent antimitotic effect on experimental tumors

[13]. It is only now with the advent of more sophisticated scientific techniques that progress has been made towards unraveling the biological mechanism(s) of ITC-mediated chemoprevention. To date, mechanistic studies have shown that ITCs, particularly the w-methylsulphinylalkyl ITCs such as sulforaphane, are potent phase II enzyme inducers up-regulating gene expression and enzyme activity in a variety of mammalian tissues. Glutathione-S-transferases (GSTs) [EC 2.5.1.18], quinone reductase (QR) [NAD (P) H: (quinone acceptor) oxidoreductase, EC 1.6.99.2] and UDP-glucuronosyl transferases [EC 2.4.1.17] have been observed to increase *in vivo* and *in vitro* in numerous studies, using cruciferous vegetable extracts and ITC [14]. The induction of phase II detoxification enzymes is believed to give these compounds their chemoprotective attributes. Indeed, consumption of watercress has been shown to alter the metabolism of the tobacco-specific carcinogen, NNK, in humans by the proposed stimulation of UDP-glucuronosyl transferases [15]. However, recent investigations have suggested that the induction of apoptosis may also be a significant contributing factor. Implicated in the apoptotic response of cancer cells to ITCs are the mitogen-activated protein kinases, c-jun N-terminal kinase, extracellular signal-regulated kinase and p38 as well as the cysteine proteases, known as the caspases. In addition, mitochondrial perturbations regulated by the anti- and pro-apoptotic members of the Bcl-2 family have also been described as playing an important role in ITC-mediated apoptosis. In the current chapter we will highlight some of the fascinating pieces of evidence implicating ITCs as potent chemopreventative agents while also describing the cellular targets and signaling pathways underlying ITC-induced apoptosis.

Glucosinolates – The Source of Bioactive ITCs

The Brassicaceae comprise a large number of vegetables including cabbage, cauliflower, Brussels sprouts, broccoli and watercress. A characteristic of individual members is their ability to synthesize phytochemicals known as glucosinolates (GSLs; Table 1). To date, 116 GSLs have been identified [16]. All share a common feature of a *b*-D-thioglucose moiety, sulfonated oxime residue and a variable side chain. The current chemical classification of GSLs recognizes three groups comprising aliphatic, aromatic and indolyl GSLs based on the amino acid from which they are derived. Chemical diversity arises from extensive modification involving elongation, hydroxylation, methylation and oxidation reactions of the precursor amino acids, leading to the formation of several structural analogues based on the initial chain elongated molecule. The homologous series of chain-elongated GSLs can be observed for the w-methylsulfinylalkyl (3-11 carbon), w-methylthioalkyl (3-8 carbon) and b-phenylalkyl derivatives (Table 1).

The identification and chemical distribution of GSLs in plant taxa, particularly the Brassicaceae, has been the focus of much work, with detailed studies by Kjaer(1974), Fenwick et al.(1983), Daxenbichler et al.(1991) and Rosa et al.(1997) [17-20]. Consequently, during the latter end of the last century significant developments were made in our understanding of the genetic regulation of GSLs biosynthesis. This has led to the development of new plant breeding programmes with possibilities of generating new crop varieties with increased chemoprotective properties [21-22].

The Glucosinolates - Myrosinase System and the Generation of Bioactive ITCs

Associated with cruciferous vegetables are a group of thioglucosidases known as myrosinases [EC 3:2:2:1]. Myrosinase enzymes are compartmentalised within myrosin cells, which are distributed throughout all parts of the plant. The separation of the myrosinase-glucosinolate system means that GSLs degradation only occurs during tissue damage, for example chewing or cutting. The catalytic degradation of the GSLs results in the production of several compounds such as nitriles, thiocyanates and isothiocyanates [18] (Figure 1). The range of compounds formed by myrosinase-mediated catalysis of GSLs appears to be influenced by several biochemical and physiological factors, including GSLs side chain structure, pH, ascorbic acid concentration and Fe^{2+} ions levels [23]. With regards to the generation of ITCs in cruciferous plant tissues, myrosinase cleaves the glucose residue of the core GSLs molecule, producing an unstable intermediate, the thiohydroximate-O-sulphate that undergoes a Lossen rearrangement to form the stable ITC. Under adverse physiological conditions such as acidic pH or in the presence of Fe^{2+} ions, the main hydrolytic products are the relatively inert nitriles. In addition to the plant

Table 1. Chemical diversity in glucosinolates

Structure of R-groups	Chemical Name	Carbon chain length	Trivial Name	Common ITC	Common and species name
Methionine-derived aliphatic glucosinolates					
CH_2=CH-$[CH_2]_n$-	Alkenyl	n = 1 - 4	Sinigrin	Allyl	Cabbage, *Brassica oleracea,*
				3-Butenyl	Chinese mustard, *B. juncea*
CH_2=CH-CHOH-CH_2-	Hydroxyalkenyl	n = 1, 2, 3	Progoitrin	4-Hydroxyl-3-butenyl	Turnip and Swede, *B. napus*
CH_3-S-$[CH_2]_n$-	Methylthioalkyl	n = 2 - 9	Glucoerucin	4-Methylthiobutyl	Rocket; *Eruca sativa*
CH_3-SO-$[CH_2]_n$-	Methylsulphinylalkyl	n = 1 -11	Glucoraphanin	4-Methylsulfinylbutyl, 8-Methylsulfinyloctyl,	Broccoli, *B. oleracea,* Watercress, *Rorripa. nasturtium aquaticum,*
				6-Methylsulphinylhexyl	Wasabi, *Wasabi japonica*
CH_3-SO_2-$[CH_2]_n$-	Methylsulfonylalkyl	n = 1, 2, 9	Glucoerysolin	9-Methylsulfonylnonyl	*Amaracea*
Aromatic glucosinolates					
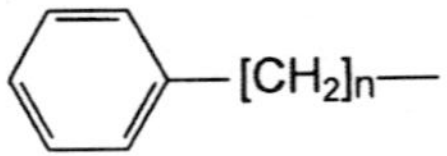	Phenylalkyl	n = 1 – 4	Glucotropaelin	Benzyl	Garden cress, *Lepidium sativa,* Papaya, *Carica papaya*
			Gluconasturtiin	β-Phenylethyl	Watercress, *R. nasturtium aquaticum*

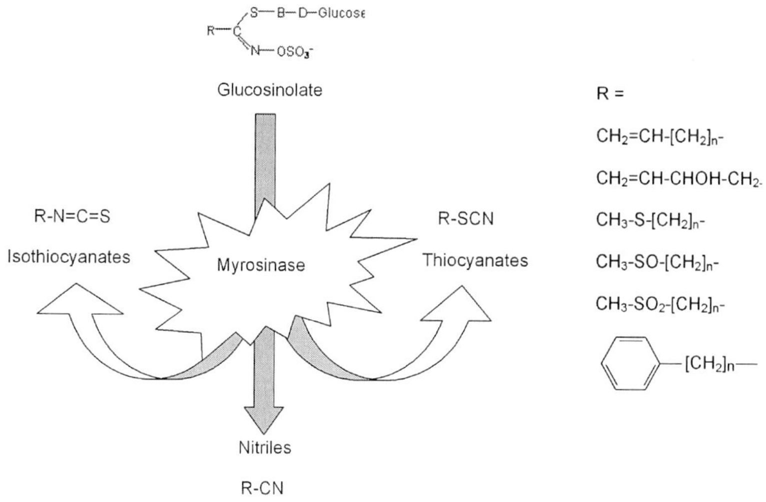

Figure 1. Glucosinolate mediated hydrolysis by myrosinase can lead to the formation of isothiocyanates, nitriles and thiocyanates during plant tissue damage.

myrosinase system, intestinal gut bacteria can also convert GSLs to bioactive ITCs along the gastrointestinal tract. Several enteric gut bacteria including *Enterobacter cloacae* and *Bacteriodes thetaiotaomicron* are able to metabolize sinigrin to allyl isothiocyanate (AITC) [24-26]. It is apparent that the GSL-myrosinase system is pivotal in the bioavailability of ITC in the human diet.

Evidence for the Chemoprotective Role of Crucifers and Associated ITCs

Results from human studies have highlighted a role of cruciferous vegetable consumption with a reduction in cancer incidence. Approximately 200 epidemiological investigations have examined the relationship between high consumption of fruits and vegetables and their protective effect against cancer. Statistical analysis of 128 out of 156 dietary studies revealed that a reduced risk of cancers of the colon, lung, stomach and bladder was associated with high intake of fruits and vegetables. Moreover, the findings from many of these studies revealed that one botanical group, namely the crucifers, were associated with the reduction of cancer incidence in humans. Verhoeven *et al.*(1976) analyzed the data of 7 cohort studies and 87 case control studies, and found a relationship between consumption of cruciferous vegetables like broccoli, cabbage and cauliflower with a reduction in lung cancer risk [3]. Lin *et al.*(2002) observed a correlation between polymorphisms in the phase II detoxification enzyme GSTT1, broccoli consumption and a reduction in colorectal adenomas [27]. Analysis of 457 cases and 505 control subjects indicated that GSTT1-positive subjects with no broccoli intake had a higher prevalence of colorectal adenomas. In contrast, a combination of both GSTM1 and GSTT1 null genotypes with high broccoli intake showed the lowest incidence of adenoma when compared to GSTM1 or GSTT1 positive subjects. Several more studies using food frequency questionnaires and urinary quantification of ITC metabolites demonstrate an inverse correlation between cruciferous vegetable consumption and lung cancer. London *et al.*(2000) described that smokers who were homozygous null for GSTM1 and GSTT1 were at a lower risk of lung cancer as compared to either GSTM1 or GSTT1 positive individuals [28]. Both GSTM1 and GSTT1 are involved in the detoxification of both the carcinogenic polycyclic aromatic hydrocarbons (PAH) and dietary ITCs. The protective effects in null individuals were attributed to a reduction in the rate of elimination of ingested ITCs. The associated reduction of lung cancer by assessing human GST polymorphisms and ITC exposure has also been reported by Zhao et al [29]. Evaluation of the role of genetic polymorphisms in two major GST isoenzymes GSTM1 and GSTT1 in Chinese women showed a positive association between ITCs exposure and a reduction in the incidence of lung cancers. Data indicated that individuals null for either GSTM1 and or GSTT1, having a high weekly intake of ITCs, showed a significant inverse association to the risk of lung cancer. Likewise, associations have also been observed for prostate cancer risk and consumption of cruciferous vegetables [30]. Of the 12 published studies, six suggested an association between cruciferous vegetable consumption and reduced risk of prostate cancer. From the available literature, it is apparent that cruciferous vegetables, the major source of ITCs in the human diet, are sources of potent chemoprotective phytochemicals. The epidemiological findings have been further supported by datum obtained from model animal studies in which subjects are exposed to a combination of known carcinogens along with dietary derived ITCs.

To date, the majority of the scientific investigations have focused on the inhibition of tumor formation induced by nitrosamines and polycyclic aromatic hydrocarbons found in tobacco smoke. Of the approximate 4000 compounds identified 43 have the potential to induce tumor formation in animal models [31]. Two of the most effective carcinogens identified are the polycyclic aromatic hydrocarbons benzo[a]pyrene (B[a]P) and the nitrosamine 4-(methylnitrosamino)-1-(3 pyridyl)-1-butanone (NNK). Both compounds promote tumor formation in animal models. B[a]P and NNK are also suggested to be potent initiating factors in the induction of tobacco-related human cancers. Dose-dependent inhibition by synthetic ITCs and their *N*-acetylcysteine conjugates against NNK- induced tumor formation in rodents has been widely addressed [32-33]. α-phenylethyl isothiocyanate (PEITC) and Benzyl isothiocyanate (BITC) have been the focus of much of this attention. Both are prominent ITCs in watercress, garden cress and papaya. PEITC can inhibit tumor formation induced by NNK in both rat and mouse models but, is ineffective at inhibiting tumor formation induced by B[a]P [34]. In contrast BITC can inhibit tumor formation induced by B[a]P while having no inhibitory effect on NNK. These contrasting differences are currently under investigation in several laboratories, and suggest that for effective chemoprevention, a combination of ITCs or other phytochemicals maybe necessary. Additional research has also shown that PEITC and BITC can inhibit DMBA-induced mammary tumors in rats, with PEITC being additionally effective in the inhibition of N-nitrosbenzylmethylamine (NBMA)-induced esophageal tumors [35-36]. Likewise, sulphoraphane (SF) can inhibit DMBA-induced tumor formation in rat mammary tissues while also showing a protective effect against azoxymethane (AZO)-induced aberrant crypt foci in rats [37]. Fisher rats F344 exposed to the carcinogen AZO develop aberrant crypt foci in colonic tissues, the formation of which is believed to be associated with the development of colon cancer. In rats fed AZO along with SF, PEITC or their respective *N*-acetylcysteine mercapturic acid metabolites, a significant reduction in the formation of aberrant crypt foci during the post-initiation phase was observed [38]. More recent data has addressed the possible role of ITC-induced apoptosis in preventing lung cancer. In a study by Yang *et al.* (2002) a high incidence of tumors in A/J mice could be induced by a single dose of the tobacco carcinogen B[a]P [39]. Administration of N-acetylcysteine conjugates of either BITC or

PEITC during the post-initiation phase showed a significant reduction in tumor multiplicity. Associated with the reduction was an increase in the number of apoptotic cells in BITC and PEITC-treated mice and the activation of the MAP kinases, extracellular signal-regulated kinase (ERK1/2), phosphorylated p38 and increases in activator protein 1 (AP1) and p53 [39]. Smith *et al.*(1998) has reported the induction of cell cycle arrest and apoptosis by sinigrin in colorectal crypt foci of rats treated with 1, 2-dimethylhydrazine (DMH) [40]. Follow-up studies [41] showed that brussels sprout extracts, a natural source of ITCs, also gave similar results, confirming that ITCs derived from the diet can induce apoptosis [40-41]. As a consequence of the positive findings implicating ITCs as potential apoptotic agents capable of inducing apoptosis in cancer cells, much research has now been conducted to determine the molecular mechanism(s) mediating the apoptotic effects of ITCs in cell culture models.

Apoptosis – A Contributing Factor in ITC-Mediated Chemoprotection?

Apoptosis, also known as programmed cell death, is a means by which living organisms control abnormalities in cells that occur either by genetic or environmental cues. Characteristic changes including cell shrinkage, chromatin condensation, plasma membrane blebbing, DNA fragmentation and finally cell breakdown with the release of apoptotic bodies are often observed during apoptosis. The initiation of apoptosis can occur via two major pathways. Firstly, interaction of extracellular ligands with membrane bound receptors leading to the initiation of an intracellular signal that promotes apoptosis. Secondly, apoptosis can be initiated via mitochondria with the release of apoptotic-initiating factors like cytochrome c, smac/diablo and apoptosis-inducing factor (AIF) [42]. Pivotal in the response of cancer cells to apoptotic stimuli are the caspases. These are intracellular cysteine-containing proteases that cleave their substrates after an aspartate residue in a tetrapeptide sequence-specific manner. Caspases are found as pro-enzymes that require activation to their proteolytic forms either through cell surface death receptors, mitochondria or a convergence of both signaling pathways. Apoptotic cues include cytochrome c release from mitochondria leading to the activation and formation of the apoptotic protease-activating factor-1 (APAF-1) - caspase-9 holoenzyme complex. Subsequently the formation of this complex leads to caspase-3 activation. Caspase-3 proteolytically degrades numerous cellular targets including poly (ADP-ribose) polymerase (PARP), Protein Kinase C delta, retinoblastoma protein, lamin, alpha-fodrin, DNAse and DNA fragmentation factor leading to the induction of apoptosis.

With regards to ITCs Chen *et al.*(1998) [43] and Yu *et al.* (1998) [44] were the first investigators to show that ITC-mediated apoptosis was predominantly caspase dependant [43-44]. Co-treatment with the pan-caspase inhibitor Z-VAD-FMK and the caspase-3 inhibitor Ac-DEVD-CHO attenuated the time-dependent induction of caspase-3 activity, PARP cleavage and DNA fragmentation in Jurkat and HeLa cells. Consequently, caspase-3 mediated proteolysis in ITC-mediated apoptosis has been determined in several cell lines. We recently showed that Z-VAD-FMK and Ac-DEVD-CHO can also abrogate caspase-3 activity in human hepatoma HepG2 cells, as measured by caspase-3 activity, PARP cleavage, DNA fragmentation and lactate dehydrogenase leakage [45]. Additional caspases have also been shown to be involved in ITC-mediated apoptosis. Using the caspase-8 specific inhibitor, Z-IETD-AFC, Xu and Thornalley(2000) [46] reported that caspase-8 inhibition can prevent PEITC-induced caspase-8 activity and DNA fragmentation as determined using SubG1 analysis by flow cytometry in HL60 cells. Pro-caspase-8 and -9 cleavage and inhibition of cytoplasmic histone-associated DNA fragments using the inhibitors, Z-IETD-FMK and z-LEHD-FMK, has also been shown in PC-3 cells treated with SF, respectively [47]. Likewise, caspase-9 activity and cleavage have been observed in human medulloblastoma DAOY cells, hepatic rat RL34 and human HepG2 cells as well as in the human bladder cancer cell lines, T24 and UM-UC-3 [48-50].

From the available literature, one of the key intracellular effects of ITCs is their action towards mitochondria. Mitochondria play an integral role in caspase activation and apoptosis. One characteristic effect on mitochondria during the induction of apoptosis is the collapse of the mitochondrial transmembrane potential ($\Delta\psi_m$) leading to the release of apoptotic signaling factors into the cytoplasm. The biochemical and physiological triggers often include a decrease in cellular GSH content, generation of reactive oxygen species (ROS), cytochrome c release and caspase activation. Furthermore, the intricate cross-talk between cellular signal transduction pathways, members of the MAP kinases, and the expression and or activation of anti- and pro-apoptotic factors such as p53 and members of the Bcl-2 family can inhibit or potentate the apoptotic response. In the mitochondrial pathway pro-apoptotic Bcl-2 family proteins like bid, bad and bax promote mitochondrial dysfunction [42]. In contrast, anti-apoptotic members including Bcl-2 and Bcl_xL can prevent mitochondrial perturbations and preserve cell viability. Bax and Bid are recognized as being able to interact with components of the mitochondrial permeability transition pore (MPT) or directly form channels in mitochondrial membranes. However, the mechanisms by which Bcl-2 family members and the MPT control mitochondrial integrity remain elusive.

Information regarding the role of anti-apoptotic Bcl-2 family proteins in ITC-mediated apoptosis has rarely been investigated. Chen *et al.*(1998) observed that human embryonic kidney 293 cells over-expressing Bcl-2 and Bcl-xL suppressed PEITC-induced apoptosis [51]. Similarly, over-expression of Bcl-2 in human Jurkat T lymphocyte cells also inhibited PEITC-induced apoptosis [52]. However, recently Xiao *et al.*(2004) recently observed that over-expression of Bcl-2 in human PC-3 human prostate cancer cells failed to prevent PEITC-induced apoptosis [53].

Additional evidence for a role of mitochondria in ITC-induced apoptosis comes from several early *in vitro* studies. Work conducted by Miko and Chance(1975) demonstrated that synthetic ITC analogues, p-bromophenyl isothiocyanate, 4,4'-diisothiocyanatebiphenyl and beta-naphtylethyl isothiocyanate, were uncoupling agents in isolated rat mitochondria, although a direct role in the induction of apoptosis was not sought [54]. More comprehensive investigations have addressed the role of mitochondria in ITC-mediated apoptosis. Nakamura *et al.*(2002) using isolated rat mitochondria and rat liver epithelial RL34 cells showed that BITC (20μM) inhibited mitochondrial respiration, induced mitochondrial depolarization and mitochondrial swelling, a characteristic associated with the activation of the MPT pore and the release of cytochrome c [48]. Likewise, mitochondrial depolarization and caspase activation have been observed in HL60/S cells exposed to AITC and BITC. In our own studies using human hepatoma HepG2 cells and isolated rat mitochondria we demonstrated that PEITC (20 μM) mediates a time dependent loss of ($\Delta\psi_m$), release of mitochondrial cytochrome c, reduction in oxygen consumption and the activation of caspase-3 and -9 [49]. PEITC-induced caspase activation appeared to be associated with the conformational change and translocation of the pro-apoptotic protein Bax to mitochondria. Bax can form pores in artificial lipid membranes, induce cytochrome c release from isolated mitochondria, while also interacting with components of the MPT namely the voltage -dependent anion channel (VDAC) and the adenine nucleotide transporter (ANT) [55-56]. These interactions ultimately lead to the activation of the initiator caspase-9 and subsequent activation of other down stream effector caspases such as caspases-3, 6 and 7. Changes in Bax protein levels have been observed in the apoptotic response of human pancreatic and leukemia cancer cells, as well as human HT29 colon cancer and Jurkat T-leukemia cells to ITCs. Indeed, 4-methylthiobutyl isothiocyanate (MTBITC) was shown to increase Bax protein levels during the apoptotic response in Jurkat T-leukemia cells [57]. In contrast, Kuang and Chen(2004) found no correlation between Bax protein levels in human non-small cell lung carcinoma A549 cells when exposed to apoptotic concentrations of PEITC, BITC or AITC [58]. These data suggest that the role of Bax in ITC-mediated apoptosis may be cell type specific and thus requires further investigation. As for additional pro-apoptotic Bcl-2 proteins, Xu and Thornalley(2001) reported that both PEITC and AITC induced the cleavage of p22 BID protein to p15, p13 and p11 fragments in human leukemia HL60 cells *in vitro* [59]. A summary of ITC-mediated apoptotic effects is represented in Figure 2.

ITC-Mediated Stress Signaling Cascades and Their Role in Apoptosis

Exposure of cells to chemotherapeutic drugs, natural products and oxidants can modulate a variety of stress signaling events that can potentate stress-related gene transcription. Pivotal in the control of numerous stress cascades are members of the mitogen-activated protein kinases (MAPK). MAPK are proline-directed serine/threonine kinases that mediate phosphorylation cascades involved in both cell survival and apoptosis. Three MAPKinase families are recognized in mammals including the c-Jun N-terminal kinase (JNK), ERK1/2 and p38 kinase. However, their role in apoptosis is complex and highly controversial.

All three kinases have been studied to determine their role in ITC-mediated phase II enzymatic induction. However, only a few reports addressed their role in ITC-induced apoptosis. Initial work by Yu *et al.*(1998) [44] revealed PEITC to be a potent inducer of JNK-1 in human ovarian HeLa cells and that sustained activation of JNK was associated with the induction of phase II detoxification enzymes. More recently, sustained JNK-1 activation has also been implicated in the apoptotic signaling response of human cancer cells to ITCs. Suppression of the JNK signaling pathway using dominant-negative mutant of JNK-1, MEKK1 or using the JNK-specific inhibitor SP600125 prevents PEITC-induced apoptosis in some human cancer cell lines. Hu *et al.*(2003) [60] reported that pretreatment with SP600125 prevents cytochrome c release and caspase activation induced by PEITC in human adenocarcinoma HT29 colon cells. In human PC-3 prostate cancer cells, PEITC treatment induces ERK1/2 and p38 but not JNK-1 activation at apoptotic concentrations (10 μM). However, only in the presence of the mitogen-activated protein kinase 1 inhibitor, PD98059, and not the p38 inhibitor, SB202190, could apoptosis be abolished. As yet the mechanism(s) of how ERK 1/2 regulates PEITC-induced apoptosis in PC-3 has not been determined [61].

In contrast, during BITC-induced apoptosis in human Jurkat T-cell leukemia, HL-60 and HeLa cells, JNK and p38 MAPK, but not extracellular signal-regulated kinase, are activated. In experiments using the JNK-specific inhibitor SP600125 or the p38 MAPK inhibitor SB202190, BITC-induced apoptosis could be prevented [62]. These data implicate both kinases as having a regulatory role in BITC-induced apoptosis. Indeed, further investigation showed that each kinase contributed to the inactivation of the anti-apoptotic protein Bcl-2 via phosphorylation, this ultimately leading to changes in the Bax Bcl-2 ratio promoting apoptosis.

CONCLUSIONS

ITCs are prominent phytochemical constituents within the damaged tissues of cruciferous vegetables. To date, they appear to be some of the most effective chemoprotective chemicals known and are readily available in the human diet. Given the extensive scientific evidence obtained from epidemiological, *in vivo* animal model and *in vitro* cell culture model

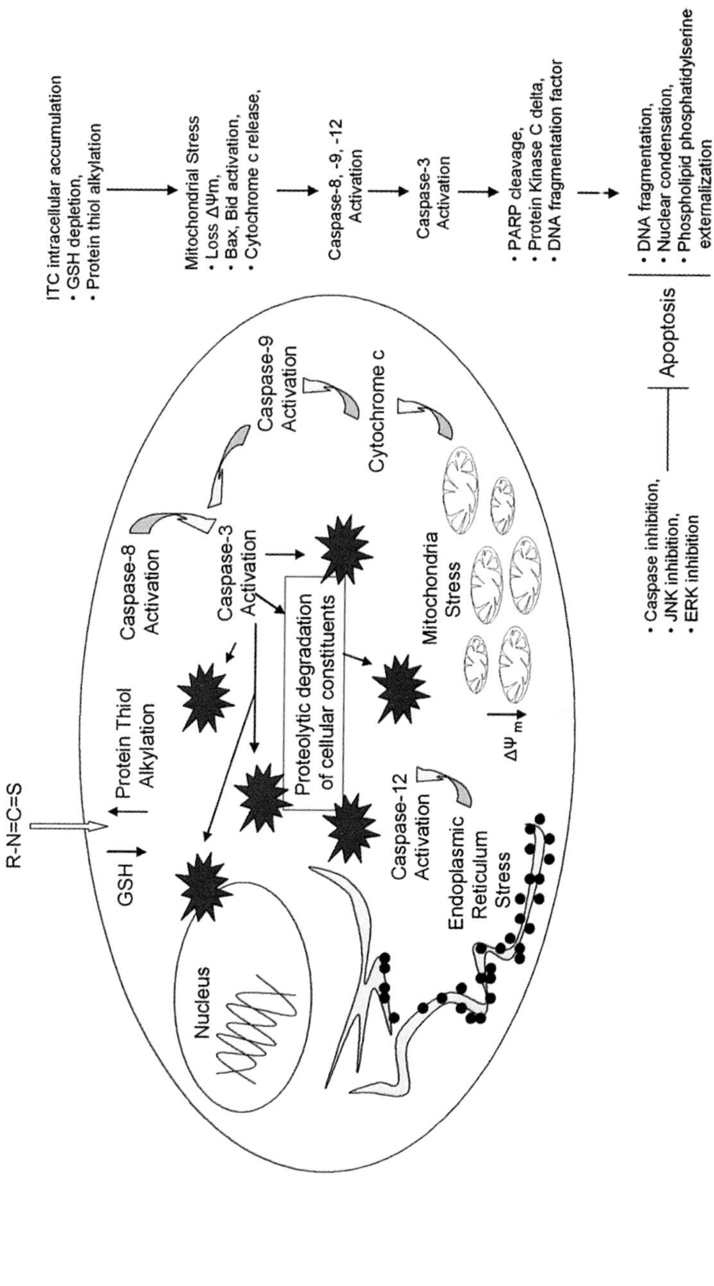

Figure 2. Schematic representation of ITC mediated apoptotic signaling in mammalian cells.

studies, we are now beginning to understand the molecular mechanisms involved in their protective effects against cancers. One of the most significant findings is the ability of ITCs to promote apoptosis in a range of cancer cells.

REFERENCES

[1] Block G, Patterson B, Subar A. Fruit, vegetables, and cancer prevention: a review of the epidemiological evidence. Nutr Cancer 1992; 18: 1-29

[2] Steinmetz KA, Potter JD. Vegetables, fruit, and cancer. I. Epidemiology. Cancer Causes Control 1991; 2: 325-57

[3] Verhoeven DT, Goldbohm RA, van Poppel G, Verhagen H, van den Brandt PA. Epidemiological studies on brassica vegetables and cancer risk. Cancer Epidemiol Biomarkers Prev 1996; 5: 733-48

[4] Bones AM, Rossiter JT. The myrosinase-glucosinolate system, its organisation and biochemistry. Physiol Plantarum 1996; 97: 194-208

[5] Ono H, Tesaki S, Tanabe S, Watanabe M. 6-methylsulphinylhexyl isothiocyanate and its homologues as food-originated compounds with antibacterial activity against *Eschericia coli* and *Staphylococcus aureus*. Biosci Biotechnol Biochem 1998; 62: 363-65

[6] Lin CM, Preston JF, Wei CI. Antibacterial mechanism of allyl isothiocyanate. J Food Prot 2000; 63: 727-34

[7] Fahey JW, Haristoy X, Dolan PM, et al. Sulforaphane inhibits extracellular, intracellular, and antibiotic-resistant strains of *Helicobacter pylori* and prevents benzo[a]pyrene-induced stomach tumors. Proc Natl Acad Sci U S A. 2002; 99: 7610-5

[8] Drobinca L, Zemanova M, Nemec K, et al. Antifungal activity of isothiocyanates and related compounds. Appl Microbiol 1967; 15: 701-9

[9] Albert-Puleo M. Physiological effects of cabbage with reference to its potential as a dietary cancer-inhibitor and its use in ancient medicine. J Ethnopharmacol 1983; 9: 261-72

[10] Burckhalter JH, Bariana DS. Synthesis of isothiocyanates as potential antineoplastic agents. J Med Chem 1963; 35: 203-5

[11] Horakova K, Drobnica L, Nemec P, Kristian P, Antos K, Martvon A. Cytotoxic and cancerostatic activity of isothiocyanates and related compounds. 3. The effect of stilbene, azobenzene and polycondensed aromatic hydrocarbon ITC derivatives on HeLa cells. Neoplasma 1969; 16: 231-7

[12] Horakova K, Drobnica L, Nemec P, Kristian P, Antos K. Cytotoxic and cancerostatic activity of isothiocyanates and related compounds. II. Activity of substituted phenylisothiocyanates on HeLa-cells. Neoplasma 1968; 15: 177-82

[13] Cruz A. Remarkable antimitotic action of watercress (*Nasturtium officinales*) on some experimental tumors. Hospital (Rio J) 1970; 77: 943-52

[14] Van Poppel G, Verhoeven DT, Verhagen H, Goldbohm RA. Brassica vegetables and cancer prevention. Epidemiology and mechanisms. Adv Exp Med Biol 1999; 472: 159-68

[15] Hecht SS, Carmella SG, Murphy SE. Effects of watercress on urinary metabolites of nicotine in smokers. Cancer Epidemiol. Biomarkers Prev 1999; 8: 907–13

[16] Fahey JW, Zalcmann AT, Talalay P The chemical diversity and distribution of glucosinolates and isothiocyanates among plants. Phytochemistry. 2001; 56: 5-51

[17] Kjaer A. The natural distribution of glucosinolates: a uniform class of sulphur containing glycosides. In: Bendz G, Santesson J, ed. Chemistry in Botanical Classification. London: Academic Press 1974: 229-34

[18] Fenwick GR, Heaney RK, Mullin WJ. Glucosinolates and their breakdown products in food and food plants. Crit Rev Food Sci Nutr 1983; 18: 123-201

[19] Daxenbichler ME, Spencer GF, Carlson DG, Rose GB, Brinker AM, Powell RG. Glucosinolate composition of seeds from 297 speices of wild plant. Phytochemistry 1991; 30: 2623-38

[20] Rosa E, Heaney RC, Fenwick R, Portas CAM. Glucosinolates in crop plants. Horticultural Reviews 1997; 19: 99-215

[21] Faulkner K, Mithen R, Williamson G. Selective increase of the potential anticarcinogen 4-methylsulphinylbutyl glucosinolate in broccoli. Carcinogenesis 1998; 19: 605-9

[22] Mithen R, Faulkner K, Magrath R, Rose P, Williamson G, Marquez J. Development of isothiocyanate-enriched broccoli, and its enhanced ability to induce phase 2 detoxification enzymes in mammalian cells. Theor Appl Genetics 2003; 106: 727-34

[23] Rask L, Andreasson E, Ekbom B, Eriksson S, Pontoppidan B, Meijer J. Myrosinase: gene family evolution and herbivore defence in Brassicaceae. Plant Mol Biol 2000; 42: 93–113

[24] Tani N, Ohtsuru M, Hata T. Purification and general characteristics of bacterial myrosinase produced by *Enterobacter cloacae.* Agric Biol Chem 1974; 38: 1623-30

[25] Krul C, Humblot C, Philippe C, Vermeulen M, Van Nuenen M, Havenaar R, Rabot S. Metabolism of sinigrin (2-propenyl glucosinolate) by the human colonic microflora in a dynamic in vitro large-intestinal model. Carcinogenesis 2002; 23: 1009-16

[26] Elfoul L, Rabot S, Khelifa N, Quinsac A, Duguay A, Rimbault A. Formation of allyl isothiocyanate from sinigrin in the digestive tract of rats monoassociated with a human colonic strain of *Bacteroides thetaiotaomicron*. FEMS Microbiol Lett 2001; 197: 99-103

[27] Lin HJ, Zhou H, Dai A. Glutathione transferase GSTT1, broccoli, and prevalence of colorectal adenomas. Pharmacogenetics 2002; 12: 175-9

[28] London SJ, Yuan JM, Chung FL. Isothiocyanates, glutathione S-transferase M1 and T1 polymorphisms, and lung-cancer risk: a prospective study of men in Shanghai, China. Lancet 2000; 356: 724-9
[29] Zhao B, Seow A, Lee EJ. Dietary isothiocyanates, glutathione S-transferase -M1, -T1 polymorphisms and lung cancer risk among Chinese women in Singapore. Cancer Epidemiol Biomarkers Prev 2001; 10: 1063-7
[30] Kristal AR, Lampe JW. Brassica vegetables and prostate cancer risk: a review of the epidemiological evidence. *Nutr Cancer* 2002; 42: 1-9
[31] Hecht SS. Chemoprevention by isothiocyanates. J Cell Biochem (Suppl) 1995; 22: 195-209
[32] Morse MA, Eklind KI, Amin SG, Hecht SS, Chung FL. Effects of alkyl chain length on the inhibition of NNK-induced lung neoplasia in A/J mice by arylalkyl isothiocyanates. Carcinogenesis 1989; 10: 1757-9
[33] Morse MA. Inhibition of NNK-induced lung tumorigenesis by modulators of NNK activation. Exp Lung Res 1998; 4: 595-604
[34] Hecht SS, Kenney PM, Wang M, Trushin N, Upadhyaya P. Effects of phenethyl isothiocyanate and benzyl isothiocyanate, individually and in combination, on lung tumorigenesis induced in A/J mice by benzo[a]pyrene and 4-(methylnitrosamino)-1-(3-pyridyl)-1-butanone. Cancer Lett 2000; 150: 49-56
[35] Sticha KR, Kenney PM, Boysen G. Effects of benzyl isothiocyanate and phenethyl isothiocyanate on DNA adduct formation by a mixture of benzo[a]pyrene and 4-(methylnitrosamino)-1-(3-pyridyl)-1-butanone in A/J mouse lung. Carcinogenesis 2002; 23: 1433-9
[36] Futakuchi M, Hirose M, Miki T, Tanaka H, Ozaki M, Shirai T. Inhibition of DMBA-initiated rat mammary tumour development by 1-O-hexyl-2,3,5-trimethylhydroquinone, phenylethyl isothiocyanate, and novel synthetic ascorbic acid derivatives. Eur J Cancer Prev 1998; 2: 153-9
[37] Zhang Y. Cancer-preventive isothiocyanates: measurement of human exposure and mechanism of action. Mutat Res 2004; 555: 173-90
[38] Johnson IT. Anticarcinogenic effects of diet-related apoptosis in the colorectal mucosa. Food Chem Toxicol 2002; 40: 1171-8
[39] Yang YM, Conaway CC, Chiao JW, et al. Inhibition of benzo(a)pyrene-induced lung tumorigenesis in A/J mice by dietary N-acetylcysteine conjugates of benzyl and phenethyl isothiocyanates during the postinitiation phase is associated with activation of mitogen-activated protein kinases and p53 activity and induction of apoptosis. Cancer Res 2002; 62: 2-7
[40] Smith TK, Lund EK, Johnson IT. Inhibition of dimethylhydrazine-induced aberrant crypt foci and induction of apoptosis in rat colon following oral administration of the glucosinolate sinigrin. Carcinogenesis 1998; 19: 267-73
[41] Smith TK, Mithen R, Johnson IT. Effects of Brassica vegetable juice on the induction of apoptosis and aberrant crypt foci in rat colonic mucosal crypts in vivo. Carcinogenesis 2003; 24: 491-5

[42] Gupta S. Molecular steps of death receptor and mitochondrial pathways of apoptosis. Life Sci 2001; 69: 2957-64
[43] Chen YR, Wang W, Kong AN, Tan TH. Molecular mechanisms of c-Jun N-terminal kinase-mediated apoptosis induced by anticarcinogenic isothiocyanates. J Biol Chem 1998; 273: 1769-75
[44] Yu R, Mandlekar S, Harvey KJ, Ucker DS, Kong AN. Chemopreventive isothiocyanates induce apoptosis and caspase-3-like protease activity. Cancer Res 1998; 58: 402-8
[45] Rose P, Whiteman M, Huang SH, Halliwell B, Ong CN. beta-Phenylethyl isothiocyanate-mediated apoptosis in hepatoma HepG2 cells. Cell Mol Life Sci 2003; 60: 1489-503
[46] Xu K, Thornalley PJ. Studies on the mechanism of the inhibition of human leukaemia cell growth by dietary isothiocyanates and their cysteine adducts in vitro. Biochem Pharmacol 2000; 60: 221-31
[47] Singh AV, Xiao D, Lew KL, Dhir R, Singh SV. Sulforaphane induces caspase-mediated apoptosis in cultured PC-3 human prostate cancer cells and retards growth of PC-3 xenografts in vivo. Carcinogenesis 2004; 25: 83-90
[48] Nakamura Y, Kawakami M, Yoshihiro A. Involvement of the mitochondrial death pathway in chemopreventive benzyl isothiocyanate-induced apoptosis. J Biol Chem 2002; 277: 8492-9
[49] Rose P, Armstrong JS, Chua YL, Ong CN, Whiteman M. Beta-phenylethyl isothiocyanate mediated apoptosis; contribution of Bax and the mitochondrial death pathway. Int J Biochem Cell Biol 2005; 37: 100-19
[50] Tang L, Zhang Y. Dietary isothiocyanates inhibit the growth of human bladder carcinoma cells. J Nutr 2004; 134: 2004-10
[51] Chen YR, Wang W, Kong AN, Tan TH. Molecular mechanisms of c-Jun N-terminal kinase-mediated apoptosis induced by anticarcinogenic isothiocyanates. J Biol Chem 1998; 273: 1769-75
[52] Pullar JM, Thomson SJ, King MJ, Turnbull CI, Midwinter RG, Hampton MB. The chemopreventive agent phenethyl isothiocyanate sensitizes cells to Fas-mediated apoptosis. Carcinogenesis 2004; 25: 765-72
[53] Xiao D, Johnson CS, Trump DL, Singh SV. Proteasome-mediated degradation of cell division cycle 25C and cyclin-dependent kinase 1 in phenethyl isothiocyanate-induced G2-M-phase cell cycle arrest in PC-3 human prostate cancer cells. Mol Cancer Ther 2004; 3: 567-75
[54] Miko M, Chance B. Isothiocyanates. A new class of uncouplers. Biochim Biophys Acta 1975; 396: 165-74
[55] Epand RF, Martinou JC, Montessuit S, Epand RM, Yip CM. Direct evidence for membrane pore formation by the apoptotic protein Bax. Biochem Biophy Res Comm 2002; 298: 744-9
[56] Marzo I, Brenner C, Zamzami N, Jurgensmeier JM, Susin SA. Bax and adenine nucleotide translocator cooperate in the mitochondrial control of apoptosis. Science 1998; 281: 2027-31

[57] Fimognari C, Nusse M, Iori R, Cantelli-Forti G, Hrelia P. The new isothiocyanate 4-(methylthio)butylisothiocyanate selectively affects cell-cycle progression and apoptosis induction of human leukemia cells. Invest New Drugs 2004; 22: 119-29

[58] Kuang YF, Chen YH. Induction of apoptosis in a non-small cell human lung cancer cell line by isothiocyanates is associated with P53 and P21. Food Chem Toxicol 2004; 42: 1711-8

[59] Xu K, Thornalley PJ. Signal transduction activated by the cancer chemopreventive isothiocyanates: cleavage of BID protein, tyrosine phosphorylation and activation of JNK. Br J Cancer 2001; 84: 670-3

[60] Hu R, Kim BR, Chen C, Hebbar V, Kong AN. The roles of JNK and apoptotic signaling pathways in PEITC-mediated responses in human HT-29 colon adenocarcinoma cells. Carcinogenesis 2003; 24: 1361-7

[61] Xiao D, Singh SV. Phenethyl isothiocyanate-induced apoptosis in p53-deficient PC-3 human prostate cancer cell line is mediated by extracellular signal-regulated kinases. Cancer Res 2002; 62: 3615-9

[62] Miyoshi N, Uchida K, Osawa T, Nakamura Y. A link between benzyl isothiocyanate-induced cell cycle arrest and apoptosis: involvement of mitogen-activated protein kinases in the Bcl-2 phosphorylation. Cancer Res 2004; 64: 2134-42

Tanshinone I and Tanshinone IIA from *Salvia Miltiorrhiza* Inhibit Growth of K1735M2 Murine Melanoma Cells via Different Pathways

Zhenlong Wu[a,b], Ying Yang[c], Ruolin Yang[b], Guoliang Xia[d], Liping Xie[b] and Rongqing Zhang[b,*]
[a] Department of Biological Sciences, National University of Singapore, Singapore
[b] Department of Biological Science and Biotechnology, Tsinghua University, Beijing 100084
[c] Division of Animal Nutrition and Feed Science, China Agricultural University, Beijing, 100094
[d] Department of Animal Physiology, China Agricultural University, Beijing, 100094, P. R. China
**Corresponding email: rqzhang@mail.tsinghua.edu.cn*

INTRODUCTION

Dan shen, a Chinese traditional medicine prepared from the roots of a Chinese traditional plant, *Salvia miltiorrhiza*, has been widely used in China for many years for the treatment of various cardiovascular diseases including angina pectoris, myocardial infarction and stroke [1-11], and also the amelioration carbon tetrachloride-induced hepatic injury and hepatic fibrosis [12-15]. Substances isolated from *Salvia miltiorrhiza* have been reported to exhibit antioxidant, antibacterial, anti-dermatophytic, anti-inflammatory and anti-platelet activities [16-27]. In addition, several lines of research reported that *Salvia miltiorrhiza* or compound extracted from it showed marked anti-tumor effect in various cancer cell lines [19,28-34]. *In vitro* studies demonstrated Salvia miltiorrhiza possessed cytotoxic activities against various types of cell lines derived from human carcinomas of lung, colon, cervix and larynx [35,36]. *Salvia miltiorrhiza* could markedly prolong the survival period of Ehrlich carcinoma-bearing mice with ascites [37]. In a recent study, water- soluble extracts from *Salvia miltiorrhiza* exhibited potent effect against HIV integrase *in vitro* and viral replication *in vivo* [38].

Cancer remains a major threat to public health. Though basic cancer research has produced remarkable advance in the last decade, chemotherapy is still considered the most effective method of cancer treatment. However, most anti-tumor drugs are not sufficiently tumor-selective and sometimes cause hematopoietic disorders. Resistance to the chemotherapeutic regimen also occurs [39]. Hence, herbal medicine has attracted a great deal of attention as an alternative cancer therapy that causes less toxicity with cost benefit. In this study, we report the inhibitory effects of tanshinone I and tanshinone IIA on proliferation of murine melanoma cells and attempt to elucidate the possible mechanism(s).

MATERIALS AND METHODS

Plant material

Dan Shen, the root of *Salvia miltiorrhiza* Bunge, was obtained from Tong Ren Tang (Beijing, China) and identified by a pharmacologist.

Extraction and isolation

The dried and powdered roots (1.0 kg) of *Salvia miltiorrhiza* were extracted three times with MeOH for 24 h each time at room temperature. The combined MeOH extracts were concentrated under reduced pressure to yield brown syrup (50g).The concentrated extract was suspended in water and partitioned successfully with CH_2Cl_2 and EtOAC to yield 20g of CH_2Cl_2 soluble fraction and 10 g of EtOAc fraction. The soluble fraction was applied to a column packed with silica gel and eluted with n-hexane-EtOAc gradient solvent system to yield sub-fractions 1-8. Among them, subfractions 4 and 7 were re-crystallized and identified by comparing spectral data of UV, IR, ^{1}H-NMR and ^{13}C-NMR studies with those of tanshinone I and tanshinone IIA as reported earlier [40,41]. The two compounds were dissolved in dimethyl sulfoxide (DMSO; Sigma Aldrich Chemicals, USA) and stored at 4°C before each experiment.

Cell cultures

The K1735M2 murine melanoma cells were obtained from Dr. I. F. Fidler (Anderson Cancer Center, Houston, TX) and maintained in minimal essential medium supplemented with 10% FBS, 100 units/ml penicillin and 100mg/ml streptomycin (all from GIBCO BRL Life Technologies, Inc.) at 37°C in 5% CO_2 humidified atmosphere. Only single cell suspension of more than 90% viability, as determined by trypan blue exclusion, was used in the study.

MTT assay

Viable cells treated with different concentration of tanshinone I or tanshinone IIA were evaluated by the modified MTT assay. Briefly, the K1735M2 cells (1×10^4) were seeded into 96-well plates containing 100 μl of the growth medium. Cells were permitted to adhere for 24 hours, and then the medium was replaced with tanshinone I or tanshinone IIA dissolved in medium. The micro-titre plate was incubated for the desired period of time. At the end of the incubation, 20 μl of MTT (5mg/ml) was added into each well and the mixture was incubated at 37°C for a further 4 hs, after which 100 μl of DMSO was added to dissolve the formazan crystals formed. Optical density at 490 nm was read on a scanning multi-well spectrophometer (Bio-Rad 550). All experiments were performed in triplicate.

Determination of cytotoxicity

Lactate dehydrogenase (LDH) activity was determined with an Abbott VP Biochemical Analyzer using commercial test kit (Abbott Laboratories) as described previously [30]. The percentage of LDH leakage from the cells was used as the index of cell viability which was calculated according to the formula: %LDH leakage= (LDH activity present in the medium after incubation/total LDH activity in cells) $\times100$.

Cell cycle assay by flow cytometry

K1735M2 cells were cultured in the presence of tanshinone I or tanshinone IIA at various concentrations for 72 hr, and then approximately 10^6 cells were harvested with trypsin/EDTA. After washing twice with PBS, the cells were fixed by 70% ethanol and treated with RNase (100μg/ml) for about 30 min at 37°C. The cells were stained with propidium iodide and analyzed by a flow cytometer (Partec, CCA-I)

Statistics

All data are presented as means±SD from three independent experiments. Statistical differences were evaluated using the Student t-test and were considered significant at $p<0.05$.

RESULTS

In this study, the effects of tanshinone I and tanshinone IIA on the growth of K1735M2 melanoma cells were assessed (Figure 1). The proliferation of the K1735M2 cells was inhibited by the two compounds. However, tanshinone IIA was more potent than tanshinone I in inhibiting cell growth. Fifty percent inhibition of cell growth was achieved by 0.6μg/ml tanshinone IIA and 3μg/ml tanshinone I, respectively. Moreover, the anti-proliferative effects were also found to be dose- and time- dependent (Figures 1 and 2).

Figures 3 and 4 show the cytotoxic effects of tanshinone I or tanshinone IIA measured as a percentage of LDH leakage [30]. After 24 h of incubation, cells with lowest concentration of the two compounds

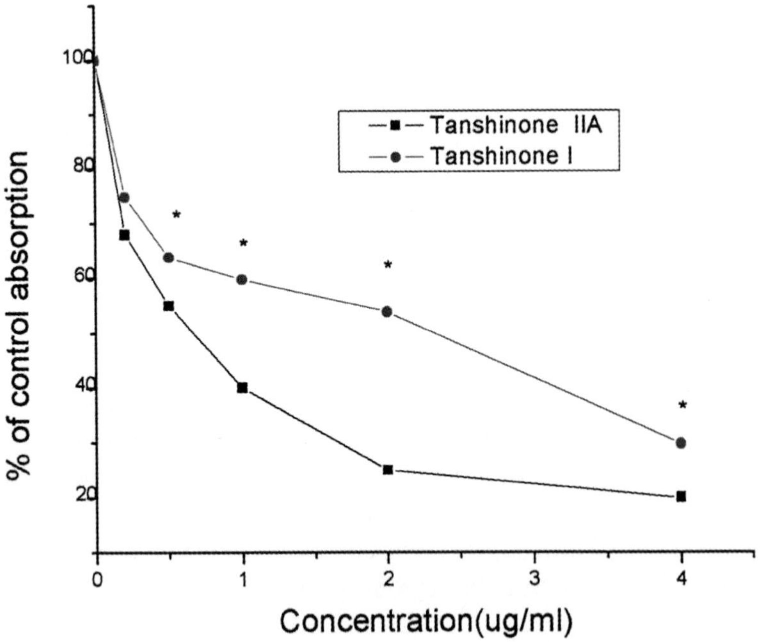

Figure 1. Dose-dependent inhibitory effect of tanshinone I or tanshinone IIA on K1735M2 cells.
Cells were treated with various concentrations of tanshinone I or tanshinone IIA for 24h; MTT assay was performed according to the procedure described in the Materials and Methods. * $P<0.01$ vs untreated control.

showed no cytotoxicity. However, with increase in the dose and time, dose- and time-dependent increases in LDH leakage were observed.

To test whether an alteration of cell cycle distribution contributed to the inhibitory effects, cells were treated with tanshinone I or IIA for 12, 24, 36h and their cell cycle were analyzed. As shown in Table 1, compared with control, the cells treated with tanshinone I showed a significant increase to 78% in the G1 phase after 36 h, and a decrease in the percentage of cells in the S phase at this period. However, neither the cells treated with tanshinone IIA nor the untreated control cells showed a significant change in cell cycle distribution.

DISCUSSION

In the present study, the anti-proliferative effect, cytotoxicity effect and cell cycle distribution of K1735M2 murine melanoma cells treated with tanshimone I and IIA were evaluated.

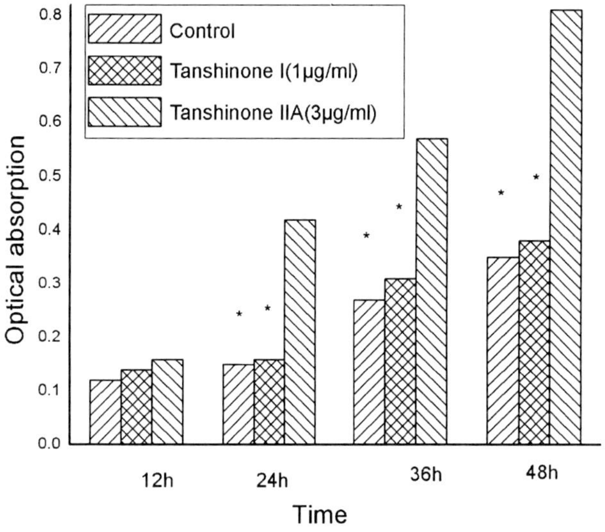

Figure 2. Time-dependent inhibitory effect of tanshinone I or tanshinone IIA on K1735M2 cells.
Cells were treated with indicated concentration of tanshinone I (3μg/ml) or tanshinone IIA (1.0μg/ml) for 12, 24, 36, 48h. MTT assay was performed according to the procedure described in Materials and Methods. * $P<0.01$ vs untreated control.

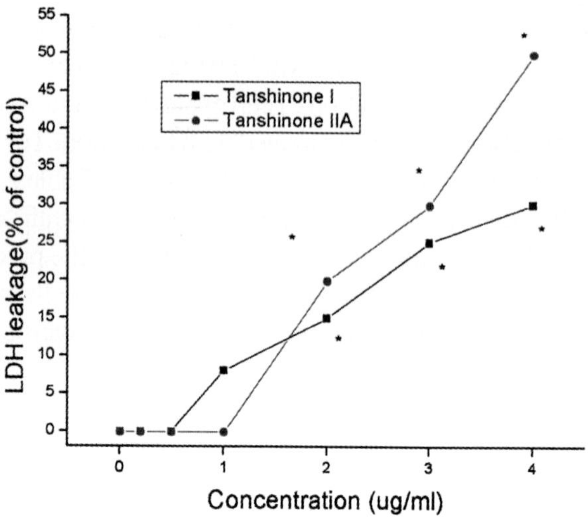

Figure 3. Dose-dependent cytotoxic effect of tanshinone I or tanshinone IIA on K1735M2 cells.
Cytotoxicity was determined by LDH leakage after treatment with various concentrations of tanshinone I or tanshinone IIA for 24 h.
* $P<0.05$ compared with that of control.

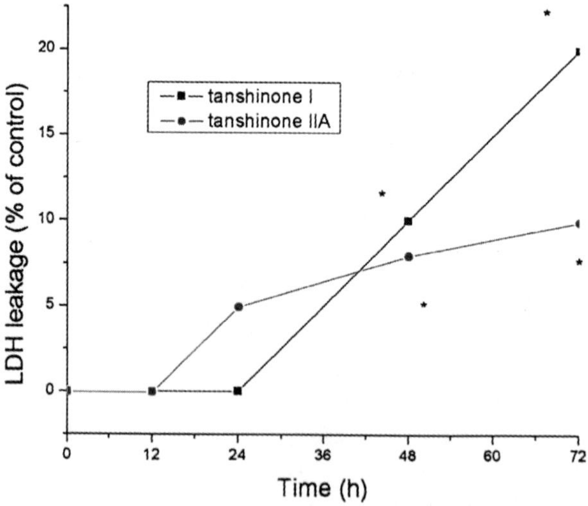

Figure 4. Time-dependent cytotoxic effect of tanshinone I or tanshinone IIA on K1735M2 cells.
Cytotoxicity was determined by LDH leakage after treatment with 1.5μg/ml tanshinone I or 0.5μg/ml tanshinone IIA for different time periods.
* $P<0.05$ compared with that of control.

Table 1. Cell cycle analysis of K1735M2 cells after treatment with 1.5μg/ml tanshinone I or 0.5μg/ml tanshinone IIA

Time (H)	Cell Cycle	Control	Tanshinone I	Tanshinone II A
12	G1	60.1±2.5	58.3±3.0	61.2±1.0
	S	22.4±1.6	25.4±2.0	27.1±1.3
	G2/M	17.5±2.0	16.2±1.5	11.2±0.5
24	G1	58.5±1.7	68.1±2.2	60.5±1.6
	S	20.7±1.1	24.6±0.8	23.8±1.7
	G2/M	20.1±2.0	7.5±0.3	15.5±2.1
36	G1	62.1±1.9	78.1±0.9	60.1±0.9
	S	20.8±1.2	15.1±0.4	25.6±0.8
	G2/M	17.0±0.5	16.6±1.1	14.3±1.0

Anti-proliferative effect has been one of the basic indexes in searching for new potential anti-tumor drugs from natural product or synthetic compound [42]. *Salvia miltiorrhiza* has been found to have anti-proliferative effect in several tumor cells *in vitro* or *in vivo* [19,30,35], but the mechanism is not determined. In our study, tanshinone I and tanshinone IIA inhibited proliferation of K1735M2 in a dose- and time-dependent manner which is in accord with the previous results [36]. Viability of cells treated with tanshinone at high concentration significantly decreased, demonstrating that growth inhibition was caused by the cytotoxic effect of tanshinone (Figures 3 and 4).

In order to confirm whether cell cycle arrest was involved in the anti-proliferative effect, K1735M2 cells treated with 0.5μg/ml tanshinone I or 1.5μg/ml tanshinone IIA were analyzed by flow cytometry. The results revealed G1 arrest in tanshinone I-treated cells. A previous report showed that growth arrest is associated with cyclin-dependent kinases which are a family of serine theronine protein kinases that regulate cell cycle distribution. Whether cyclin-dependent kinases are involved in the G1 arrest induced by tanshinone I need more research. Cells treated with tanshinone IIA did not show cell cycle arrest, so the inhibition may result from other signal pathway. Accumulating evidence indicates that chemotherapeutic agents induce tumor regression through inhibition of proliferation and/or activation of apoptosis. Tanshinone I inhibited the growth of K1735M2 significantly and induced cell cycle arrest. Whether the inhibitory effect resulted from apoptosis or other mechanism(s) needs more study.

In the present study, both tanshinone I and tanshinone IIA showed significant inhibitory effects on the growth of K1735M2 murine melanoma cells. Cells treated with tanshinone I showed cell cycle arrest at G1 phase, but the cell cycle was not affected in tanshinone IIA-treated cell. In order to search for more novel potential anti-tumor agents from natural products, more research should be done to elucidate the mechanisms by which this effect is caused. This is currently being conducted in our group.

REFERENCES

[1] Zhou W, Ruigrok T. Protective effects of Dan Shen during myocardial ischemia and reperfusion: an isolated rat heart study. Am J Chin Med 1990; 18: 19-24

[2] Tian G, Zhang Y, Zhang T, Yang F, Ito Y. Separation of tanshinones from *Salvia miltiorrhiz*a Bunge by high-speed counter-current chromatography using stepwise elution. J Chromatogr A 2000; 904: 107-11

[3] Cheng YY, Fong SM, Chang HM. Protective action of *Salvia miltiorrhiza* aqueous extract on chemically induced acute myocardial ischemia in rats. Zhong Xi Yi Jie He Za Zhi 1990; 10: 609-11, 582

[4] Chen YL, Yang SP, Shiao MS, Chen JW, Lin SJ. *Salvia miltiorrhiza* inhibits intimal hyperplasia and monocyte chemotactic protein-1 expression after balloon injury in cholesterol-fed rabbits. J Cell Biochem 2001; 83: 484-93

[5] Fung KP, Zeng LH, Wu J, Wong HN, Lee CM, Hon PM, Chang HM, Wu TW. Demonstration of the myocardial salvage effect of lithospermic acid B isolated from the aqueous extract of *Salvia miltiorrhiza*. Life Sci 1993; 52: 239-44

[6] Huang YT, Lee TY, Lin HC, Chou TY, Yang YY, Hong CY. Hemodynamic effects of *Salvia miltiorrhiza* on cirrhotic rats. Can J Physiol Pharmacol 2001; 79: 566-72

[7] Ji XY, Tan BKH, Zhu YZ. *Salvia miltiorrhiza* and ischemic diseases. Acta Pharmacol Sin 2000; 21: 1089-94

[8] Lei XL, Chiou GC. Cardiovascular pharmacology of *Panax notoginseng* (Burk) F.H. Chen and *Salvia miltiorrhiza*. Am J Chin Med 1986; 14: 145-52

[9] Lo CJ, Lin JG, Kuo JS, Chiang SY, Chen SC, Liao ET, Hsieh CL. Effect of *Salvia miltiorrhiza* Bunge on cerebral infarct in ischemia-reperfusion injured rats. Am J Chin Med 2003; 31: 191-200

[10] Maki T, Kawahara Y, Tanonaka K, Yagi A, Takeo S. Effects of tanshinone VI on the hypertrophy of cardiac myocytes and fibrosis of cardiac fibroblasts of neonatal rats. Planta Med 2002; 68: 1103-7

[11] Sugiyama A, Zhu BM, Takahara A, Satoh Y, Hashimoto K. Cardiac effects

of *Salvia miltiorrhiza*/dalbergia odorifera mixture, an intravenously applicable Chinese medicine widely used for patients with ischemic heart disease in China. Circ J 2002; 66: 182-4

[12] Lee TY, Mai LM, Wang GJ, Chiu JH, Lin YL, Lin HC. Protective mechanism of *Salvia miltiorrhiza* on carbon tetrachloride-induced acute hepatotoxicity in rats. J Pharmacol Sci 2003; 91: 202-10

[13] Lee TY, Wang GJ, Chiu JH, Lin HC. Long-term administration of *Salvia miltiorrhiza* ameliorates carbon tetrachloride-induced hepatic fibrosis in rats. J Pharm Pharmacol 2003; 55: 1561-8

[14] Nan JX, Park EJ, Kang HC, Park PH, Kim JY, Sohn DH. Anti-fibrotic effects of a hot-water extract from *Salvia miltiorrhiza* roots on liver fibrosis induced by biliary obstruction in rats. J Pharm Pharmacol 2001; 53: 197-204

[15] Wasser S, Ho JM, Ang HK, Tan CE. *Salvia miltiorrhiza* reduces experimentally-induced hepatic fibrosis in rats. J Hepatol 1998; 29: 760-71

[16] Lee DS, Lee SH, Noh JG, Hong SD. Antibacterial activities of cryptotanshinone and dihydrotanshinone I from a medicinal herb, *Salvia miltiorrhiza* Bunge. Biosci Biotechnol Biochem 1999; 63: 2236-9

[17] Cao CM, Xia Q, Zhang X, Xu WH, Jiang HD, Chen JZ. *Salvia miltiorrhiza* attenuates the changes in contraction and intracellular calcium induced by anoxia and reoxygenation in rat cardiomyocytes. Life Sci 2003; 72: 2451-63

[18] Cao EH, Liu XQ, Wang JJ, Xu NF. Effect of natural antioxidant tanshinone II-A on DNA damage by lipid peroxidation in liver cells. Free Radic Biol Med 1996; 20: 801-6

[19] Liu J, Yang CF, Wasser S, Shen HM, Tan CE, Ong CN. Protection of *Salvia miltiorrhiza* against aflatoxin-B1-induced hepatocarcinogenesis in Fischer 344 rats: dual mechanisms involved. Life Sci 2001; 69: 309-26

[20] Choi JS, Kang HS, Jung HA, Jung JH, Kang SS. A new cyclic phenyllactamide from *Salvia miltiorrhiza*. Fitoterapia 2001; 72: 30-4

[21] Kang DG, Oh H, Sohn EJ, Hur TY, Lee KC, Kim KJ, Kim TY, Lee HS. Lithospermic acid B isolated from *Salvia miltiorrhiza* ameliorates ischemia/reperfusion-induced renal injury in rats. Life Sci 2004; 75: 1801-16

[22] Kim SY, Moon TC, Chang HW, Son KH, Kang SS, Kim HP. Effects of tanshinone I isolated from *Salvia miltiorrhiza* Bunge on arachidonic acid metabolism and *in vivo* inflammatory responses. Phytother Res 2002; 16: 616-20

[23] Lin HC, Ding HY, Chang WL. Two new fatty diterpenoids from *Salvia miltiorrhiza*. J Nat Prod 2001; 64: 648-50

[24] Liu J, Yang CF, Lee BL, Shen HM, Ang SG, Ong CN. Effect of *Salvia miltiorrhiza* on aflatoxin B1-induced oxidative stress in cultured rat hepatocytes. Free Radic Res 1999; 31: 559-68

[25] Niu XL, Ichimori K, Yang X, Hirota Y, Hoshiai K, Li M, Nakazawa H. Tanshinone II-A inhibits low density lipoprotein oxidation *in vitro*. Free Radic Res 2000; 33: 305-12

[26] Onitsuka M, Fujiu M, Shinma N, Maruyama HB. New platelet aggregation

inhibitors from Tan-Shen; radix of *Salvia miltiorrhiza* Bunge. Chem Pharm Bull (Tokyo) 1983; 31: 1670-5

[27] Wang N, Luo HW, Niwa M, Ji J. A new platelet aggregation inhibitor from Salvia miltiorrhiza. Planta Med 1989; 55: 390-1

[28] Liu J, Shen HM, Ong CN. Role of intracellular thiol depletion, mitochondrial dysfunction and reactive oxygen species in *Salvia miltiorrhiza* -induced apoptosis in human hepatoma HepG2 cells. Life Sci 2001; 69: 1833-50

[29] Ji ZN, Liu GQ. Inhibition of serum deprivation-induced PC12 cell apoptosis by tanshinone II A. Acta Pharmacol Sin 2001; 22: 459-62

[30] Liu J, Shen HM, Ong CN. *Salvia miltiorrhiza* inhibits cell growth and induces apoptosis in human hepatoma HepG(2) cells. Cancer Lett 2000; 153: 85-93

[31] Mosaddik MA. In vitro cytotoxicity of tanshinones isolated from *Salvia miltiorrhiza* Bunge against P388 lymphocytic leukemia cells. Phytomedicine 2003; 10:682-5

[32] Sung HJ, Choi SM, Yoon Y, An KS. Tanshinone IIA, an ingredient of *Salvia miltiorrhiza* BUNGE, induces apoptosis in human leukemia cell lines through the activation of caspase-3. Exp Mol Med 1999; 31:174-8

[33] Wang X, Bastow KF, Sun CM, Lin YL, Yu HJ, Don MJ, Wu TS, Nakamura S, Lee KH. Antitumor Agents. Isolation, structure elucidation, total synthesis, and anti-breast cancer activity of neo-tanshinlactone from *Salvia miltiorrhiza*. J Med Chem 2004; 47: 5816-9

[34] Yoon Y, Kim YO, Jeon WK, Park HJ, Sung HJ. Tanshinone IIA isolated from *Salvia miltiorrhiza* BUNGE induced apoptosis in HL60 human premyelocytic leukemia cell line. J Ethnopharmacol 1999; 68: 121-7

[35] Ryu SY, Lee CO, Choi SU. *In vitro* cytotoxicity of tanshinones from *Salvia miltiorrhiza*. Planta Med 1997; 63: 339-42

[36] Wu WL, Chang HY, Chen CF. Cytotoxic activities of tanshinones against human carcinoma cell lines. American Journal of Chinese Medicine 1991; 19: 207-16

[37] Chang H, But P. Pharmacology and applications of Chinese Materia Medica. World Scientific Publishing: Singapore, 1986

[38] Abd-Elazem IS, Chen HS, Bates RB, Huang RC. Isolation of two highly potent and non-toxic inhibitors of human immunodeficiency virus type 1 (HIV-1) integrase from *Salvia miltiorrhiza*. Antiviral Res 2002; 55: 91-106

[39] Sorimachi K AK, Ikehara Y, Inafuku K, Okubo A. Secretion of TNF-a, IL-8 and nitric oxide by macrophages activated with *Agaricus blazei* Murill fractions *in vitro*. Cell Structure and Function 2001; 26: 103-8

[40] Ong ES, Len SM. Evaluation of pressurized liquid extraction and pressurized hot water extraction for tanshinone I and IIA in *Salvia miltiorrhiza* using LC and LC-ESI-MS. J Chromatogr Sci 2004; 42: 211-6

[41] Li JF, Wei YX, Xu ZC, Dong C, Shuang SM. Studies on the spectroscopic behaviuor of cryptotanshinone, tanshinone IIA, and tanshinone I. Spectrochim Acta A Mol Biomol Spectrosc 2004; 60: 751-6

[42] Chen HC, Sheng CK, Li J, Chung HL, Hsu SL. A phenylacetate derivative, SK6, inhibits cell proliferation via G1 cell arrest and apoptosis. European Journal of Pharmacology 2003; 467: 31-9

New Pesticidal Compounds from Limonoids

Trevor H Yee*
Natural Products Institute, University of the West Indies, Mona, Kingston 7, Jamaica
**Corresponding email: trevor.yee@uwimona.edu.jm*

INTRODUCTION

An impressive list of limonoids (tetranortriterpenes) and protoliminoids have been extracted from the related plant families of *Rutaceae* and *Meliaceae* at the University of the West Indies (UWI), Table 1. Very little investigations, however, have been done on the pesticidal potential of these compounds, although the azadirachtin (Figure 1a and 1b) which are well researched pesticidal compounds from *Meliaceae*, are members of this group of compounds, and the pesticidal properties of other liminoids have been reported [1-5]. They have been isolated and structurally elucidated and synthesized at UWI primarily for academic purposes.

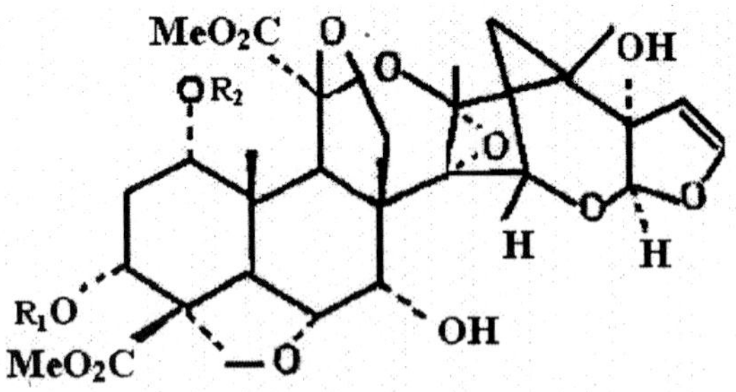

Figure 1. Structure of Azadirachtin

1a. R_1 = Ac , R_2 = Tiglo = azadarachtin a
1b. R_1 = Tiglo, R_2 = H = azadarachtin b

Table 1A. Some Limonoids and Protolimonoids Extracted from *Meliaceae*

Name of plant	Part	Extraction	Structure	Yield	Bioactivity	Ref
Cedrela odorata	Heartwood	Benzene by cold percolation	$C_{30}H_{46}O_4$ Methyl angolensate mp 205°C $[\alpha]_D$ -43.9° 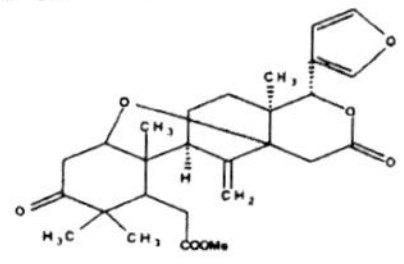	1%	Active against *Heliothis zea, Spadoptera fungiperda, Pactinophora gossypiella*	2,6
Cedrela odorata	Heartwood	Benzene by cold percolation	$C_{30}H_{46}O_4$ Gedunin mp 247-248°C $[\alpha]_D$ -45° 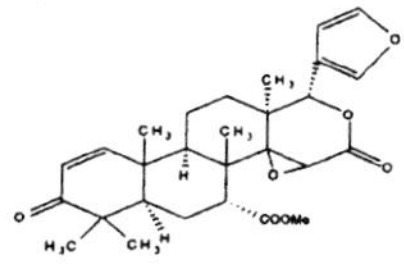		Active against a variety of insects	2,6
Cedrela odorata	Heartwood	Benzene by cold percolation	$C_{30}H_{50}O_4$ isoodoratol mp 246-248°C $[\alpha]_D$ -48.5° 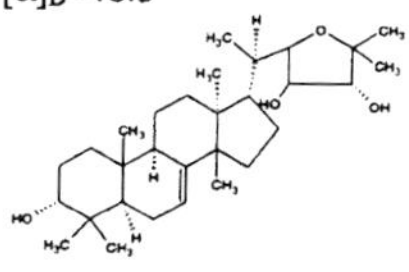			7
Cedrela odorata	Heartwood	Benzene by cold percolation	$C_{30}H_{48}O_4$ odoratone mp 236-238°C $[\alpha]_D$ -62° 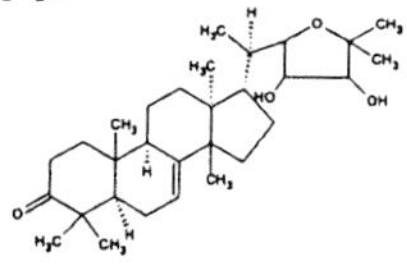			7
Cedrela odorata	Heartwood	Benzene by cold percolation	$C_{30}H_{50}O_4$ Odoratol 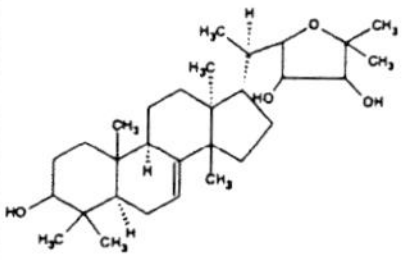			7

Trichilia havanensis	Fruits	Acetone by cold percolation	$C_{26}H_{34}O_4$ Trichilone mp 199-201°C $[\alpha]_D$ +17° 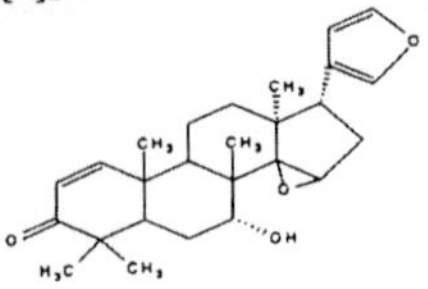			8
Trichilia havanensis	Fruits	Acetone by cold percolation	$C_{30}H_{42}O_7$ 1,7- diacetylhavanensin mp 185-187°C 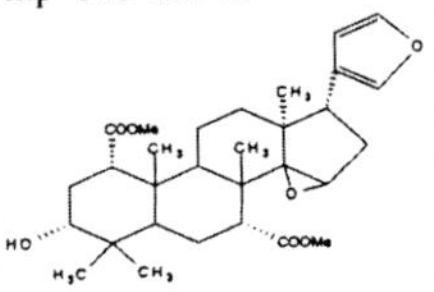			8
Trichilia havanensis	Fruits	Acetone by cold percolation	$C_{30}H_{42}O_7$ 3,7- diacetylhavanensin mp 185-187°C $[\alpha]_D$ -39° 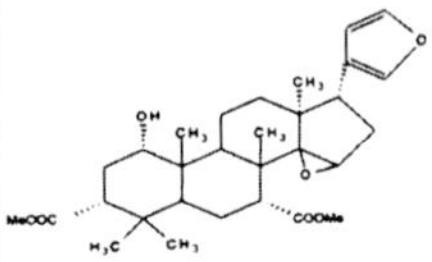			8
Trichilia havanensis	Fruits	Acetone by cold percolation	$C_{32}H_{44}O_8$ triacetylhavensin mp 188-189°C $[\alpha]_D$ -76° 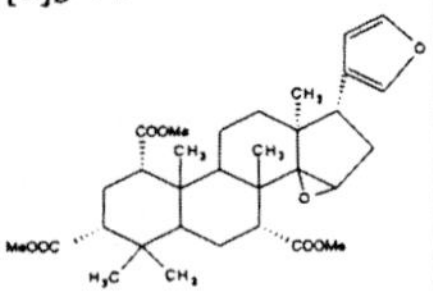			8
Trichilia havanensis	Fruits	Acetone by cold percolation	$C_{30}H_{42}O_7$ Havensin B mp 295-299°C 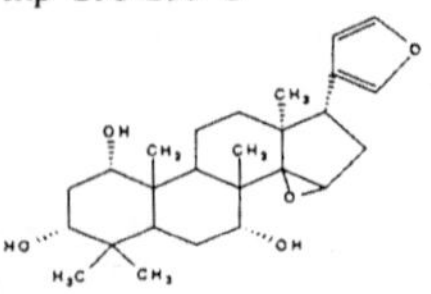			8

Entandrop-hragma cylindricum	Heartwood	Cold percolation with Benzene	$C_{30}H_{50}O_4$ m.p.216 – 217 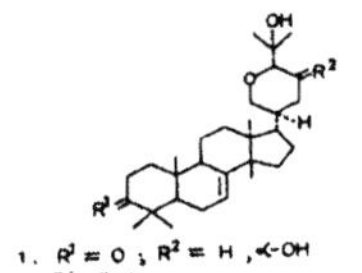 1. R¹ = O ; R² = H, α-OH Sapelin A			9
Entandroph-ragma cylindricum	Heartwood	Cold percolation with Benzene	$C_{30}H_{50}O_4$ m.p.173 -174 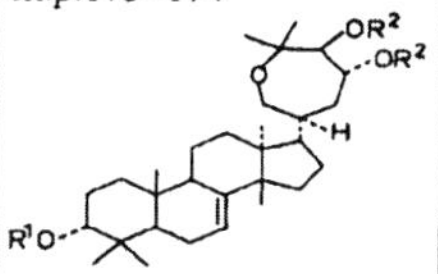 12 . R¹ = R² = H Sapelin B			9
Entangroph-ragma cylindricum	Heartwood	Cold percolation with Benzene	$C_{30}H_{46}O_5$ m.p.243 – 246 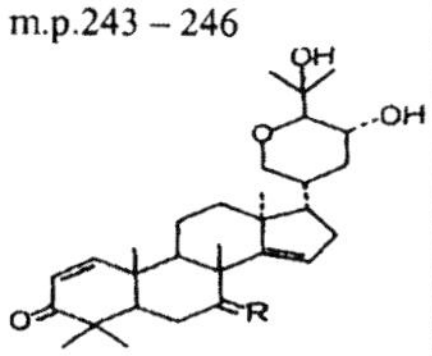 21. R = H, α-OH Sapelin C			10
Entandrop-hragma cylindricum	Heartwood	Cold percolation with Benzene	$C_{30}H_{50}O_5$ m.p 243 - 244. 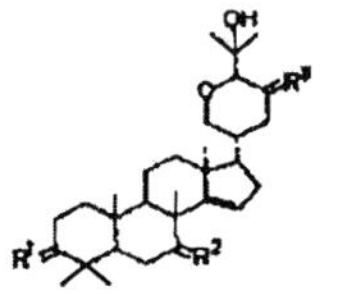 27. R¹ + R⁴ = R⁸ = H, α - OH Sapelin D			10
Entandro-phragma cylindricum	Heartwood	Cold percolation with Benzene	$C_{30}H_{46}O_5$ m.p. 239 – 241 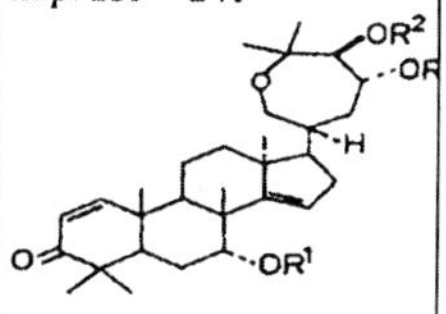 39 . R¹ = R² = H Sapelin E			10

Entandrophragma cylindricum	Heartwood	Cold percolation with Benzene	$C_{30}H_{52}O_5$ m.p. 210 – 212 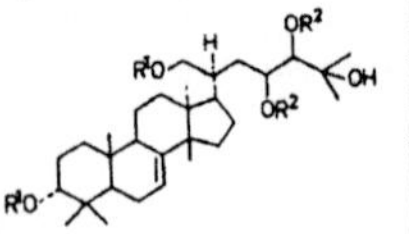 48 . $R^1 = R^2 = H$ Sapelin F			10
Trichilia hirta		Acetone by cold percolation	$C_{32}H_{36}O_{11}$ Hirtin 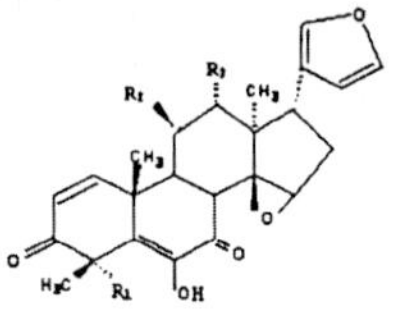 a. $R_1 = CO_2Me$, $R_2 = OAc$, $R_3 = CO_2Et$ = Hirtin b. $R_1 = CO_2Me$, $R_2 = OH$, $R_3 = CO_2Et$ = Deacetylhirtin		Anti-feedant and growth inihibition of *Peridroma saucia*	3,5, 11
Trichilia reticulata	Leaves and twigs	Hexane by cold percolation	$C_{32} H_{50} O_3$ 3β-acetoxy-cycloart-24-en-23-one mp 147-149°C $[\alpha]_D$ +17° 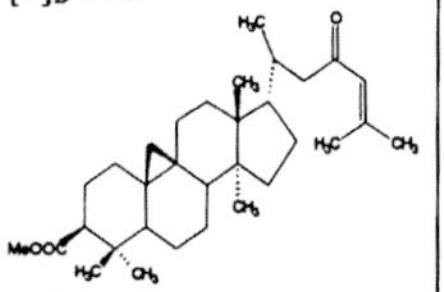	0.002%		12
Trichilia reticulata	Leaves and twigs	Hexane by cold percolation	$C_{30} H_{50} O_2$ Cycloart-23(Z)-en-3β,25-diol mp 197-199°C $[\alpha]_D$ +25° 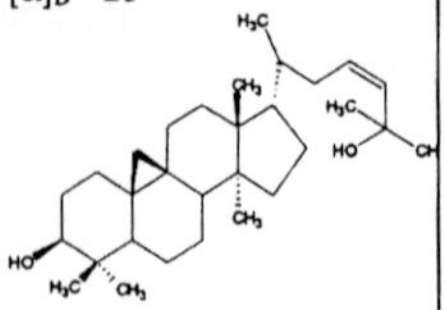	0.0003 %		12

Trichilia reticulata	Leaves and twigs	Acetone by cold percolation	$C_{30}H_{46}O_2$ Cycloart-24-en-3,23-dione mp 133-134°C $[\alpha]_D$ +19° 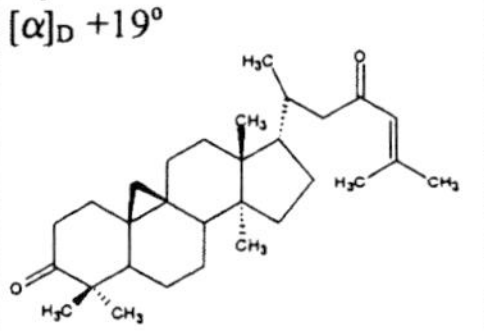	0.005%		12
Trichilia reticulata	Leaves and twigs	Hexane by cold percolation	$C_{30}H_{50}O_3$ Dihydroniloticin mp 168-169°C $[\alpha]_D$ -22° 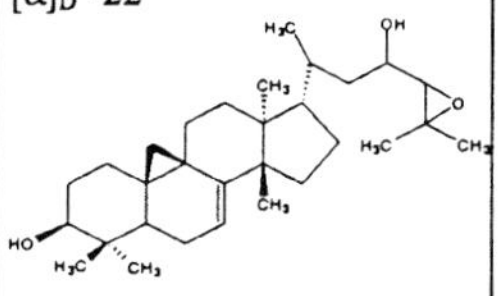	0.0002%	*Little activity against *Pseudomonas aureginosa*	12
Trichilia reticulata	Leaves and twigs	Acetone by cold percolation	$C_{30}H_{46}O_4$ Melianone $[\alpha]_D$+5° 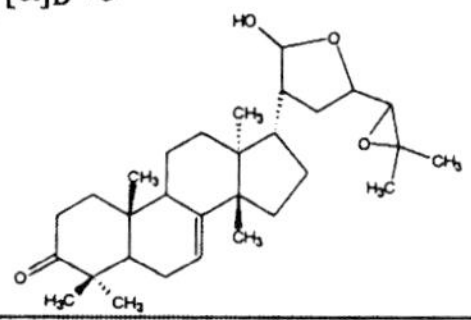	0.002%	*Activity against *Bacillus. subtilis, Candida ablicans, Staphhylococcus. aureus*	12
Trichilia reticulata	Leaves and twigs	Acetone by cold percolation	$C_{30}H_{48}O_5$ Melianodiol mp 210-215°C lit:219-222 °C 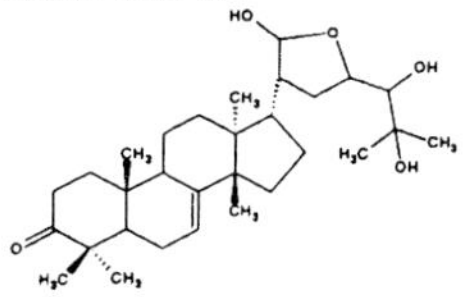	0.0005%	*Activity against *Bacillus subtilis* (more than melianone)	12
Guarea jamaicensis	Leaves and twigs	Acetone by cold percolation	$C_{36}H_{54}O_8$ Glabretal-21-acetate (oil) $[\alpha]_D$ -18° 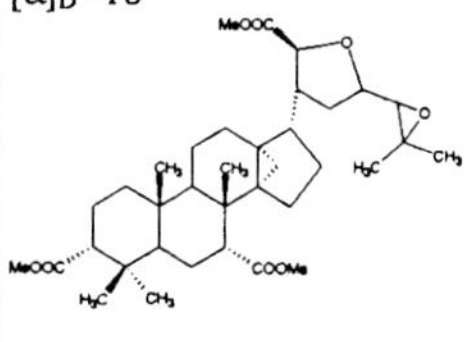	0.0001%		13

Guarea jamaicensis	Leaves and twigs	Ethanol by cold percolation	$C_{40}H_{60}O_{11}$ Compound N (oil) $[\alpha]_D$ -23° 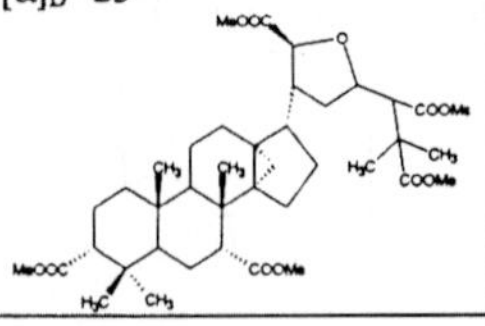	0.00004%		13
Guarea jamaicensis	Leaves and twigs	Ethanol by cold percolation	$C_{44}H_{66}O_{14}$ Compound O (oil) $[\alpha]_D$ +7° 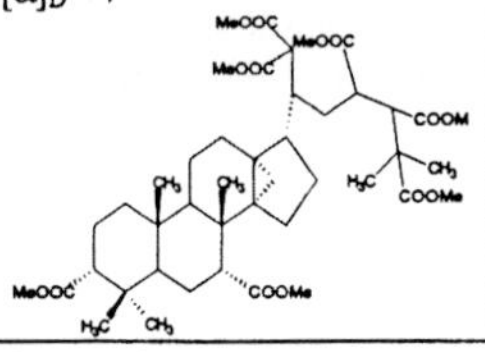	0.00004%		13
Guarea jamaicensis	Leaves and twigs	Ethanol by cold percolation	$C_{29}H_{48}O$ A-Spinasterol mp 168-169°C $[\alpha]_D$ -5° 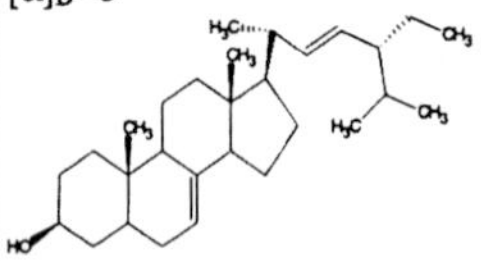	0.0002%	*No activity	13
Guarea jamaicensis	Leaves and twigs	Ethanol by cold percolation	$C_{45}H_{72}O_{10}$ Sitosteroyl-3-O-β-D-Glucopyranoside-2',3',4'-Triacetate-6'-O-Palmitate (oil) $[\alpha]_D$ -26° 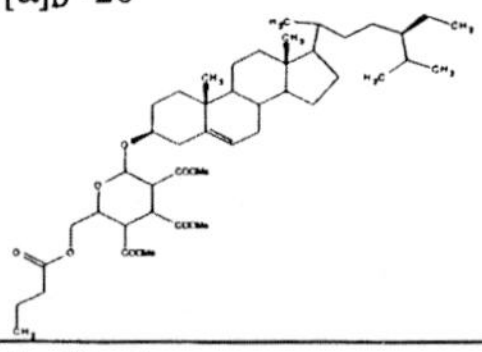	0.0001%		13
Swietenia macrophylla	Fruits	Acetone by cold percolation	$C_{38}H_{48}O_{14}$ Augustineolide (Oil) $[\alpha]_D$ -8.3° 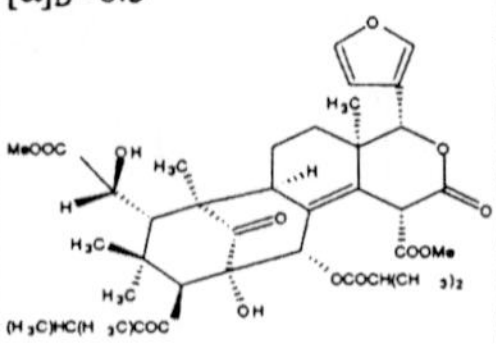	0.004%		14

Swietenia macroph-ylla	Fruits	Acetone by cold percolation	$C_{27}H_{34}O_8$ 3β,6-Dihydroxydihydrocarapin $[\alpha]_D$ 39.2° 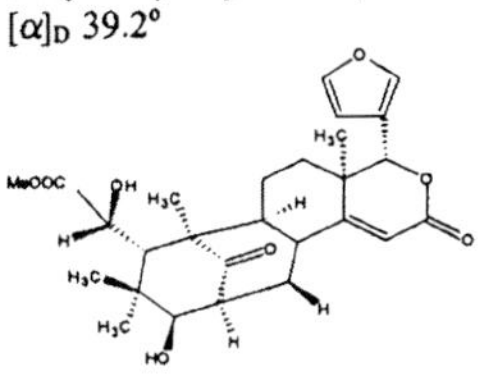	0.004%		14
Swietenia macroph-ylla	Fruits	Acetone by cold percolation	Swietenolide mp 173-174°C 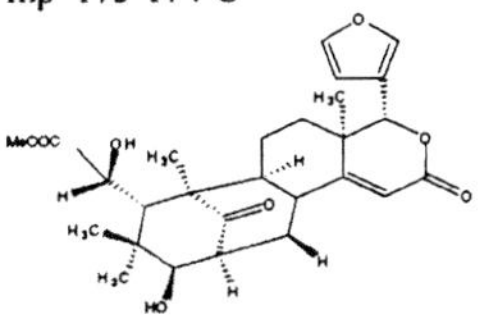		+	14
Swietenia macroph-ylla	Fruits	Acetone by cold percolation	6-O-acetylswietenolide mp 262-264°C 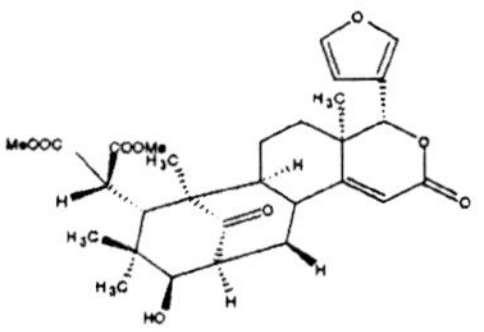	0.009%	+	14
Swietenia macroph-ylla	Fruits	Acetone by cold percolation	3,6-O,O-diacetylswietenolide mp 223-225°C 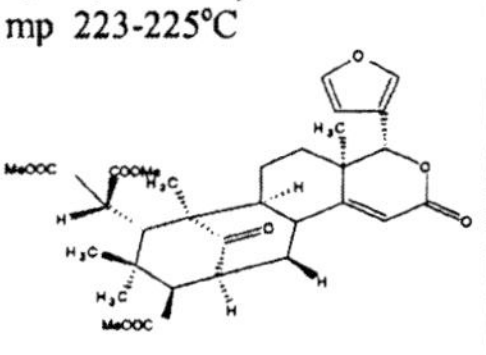	0.020%	+	14
Swietenia macroph-ylla	Fruits	Acetone by cold percolation	Swietemahonin F mp 279-281°C 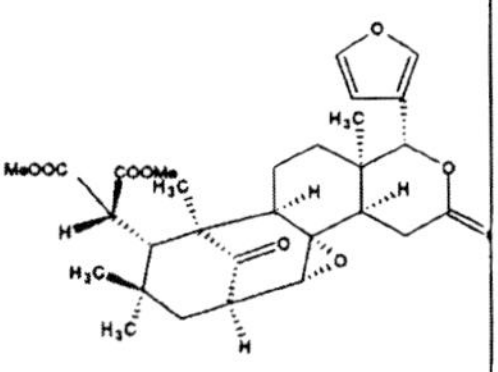	0.033%	+	14

Ruagen glabra	Fruits	Benzene by cold percolation	Xylocarpin $[\alpha]_D$ -79.8° 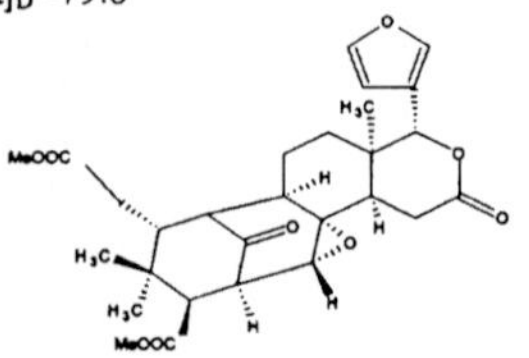	0.008%	+	15
Ruagea glabra	Fruits	Benzene by cold percolation	$C_{31}H_{40}O_9$ Ruageanin A $[\alpha]_D$ -63.9° 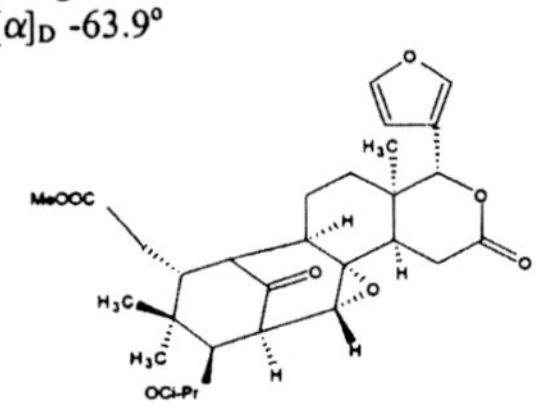	0.01%	+	15
Ruagea glabra	Fruits	Benzene by cold percolation	$C_{32}H_{40}O_{10}$ Ruageanin B mp 227-229°C $[\alpha]_D$ -17.7° 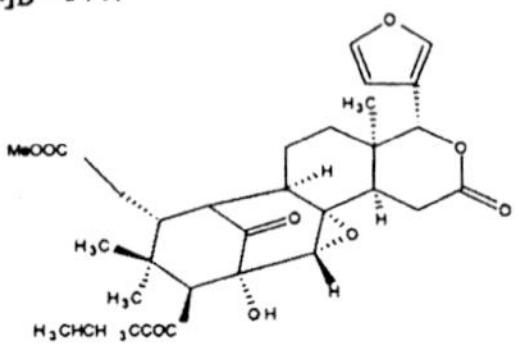	0.03%	+	15
Ruagea glabra	Fruits	Benzene by cold percolation	$C_{29}H_{36}O_{10}$ Ruageanin C mp 217.5-219°C $[\alpha]_D$ -15.9° 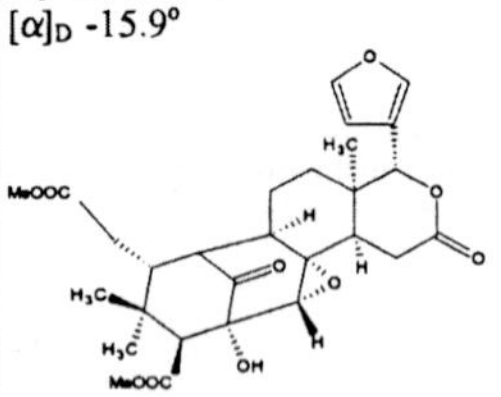	0.06%		15

Ruagea glabra	Fruits	Benzene by cold percolation	$C_{29}H_{36}O_9$ Ruageanin D $[\alpha]_D$ -15.9° 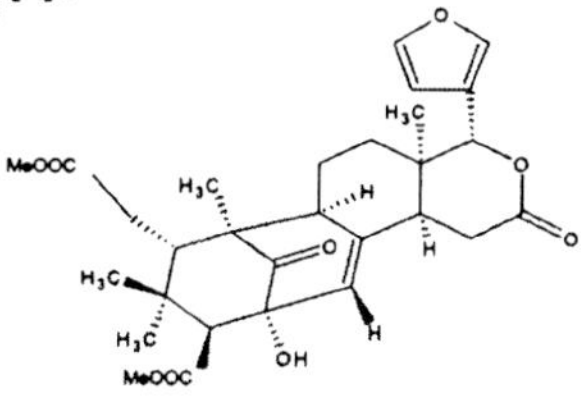	0.004%		15
Cedrela mexicana	Fruit	Benzene by cold percolation	$C_{32}H_{40}O_{12}$ 7α,11β-diacetoxydihydronomilin mp.262-264°C 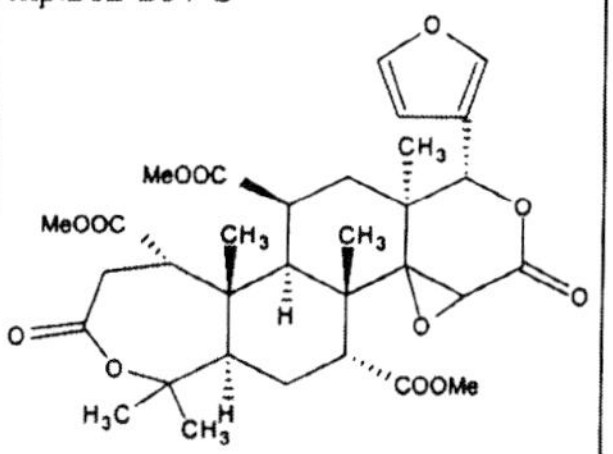			16

Table 1B. Some Limonoids and Protolimonoids Extracted from *Rutaceae*

Name of plant	Part	Extract	Struc. Formula etc	Yield	Bioactivity	Ref
Spathelia glabrescens	Bark	Hexane by cold percolation	$C_{27}H_{32}O_9$ Deacetylspathelin mp 106-108°C $[\alpha]_D$ -10.7° 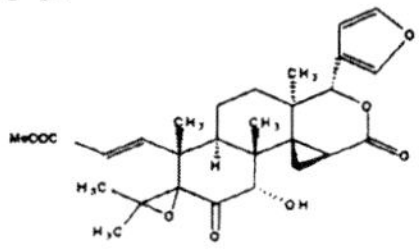	0.004 %	Little activity against E. coli and has a mortality rate of 14.5% after 2hrs. against the mosquito *Aedes aegypti* L.	17
Spathelia coccinia	Heartwood	Hexane by cold percolation	$C_{29}H_{34}O_{10}$ spathelin mp 181-182°C lit. 178.5-180 $[\alpha]_D$ -8° lit: -7° 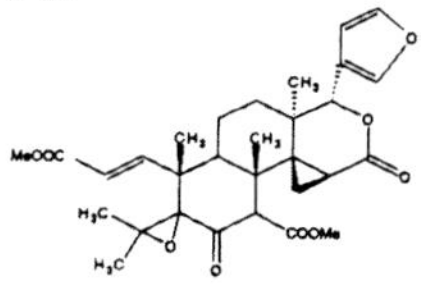	0.0002 %		18

PESTICIDAL POTENTIAL

Limonoids and Protoliminoids

There is a considerable research to be done in the evaluation of the activity of these compounds for pesticidal and other biologically active properties. Some early study into the activity has started at the St. Augustine campus of the UWI, and these results are shown in Table 2. The compounds tested so far have shown very limited anti-bacterial activity but the early work has shown anti-feedant activity against the moth *Spodoptera frugiperda* (Figure 2), which is a pest for several crops such as the crucifirus vegetables, the West Indian vegetable Calaloo, *Amaranthus sp.,* and cotton, etc.

Figure 2. *Spodoptera frugiperda*

* Bioactivity studies were conducted against A*spergillus niger, Bacillus subtilis, Escherichia coli, Candida ablicans, Stophylococcus aureus, Staphylococcus cholesteramis, Pseudomonas aureginosa and Pseudomonas seringinae* and the mosquito *Aedes aegypti*. Studies were also done on *Panagrellus redivirus and Caenorhabditis elegans* however, compounds used were found inactive even up to concentrations of 100ppm.

Table 2. Results from bioassay conducted against final instar larvae of *Spodoptera frugiperda*

Compounds	AI[a] ± SEM
swietenolide	94.1 ± 2.90
6-O-acetylswietenolide	72.2 ± 19.60
3,6-O,O-diacetylswietenolide	72.0 ± 19.60
Swietemahonin	70.2 ± 8.60
Methylangolensate	66.4 ± 10.63
Xylocarpin	77.8 ± 6.90
Ruageanin A	72.6 ± 19.60
Ruageanin B	86.3 ± 6.41

+Antifeedant bioassays were conducted against the final instar larvae of *Spodoptera frugiperda* at concentrations 1000 ppm. The compounds used are listed in the table above.

Quassinoids

The Natural Products Institute is engaged in further evaluations of the pesticidal properties of the above and other compounds. In investigating anti-feedant activities, there is an even greater scope for research. The quassinoid group of compounds from another related plant family, *Simaroubaceae*, are exceedingly bitter, with a bitter index of 1:60,000 and quoted as high as 1:250,000 [19, 20]. These compounds are known for their pesticidal activity, especially anti-feedant properties, and the investigation of this group provides additional scope for research.

In general, the quassinoids can be viewed as derivatives of liminoids, via a pathway known as the "merolimonol rearrangement" [21], which occurs by base hydrolysis of liminoids with a lactone in ring D and oxygenation at the C-7 position in ring B. The mechanism can be depicted as outlined in Figure 3 below. It can be seen from the structure of quassin (Figure 4), that this is a likely pathway for the quassinoids group of compounds, which have been isolated from the related plant family, Simaroubaceae.

An example of such a "merolimonol rearrangement" product, synthesized in the course of the attempted structural elucidation of the limonoid utilin but uninvestigated for pesticidal properties is the merolimonol rearrangement compound [22], synthesized in the process described in Figure 5.

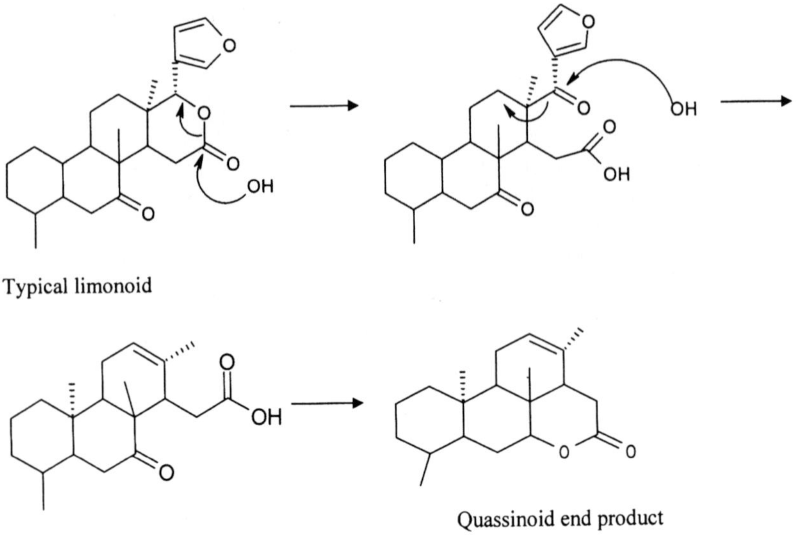

Figure 3. Depiction of the Merolimonol rearrangement of limonoids.

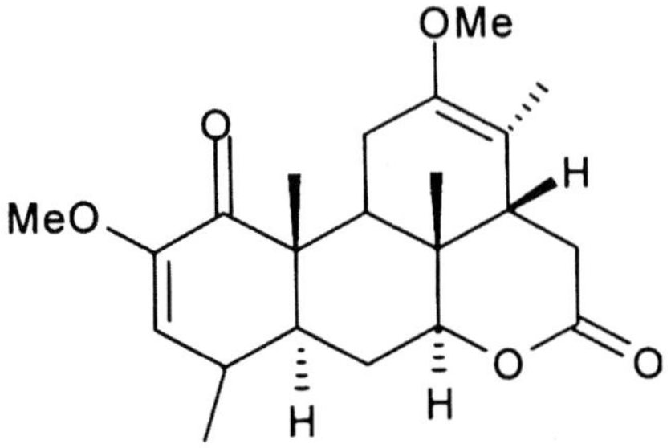

Figure 4. Quassin

Rotenoids

There is even further scope for research into the pesticidal potential of compounds that have been isolated or synthesized at the U.W.I. Rotenoids and their analogues, the naturally occurring rotenonoids, and the synthetic b-rotenonoids, are the area of synthesis of one group at the UWI, on an purely academic basis [23-25], with very little investigations as yet on their pesticidal properties (Figures 6-8). The b-rotenonoids are known to rearrange under alkaline conditions to give rotenonoids.

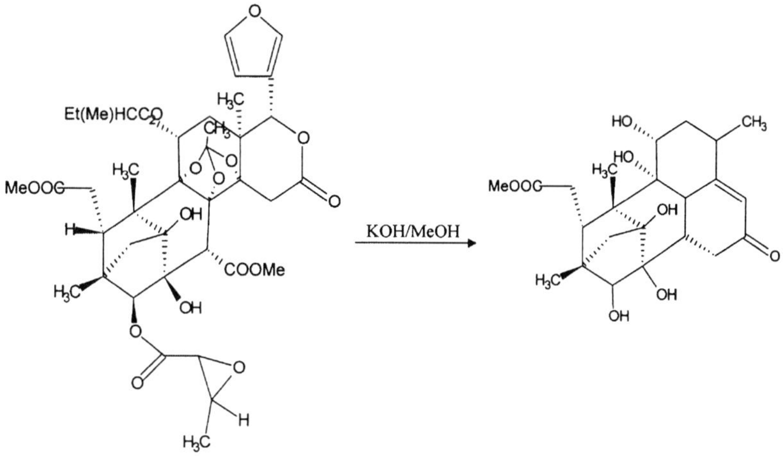

Utilin

Uninvestigated quassinoid from the merolimonol rearrangement of utilin.

Figure 5. Formation of Merolimonol rearrangement compound.

REFERENCES

[1] Nakatani M, James JC, Nakanishi K Isolation and Structures of Trichilins, Antifeedants against the Southern Army Worm. J. Am. Chem. Soc 1981; 103: 1228–30.

[2] Champagne DE, Koul O, Isman MB, Scudder GGE, Towers GHN. Biological Activity of Limonoids from the Rutales. Phytochemistry 1992; 31: 377-94

[3] Xie YS, Isman MB, Gunning P, MacKinnon S, Amason JT, Taylor DR, Sanchez P, Hasbun C, Towers GHN. Biological activity of *Trichilia* species and the limonoid hirtin against Lepidopteran larvae. Biochem Syst. Ecol 1994; 22: 129-36

[4] Zhou J-B, Okamura H, Iwagawa T, Nakatani M. Limonoid Antifeedants from *Melia toosendan*. Phytochemistry 1996; 35: 117-20

[5] Simmonds SJ, Stevenson PC, Porter EA, Veitch NC, Insect Antofeedant Activity of Three Tetranortriterpenes from *Trichilia pallida*. J. Nat. Prod. 2001; 64: 1117-20

[6] Mootoo, B.S., Ph. D. Thesis. Studies on Tropical Hardwood Extracts. 1965. University of the West Indies, Mona, Kingston 7, Jamaica

[7] Chan WR., Holder NL, Taylor DR, Snatzke G, Felhaber HW. Extractives of *Cedrela odorata* L. Part II. The Structures of the *Cedrela* tatracyclic

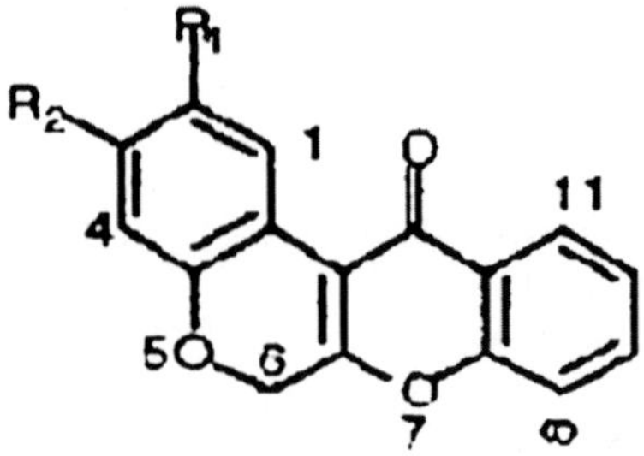

Figure 6. Typical rotenoid

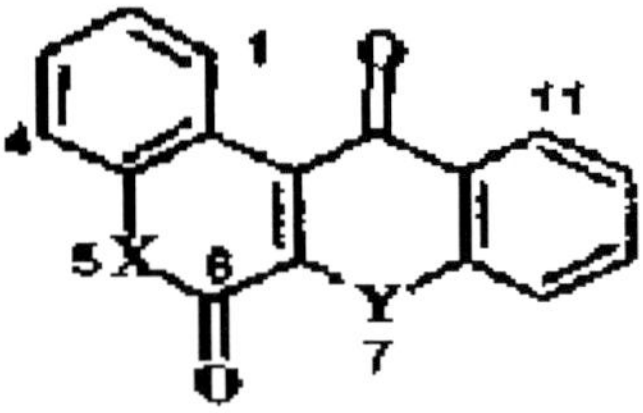

X = Y = O

Figure 7. Typical rotenonoid

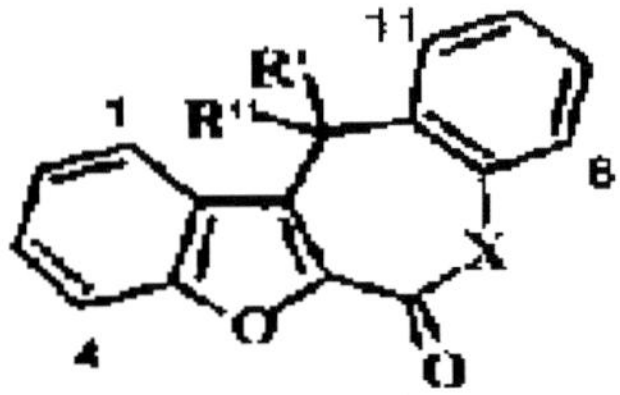

X = O; R',R'' = O

Figure 8. Typical b-rotenonoid

Triterpenes, Odoratol, Iso-odoratol, and Odoratone. J. Chem Soc. (C) 1968: 2485

[8] Chan WR, Gibbs JA, Taylor DR. Triterpenoids from *Trichilia havenensis.* Jacq. Part I. The Acetates of Havenensin and Trichilenone, New Tetracarbocyclic Tetranortriterpenes. J. Chem. Soc. Perkins Transaction. 1. 1973: 1047-50

[9] Chan WR, Taylor DR, Yee T. Triterpenoids from *Entandrophragma cylindricum*. Sprague. Part I. Structures of Sapelins A and B. J. Chem. Soc (C) 1970: 311 – 4

[10] Chan WR, Taylor D.R, Yee, T.Triterpenoids from Entandrophragma cylindricum. Sprague. Part II. The Structures of Sapelins C,D,E and F. J.Chem. Soc. (C) 1971: 2662 – 7

[11] Chan W, Taylor DR. Hirtin and Deacetylhirtin: New "Limonoids" from *Trichilia hirta*. J. Chem Soc. Chem. Commun. 1966: 206-7

[12] Harding WW, Jacobs H, Lewis PA, McLean S, Reynolds WF. Cycloatranes, Protoliminoids, a Pregnane and a new Ergostane from *Trichilia reticulata*. Natural Product Letters. 2001; 15: 253 – 60

[13] Harding WW, Jacobs H, McLean S, Reynolds WF. Identification and Complete ^{1}H and ^{13}C Spectral Assignments for the terpene fern-9(11)-en-28-oic acid. Magnetic Resonance in Chemistry. 2001; 39: 719 – 22

[14] Mootoo BS, Ali A, Motilal R, Pingal R, Ramlal A, Khan A, Reynolds WF, McLean S. Limonoids from *Swetenia macrophylla* and *S. anbrevilleana*. J. of Natural Products. 1999; 62: 1514

[15] Mootoo BS, Ramsewak R. Tetranortriterpenoids from *Ruagea glabra*. J. of Natural Products. 1996; 59: 544

[16] Marcelle GB, Mootoo BS. 7a,11b-Diacetoxydihydronomilin New Tetranortriterpene from *Cedrela mexicana*. Terahedron Leters. 1981; 22: 505

[17] Harding WW, Simpson DS, Jacobs H, McLean S, Reynolds .F. A new Squalene- derived epoxy tri-THF diol from *Spathelia glabrescens*. Tetrahedron Letters. 2001; 42: 7379 – 81

[18] Jacobs, H. personal communication.

[19] The Merck Index, Ninth Edition. 1976. Merck & Co. Inc., Rahway, New Jersey, USA, Pp. 1042

[20] The Code of Federal Regulations, U.S. National Archives and Records Aministration. 2004. Pp. 499

[21] Melera A, Schaffner K, Arigoni D, Jeger O. Zur Konstitution des Limonin I. Uber den verlauf der alkalischen Hydrolyse von Limonin und Limonol. Helv. Chim. Acta. 1957; 40: 1430

[22] Yee TH, Ph. D. Thesis. Triterpenoids from *Entanmdrophragma cylindricum* Sprague. 1971. The University of the West Indies, Mona, Kingston 7, Jamaica

[23] Jackson Y.A. Improved Synthesis of Esculetin. Heterocycles. 1995; 41: 1979 – 85

[24] Marriot KC, Anderson M, Jackson YA. Synthesis of a 2,3-dimethoxyrotenoid. Heterocycles. 2001; 55: 91 – 7

[25] Jackson YA, Marriot KC. Synthesis of 2,3-Dimethoxy-7-methyl-7,12-dihydro-6H-[1]-benzofuro-[2,3-C]-[1]-benzezepin-6,12-dione. Molecules. 2002; 7: 353 – 62

Insecticidal Properties of *Anacardium Occidentale* L.

Shamima A Parveen[a], Masaru Hashimoto[b], Nurul Islam[a], Toshikatsu Okuno[b] and M Khalequzzaman[a]

[a]*Department of Zoology, University of Rajshahi, Rajshahi-6205, Bangladesh*

[b]*Faculty of Agriculture and Life Science, Hirosaki University, 3 Bunkyo-cho, Hirosaki, Aomori 036-8561, Japan*

INTRODUCTION

The cashew tree, *A. occidentale* L. is a botanical species that is native to eastern Brazil. It was introduced into other tropical countries such as India, Africa, Indonesia and South East Asia in the 16th century [1]. Its stems exude a clear gum, Cashew gum, used in pharmaceuticals and as a substitute for Gum arabic [2, 3]. The principle function of the secondary chemicals in the Family Anacardiaceae is probably to serve as a defense against vertebrate and insect herbivores. Contact with the poisonous members of the Anacardiaceae usually causes a cell-mediated dermatitis. Per 100 g, the mature seed is reported to contain 561 calories, 5.2 g H_2O, 17.2 g protein, 45.7 g fat, 29.3 g total carbohydrate, 1.4 g fiber, 2.6 g ash, 38 mg Ca, 373 mg P, 3.8 mg Fe, 15 mg Na, 464 mg K, 60 mg β-carotene equivalent, 0.43 mg thiamine, 0.25 mg riboflavin, and 1.8 mg niacin. The apple contains 87.9% water, 0.2% protein, 0.1% fat, 11.6% carbohydrate, 0.2% ash, 0.01% Ca, 0.01% P, 0.002% Fe, 0.26% vitamin C, and 0.09% carotene [4]. The fruit-shell contains β-catechin, β-sitosterol, and *1*-epicatechin; also proanthocyanadin, leucocyanadin, and leucopelargodonidine. The dark color of the nut is due to an iron-polyphenol complex. The shell oil contains about 90% anacardic acid ($C_{22}H_{32}O_3$ and 10% cardol ($C_{32}H_{27}O_4$) mentioned by Deszcz and Kozubek [5]. It yields glycerides, linoleic, palmitic, stearic and lignoceric acids and a ksitosterol. Gum exudates contain arabinose, galactose, rhamnose and xylose.

Crude cashew nut shell liquid represents one of the major and cheapest sources of naturally occurring non-isoprenoid phenolic lipids, such as anacardic acids, cardols, cardanols, methylcardols and polymeric materials [6]. The fruit-shell juice and the nut oil are both said to be folk remedies for calluses, corns and warts, cancerous ulcers and even elephantiasis. Lans (1992) [7] reported that the stem-bark of *A. occidentale* is used to control diarrhea in ruminants and other pet animals. Anacardol and anacardic acid have shown some activity against Walker carcinosarcoma 256. Decoction of the astringent bark is given for severe diarrhea and thrush. Old leaves are applied to skin afflictions and burns [8]. Oily substance from pericarp is used for cracks on the feet. Cuna Indians used the bark in herb teas for asthma, colds and congestion [9-13]. The seed oil is believed to be alexeritic and amoebicidal and is used to treat gingivitis, malaria, and syphilitic ulcers [14]. This extentively studied plant was subjected to further investigation to evaluate its pesticidal potential.

MATERIALS AND METHODS

In this investigation, different parts of *A. occidentale,* viz. nut-shell, root-bark and stem-bark have been collected for screening to trace the presence of toxic, as well as bio-active constituents since the plant is well-known as a medicinal plant and also considered to contain toxic constituents.

Collection of plant materials

For the extraction, fresh materials of *A. occidentale* were collected from the botanical garden of Rajshahi University Campus. Fresh leaves, nut-shells, root-bark and stem-bark of *A. occidentale* were powdered and extracted separately by chloroform and methanol, successively. The powdered materials were weighed and placed in separate conical flasks with chloroform ($CHCl_3$, Merck, Germany) [500g × 1500ml × 3 filtrations with Whatman filter paper at 24 h interval in the same collection flask] to yield the first extracts (for leaves, nut-shell, root-bark and stem-bark) separately. After filtration extraction by $CHCl_3$ to collect aglycones (or the components from the inter-cellular space) was completed, the same amount of methanol (MeOH, Merck, Germany) was added in the same way to extract glycosides from leaves, root-bark and stem-bark till extraction of all possible components was done. The extracts were then transferred to a round bottom flask connected to a vacuum rotary evaporator. The output extracts were removed into glass vials and preserved in a refrigerator at 4°C with proper labeling.

Selection of test organisms

To screen for insecticidal properties of the extracts of different parts of *A. occidentale,* a test organism *T. castaneum* (Hbst.) was selected, because it is an easily cultivable and noble laboratory animal. The beetles were originally received from the Crop Protection Department of the University of Newcastle upon Tyne, U.K. and were reared in the Crop Protection and Toxicology Laboratory, Department of Zoology, University of Rajshahi, Bangladesh. Whole-wheat flour was used as the food medium for the insects. The flour was sterilized at 60°C for 24 hours in an oven. A standard mixture of whole-wheat flour with powdered dry yeast in a ratio of 19:1 was used as food medium throughout the experimental period [15-17]. Both the flour and the powdered dry yeast were sterilized at 60°C for six hours in an oven. Food was not used until at least 15 days after sterilization to allow its moisture content to equilibrate with the environment.

About 500 beetles were placed in a 500ml beaker containing food medium. The beaker was covered with a piece of cloth and kept in an incubator at 30°C ± 0.5°C. At regular intervals the eggs were collected by sieving the food medium with two sieves having 250 and 500 μ aperture separating the adults and eggs, respectively following the method of Khan and Selman (1981) [18]. Eggs were then transferred to a petri dish (90mm in diameter) with food and incubated at the same temperature. After 3-5 days, larvae were hatched. Newly hatched larvae (1 mm long) were then collected with a fine pointed camel hairbrush and shifted to fresh food medium for culture. The larvae are yellowish white in color and cylindrical in shape; they become 6-7mm long at maturation.

Most larvae had six instars as reported by Good [19]. The larval instars were determined by counting the number of exuvae (larval skin) deposited in the food medium according to [19]. Two days-old larvae were considered as first instar larvae while second, third, fourth, and fifth instar larvae were considered on fourth, seventh, tenth and thirteenth day after hatching, respectively. Depending on these days (according to larval instars), sixteen day-old larvae have been considered as matured larvae. Larval cultures were maintained in an incubator in the same procedure at

30°C ± 0.5°C without light and humidity control. A huge number of flour beetles were thus reared to get a regular supply of the newly formed adults. When sufficient adults were produced in the sub-cultures, they were collected from the food medium. For this purpose, some pieces of filter papers were kept inside the beaker on the food. Adults crawling on the paper were taken out with the help of forceps and the beetles were collected in a small beaker (100 ml) with the help of a camel hairbrush. Adults were 2.3 to 4.4 mm in length with no external sexual differentiation. The body is flattened with a broad and wide pronotum.

Test fungus as a tool for activity guided fractionation

As per the present search for pesticidal properties from *A. occidentale,* 8 crude extracts of different parts, viz. leaves, nut-shell, root-bark and stem-bark extracted in $CHCl_3$ and MeOH have been screened for antifungal activity against a plant pathogenic fungus, *Valsa ceratosperma* (Tode ex Fries). Initially it was cultured in the Plant Protection Division of the Aomori Green BioCenter (AGBC), Aomori, Japan to collect spores and brought to the Crop Protection Laboratory of the Department of Zoology, Rajshahi University. This fungus is a suitable organism for screening purposes; bioautographic assays on TLC with these were successful since most of the anti-*Valsa* compounds were found to be active against the test insect, *T. castaneum*.

Isolation and purification of bioactive compounds

In this investigation, isolation of the pure compounds from different parts viz. nut-shell, root-bark and stem-bark of *A. occidentale* was done mainly by open column chromatography, while thin layer chromatography (TLC) was used as a supporting tool. Several compounds were isolated and purified successfully; four were obtained in sufficient quantity to conduct characterization and bioassay. They were named NS-1, NS-2, RB-1 and SB-2 (Figures 1-4). The pathway of isolation is given in Figures 5-7.

Bioassay with the pure compounds

The whole work was based on activity-guided fractionation for the isolation and purification of biologically active secondary metabolites. Bioassay with the pure compounds was essential for a complete project and that was done accordingly. Four compounds, NS-1, NS-2, RB-1 and SB-2, were applied against the test insect, *T. castaneum*, by topical application, as described earlier. Test for repellency effect was also done in a similar manner.

RESULTS

All the four pure compounds NS-1, NS-2, RB-1 and SB-2 from the $CHCl_3$ extract of the nut-shells, root-bark and stem-bark of the test plant were tested and the findings are shown in Table 1. The LD_{50} values

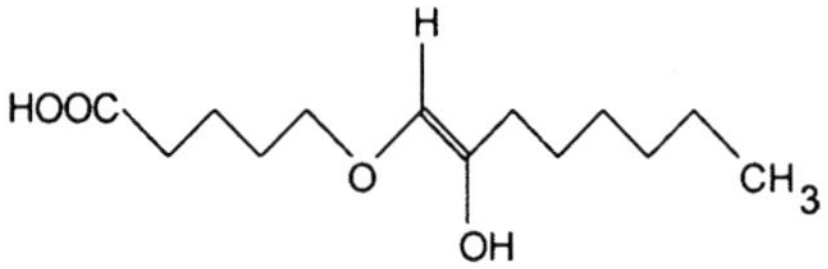

Figure 1. Compound NS-1 (α-linoleic acid)

The spectrum of NS-1 shows characteristics of an unsaturated fatty acid compared with the spectrum of NS-2; one significant difference is the lack of aromatic signals, which suggests it represents a free unsaturated fatty acid.

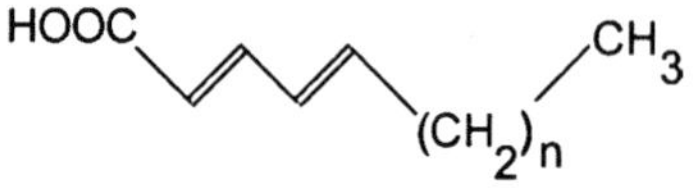

Figure 2. Structure of Compound NS-2 (anacardic acid)

Apart from the same solvent residual and TMS, the rest of the signals of the spectrum can be divided into four distinctive groups: **a**- below 1 ppm, only one signal at 0.85 ppm. The signal is a triplet, which means a methyl group attached to a methylene chain (CH_3-CH_2-...); **b**- from 1ppm to 3 ppm, all the methylene signals (-CH_2-) in the molecule, which either appear as distinctive signals, or superposed strong signals at 1.3 ppm, according to the different chemical environment of the individual protons); **c**- from 4.8 ppm to 6 ppm, signals in this region indicate olenific protons); **d**- 6.5 ppm to 7.5 ppm, where aromatic protons appear. It can be suggested from these spectral characters that the compound has two structural parts. One is an aromatic ring with certain substitution pattern; another is an aliphatic side chain with certain degree of unsaturation. The aromatic region shows the signal of three protons, the signal at 7.28 ppm is a triplet (split by two protons), while the signals at 6.82 ppm and 6.74 ppm are both doublets (split by one adjacent proton), indicating the aromatic ring is a tri-substituted ring, with three free aromatic protons adjacent to one another. This can translate into a 2-hydroxybenzoic acid unit with a side chain substitution at position 6, which agrees with the aromatic part of the structure of anacardic acid. Examination of the rest of the signals also reveals that they match those of anacardic acid. Therefore, compound NS-2 was identified as anacardic acid.

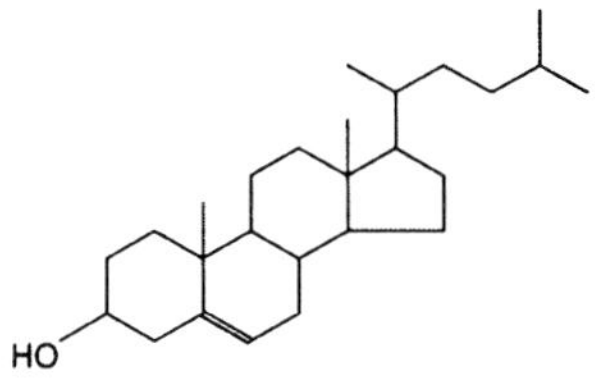

Figure 3. Structure of Compound RB-1 (β-sitosterol)

Compound RB-1 appears as white needle crystals with no UV absorption. The bright purple spot appearing with Godin reagents after the development in TLC plate suggests it to be a terpenoid or steroid. From the 1H NMR spectrum, firstly the solvent residue peak at ca. 7.27 ppm (from the residual H of $CDCl_3$) and reference peak at 0 ppm from TMS were identified. The true signals gave the following information: It has the general characteristics of a sterol. Aliphatic methylene and methine proton signals were second to and partially superposed with six methyl proton signals. Two of the methyl signals were singlets, at 1.01 ppm (Me-19) and 0.68 ppm (Me-18), which means no vicinal carbon has a proton attached (quaternary carbon). The left four methyl signals were doublets, which means they are coupled with and split by one proton carried on the vicinal carbons. The signals of one olenific proton at 5.36 ppm and one proton attached to an oxygenated carbon at 3.52 ppm were also observed. These characteristic data were in perfect match with reported values of β-sitosterol by Yu *et al.* (1989) [24].

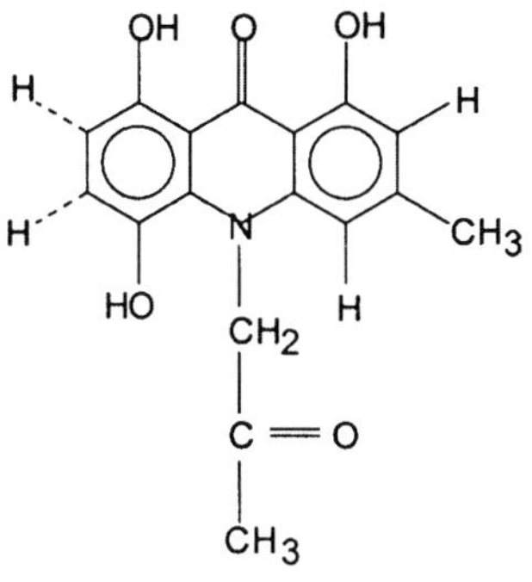

Figure 4. Structure of Compound SB-2 (unnamed)

Compound SB-2 shows the following ^{1}H-NMR spectral data: ^{1}H-NMR (400MHz, $CDCl_3$) δ: 7.65 (1H, bs, ArH), 7.35 (1H, d, J=4.0 Hz, ArH), 7.07 (1H, bs, ArH), 6.67 (1H, d, J=4.0 Hz, ArH), 3.95 (3H, S, OAc), 3.48 (2H, S, N-CH2), 2.45 (3H, S, $ArCH_3$). From the above ^{1}H-NMR spectral data of the compound SB-2, two aromatic signals appearing at δ 7.35 and 6.67 with a J value of 4.0 Hz are assigned to be vicinal aromatic protons. Moreover, broad singlets or ortho, para aromatic protons appeared at δ 7.07 and 7.65 of another aromatic ring. Methyl protons of an acetyl group appeared at δ 3.95, equivalent to three protons. Another three proton singlets appeared at δ 2.45 and can be assigned as aromatic methyl group protons. A methylene proton appearing at δ 3.48 as singlet can be assigned as N-CH_2 proton with a neighbouring OAc group.

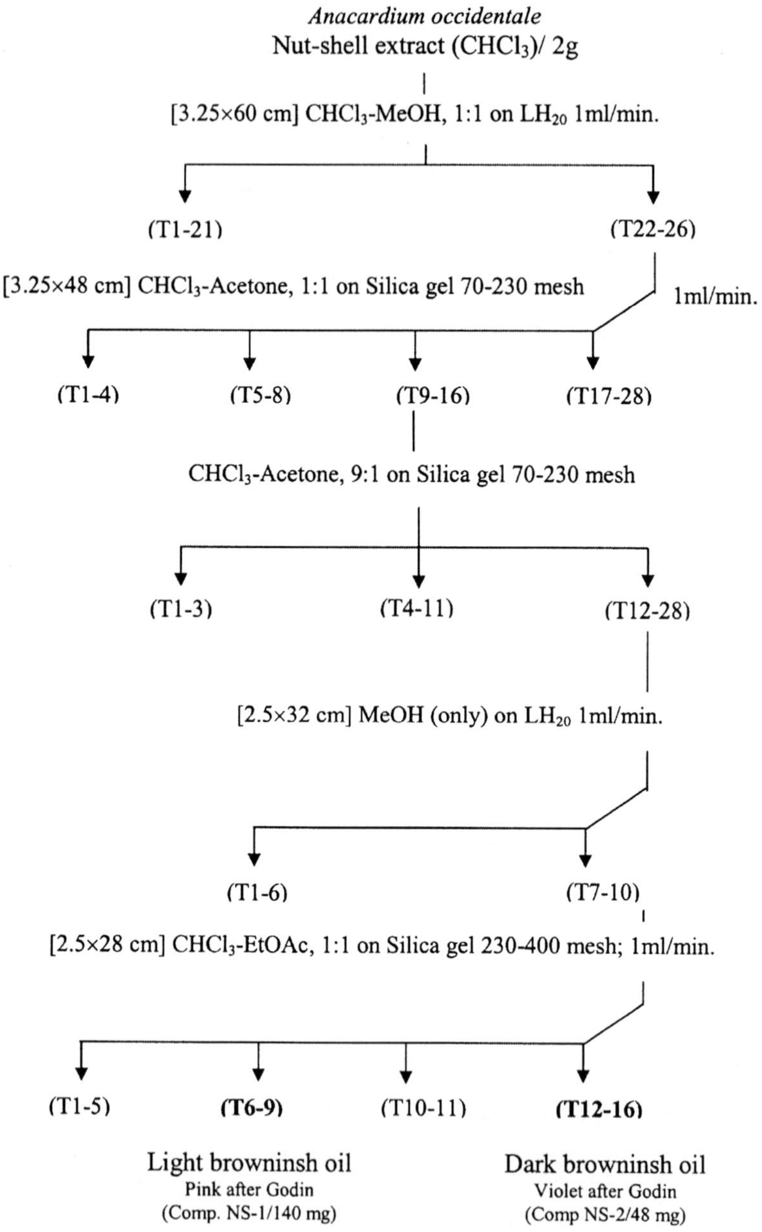

Figure 5. Flow chart of nut-shell compounds NS-1 and NS-2

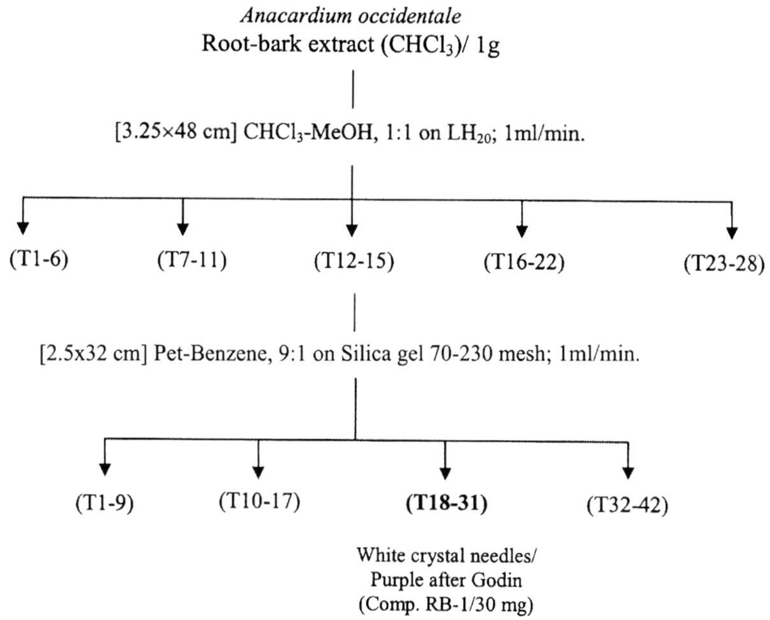

Figure 6. Flow chart of root-bark compound RB-1

were 8.918, 11.940, 13.727 and 13.318 μg/insect for 24 h of treatment for the compounds in the order given above, respectively, and 8.038, 10.545, 11.291 and 11.513 μg/insect for 48 h of exposure. The regression equations were established as Y= 3.146 + 1.950x; Y= 3.220 + 1.652x; Y= 3.458 + 1.355x; Y= 3.308 + 1.505x; Y= 3.260 + 1.921x Y= 3.300 + 1.661x; Y= 3.693 + 1.248x and Y= 3.494 + 1.420x respectively.

The results were satisfactory except RB-2, which was not possible to be applied because it was obtained in low quantity. It is necessary to mention that SB-1 was similar to RB-1 through its physical characteristics (white crystal needles, purple after Godin reagent spray [21] and R_f). Subsequent NMR spectra study confirmed that both of them are the same. The LD_{50} values found in this investigation indicate potentiality in the order of NS-1>NS-2>RB-1>SB-2.

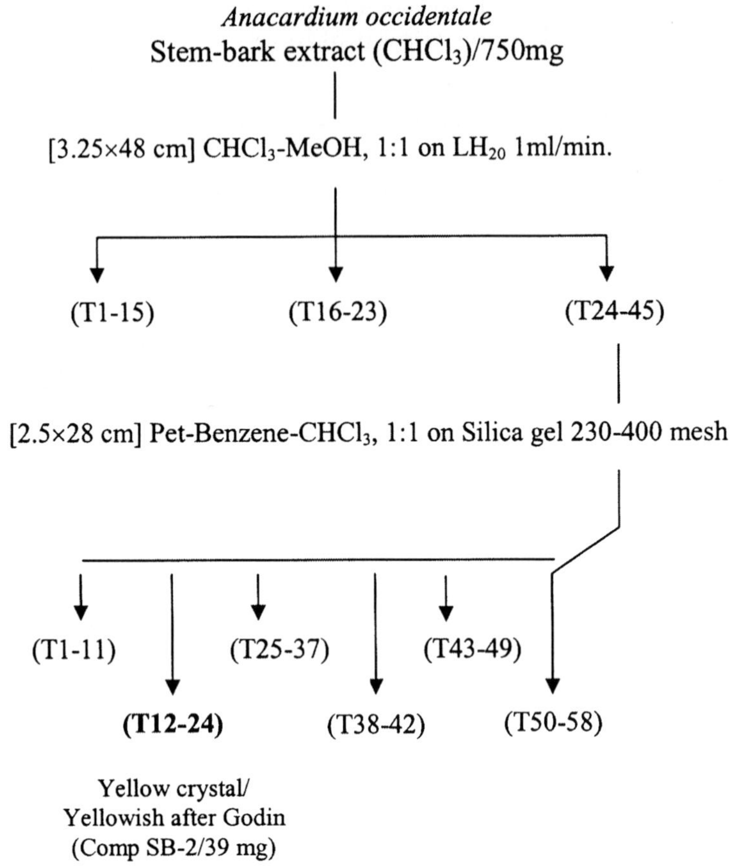

Figure 7. Flow chart of stem-bark compound SB-2

DISCUSSION

The insecticidal potential of *A. occidentale* has been well-known. Scientific explanations are now available to support this. The present investigations prove the efficacy of its constituents against the stored grain pest, *Tribolium castaneum* adults. Bioassays by previous workers using hexanolic extracts of cashew nut-shells against snails showed promising effect, with lethal concentration, LC_{90}, of sample NS-1 being 2.0 to 2.2 ppm for adult snails [11]. *A. occidentale* has good antibacterial activity

Table 1. LD_{50} values of pure compounds of *A. occidentale* against *T. castaneum* adults

Source	Compound	Regression equations		LD_{50} values μg/insect	
		After 24 hrs	After 48 hrs	24 hrs	48 hrs
Nut-shell	NS-1	Y= 3.146 + 1.950x	Y= 3.260 + 1.921x	8.918	8.232
Nut-shell	NS-2	Y= 3.220 + 1.653x	Y= 3.300 + 1.662x	11.940	10.546
Root-bark	RB-1	Y= 3.458 + 1.355x	Y= 3.693 + 1.248x	13.727	11.291
Stem-bark	SB-2	Y= 3.308 + 1.505x	Y= 3.494 + 1.420x	13.318	11.513

against *Escherichia coli* and *Pseudomonas aeruginosa* which are gram-negative bacteria [22]. The nut-shell oil of the test plant afforded mixtures of anacardic acid and cardols in varying degrees of unsaturation. Anacardic acid has been found to be effective at 10 ppm [23] and nut-shell liquid was found to be effective [24] against molluscs.

REFERENCES

[1] Mitchell JD, Mori SA. *The Cashew and Its Relatives (Anacardium: Anacardiaceae)*. New York Botanical Garden, Bronx; New York, 1987, 78 pp. ISBN 0-89327-313-9

[2] Bose S, Biswas M. *Anacardium occidentale* gum. I. Immunochemical and hydrolytic studies. Indian J Biochem 1970; 7: 68-72

[3] da Silveira NLR, Rabelo LJ, Ribeiro De Salis C, de Azevedo MR. Cashew-tree (*Anacardium occidentale* L.) exudate gum: a novel bioligand tool. Biotechnol Appl Biochem 2002; 35: 45-53

[4] Duke JA. *Handbook of Medicinal Herbs*. Florida, USA, CRC Press Inc., 1985

[5] 5. Deszcz L, Kozubek A. Higher cardol homologs (5-alkylresorcinols) in rye seedlings. Biochim et Biophys Acta 2000; 1483: 241-50

[6] Santos ML dos, Magalhães GC de. Utilisation of Cashew Nut Shell Liquid from *Anacardium occidentale* as Starting Material for Organic Synthesis:

A Novel Route to Lasiodiplodin from Cardols. J Braz Chem Soc 1999; 10: 13-20
[7] Lans C. Creole remedies: case studies of ethnoveterinary medicine in Trinidad and Tobago. Wageningen University, 2003, Dissertation no. 2992
[8] George J, Kuttan R. Mutagenic, carcinogenic and cocarcinogenic activity of cashew nut shell liquid. Cancer Lett 1997; 112: 11-6
[9] Ippen H. Contact allergy to Anacardiaceae. A review and case reports of poison ivy allergy in central Europe. Derm Beruf Umwelt 1983; 31: 140-8
[10] Fernandes L, Mesquita AM. *Anacardium occidentale* (cashew) pollen allergy in patients with allergic bronchial asthma. J Allergy Clin Immunol 1995; 95: 501-4
[11] Menezes EA, Tome ER, Nunes RN, Nunes AP, Freire, CC, Torres JC, Castro FM, Croce J. Extracts of *Anacardium occidentale* (cashew) pollen in patients with allergic bronchial asthma. J Investig Allergol Clin Immunol 2002; 12: 25-8
[12] Teuber SS, Sathe SK, Peterson WR, Roux KH. Characterization of the soluble allergenic proteins of cashew nut (*Anacardium occidentale* L.). J Agric Food Chem 2002; 50: 6543-9
[13] Wang F, Robotham JM, Teuber SS, Sathe SK, Roux KH. Ana o 2, a major cashew (*Anacardium occidentale* L.) nut allergen of the legumin family. Int Arch Allergy Immunol 2003; 132: 27-39
[14] Mota ML, Thomas G, Barbosa Filho JM. Anti-inflammatory actions of tannins isolated from the bark of *Anacardium occidentale* L. J Ethnopharmacol 1985; 13: 289-300
[15] Park T, Frank MB. The fecundity and development of the flour beetles, *Tribolium castaneum* and *Tribolium confusum* at three constant temperatures. Ecology 1948; 29: 368-75
[16] Park T. Beetles, competition and population. *Science* 1962; 138: 1369-75
[17] Zyromska-Rudzka H. Abundance and emigration of *Tribolium* in a laboratory model. Ecol Pole 1966; 14: 491-518
[18] Khan AR, Selman BJ. Some techniques for minimizing the difficulties I egg counting *Tribolium castaneum* (Herbst). Entomol Rec J Var 1981; 93: 36-7
[19] Good, N.E. The flour beetles of the genus *Tribolium*. USDA Technical Bulletin 1936; 5:27–8
[20] Yu, DQ, Yang JS, Xie JX. Spectroscopy of nuclear magnetic resonance. In Handbook of Analytical Chemistry. Chemical Industry Press, Beijing, 1989; 5: 832-4
[21] Godin P. A new spray reagent for paper chromatography of polyols and cetoses. Nature (London) 1954; 174: 134
[22] Kudi AC, Umoh JU, Eduvie LO, Gefu J. Screening of some Nigerian medicinal plants for antibacterial activity. J Ethnopharmacol 1999; 67: 225-8
[23] Consoli RA, Mendes NM, Pereira JP, Santos B de S, Lamounier MA. Effect of several extracts derived from plants on the survival of larvae of *Aedes*

fluviatilis (Lutz) (Diptera: *Culicidae*) in the laboratory. Mem Inst Oswaldo Cruz 1988; 83: 87-93

[24] Laurens A, Belot J, Delorme C. Molluscicidal activity of *Anacardium occidentale* L. (Anacardiaceae). Ann Pharm Fr 1987; 45: 471-3

Antibacterial Effect of Extracts from Persimmon Leaves

Li-Lian Ji[b,*], Qiang-Hua Zhang[a], Xu Gu[a] and Meng Wang[b]

[a]*School of Life Sciences, Huaiyin Institute of Technology, China.*
[b]*Institute of Functional Biomolecules, School of Life Sciences, Nanjing University, China*
**Corresponding email: jll2863@vip.sina.com*

INTRODUCTION

Persimmon leaf, a traditional Chinese medicine belongs to *Diospyrous kaki*, contains various nutrient and bioactive components, including flavones and flavonoids, tannins, phenolic compounds, coumarins, organic acid, carotent, vitamin C and essential amino acids [1, 2, 3, 4, 5]. As a traditional Chinese medicine, persimmon leaf is for the treatment of hypertension, various types of hemorrhage, acute and chronic inflammation, virus infection, chyluria and macula lutea [6]. Some nutrient compounds, including vitamin C, amino acid and carotine in persimmon leaf contribute to cosmetic function for softening skin and keeping skin beautiful [7]. Therefore, persimmon leaf has been using as a healthy material for functional drinks and cosmetics in China and other countries for recent 30 years. Moreover, persimmon leaf also contains antioxidant compounds and can be added to food products and cosmetics to extend their shelf life [8, 9]. However, little has been reported about antibacterial activity of persimmon leaf on food spoilage and food-borne pathogens.

The aim of this study was to investigate antibacterial effect of extract from persimmon leaves (EPL) on eight food spoilage and food-borne pathogens, including Gram-positive and Gram-negative species, and to verify whether and what extent the antibacterial activity is affected by environmental factors. Finally, the possibility for applying the EPL in orange juice was evaluated in order to assess the EPL suitability as an alternative food preservative and/or functional factor.

MATERIALS AND METHODS

Chemicals

Ethylacetate, methanol, ethanol, hydrochloric acid, and sodium hydrate were purchased from Chemical and Medicinal Company (Huaian, China). All organic solvents were of HPLC grade.

Preparation of Persimmon Leaves Extracts

Mature persimmon leaves were collected in Jiangning, China. Dry samples were obtained after the leaves were air-dried and powdered. The dry sample (100 g each) was extracted with an equal amount of ethylacetate, methanol, 75% ethanol (w/v) or water respectively so extracts from persimmon leaves (EPL) were obtained. For the EPL with organic solvent, 100 g of the persimmon leaf powder was refuxed in a water bath at 60°C for 3h. After filtration through Whatman No. 1 filter paper, the residue was re-refuxed with an additional volume of organic solvent and then filtered. The combined filtrate was then concentrated in vacuum to dryness, and then weighed. For the EPL with water solvent, persimmon leaf powder was blended with water and then boiling at 100°C for 30min, and the following operations as above with organic solvent were carried out. All EPLs were finally made up to 10ml with 0.1% Tween 80 aqueous solution in order to be tested [10, 11].

Bacterial Strains

Eight food spoilage and food-borne bacterial strains were obtained from Jiangsu Fermentation Research Institute (JSFRI) (China). They were Gram-positive species, including *Staphylococcus aureus, Bacillus subtilis, Bacillus cereus* and *Streptococcus faecalis*, and Gram-negative species, including *Escherichia coli*, *Salmonella typhimurium*, *Proteus mirabilis*, and *Pseudomonas aeruginosa*. Those strains were culture with Luria-Bertani (LB) medium on neutral plate count agar (PCA) or in broth (PCB) at 37°C [12, 13, 14].

Antibacterial Activity Assay

Antibacterial activity was tested according to agar diffusion method (whatman No.1 filter paper, 6mm in diameter). Briefly, sterile filter paper was impregnated with 10¼1 of various EPL solution respectively, and placed in the center of the agar plate, on which the test bacteria was uniformly inoculated by transferring inoculum to solidified agar plates and spreading the plates with an L-shaped glass rod. After 20min, the plate was inverted and incubated at 37°C for 24hr. The diameter of inhibitory zone shown on plate was measured using calipers and expressed in millimeters. Potassium sorbate was used as a positive control and 0.1% Tween 80 was as a negative control [10, 11].

MIC and MBC Determination

The minimum inhibitory concentration (MIC) and minimum bactericide concentration (MBC) of the EPL were determined according to broth dilution method [13]. Briefly, the selected bacteria were incubated for 24h for proliferation in LB media without added the EPL. The EPL solutions were added to sterilized tubes containing LB to reach final concentrations of 50%, 25%, 12.5%, 6.25%, 3.13%, 1.57%, 0.79%, 0.4% (EPL/ medium, w/v). Strain cultures in LB were diluted to reach a final concentration in the test tube of 10^6 colony-forming units (cfu g^{-1}). The tubes were incubated at 37°C and observed after 24h for the presence of tubidity. The MIC value was evaluated as the lowest concentration that completely inhibited the formation of visible growth. Twenty microliters of media from all of the tubes in which no growth was observed were streaked on appropriate agar plate respectively which were incubated at 37°C and checked after 24h for the presence of colonies. The MBC was determined as the lowest concentration at which fewer than five colonies were obtained. All of above experiments were replicated three times.

Bacterial Growth Curve Measurement

The antibacterial activity of the EPL was further investigated by measuring the difference in bacterial growth curves in liquid LB cultures during an incubation period of 5~30h and diluted with deionized water. The proper dilutions were plated on agar LB medium, which were incubated at 37°C. The bacterial counts were recorded after 24h. The control growth curves were obtained as above with the liquid LB cultures added potassium sorbate.

Application in Juice

In order to test application effect of EPL in Orange Juice, commercial orange juice (12Brix) was mixed with the EPL to final concentration of 0.4% (w/v) and 0.8%(w/v) with deionized water as the control. After autoclaving at 121°C for 15min, the tubes of 20ml with 12ml above mixed juice were inoculated by combined strains of eight test bacteria and kept at 30°C. The samples were taken every 24h followed by serial dilution, pour-plating onto PCA and incubation at 37°C for 24h [11].

Statistics Analysis

Analysis of variance was performed on the data of all tables and figures using Statistical Analysis System (SAS 1982), and significant differences among the means were determined by the Duncan new multiple range test.

RESULTS

Yields of the EPL

Yields of the EPL with solvent were listed in Table 1. Accordingly, 75% ethanol was shown to be one of the most suitable solvent with the highest yield of EPL (14.98 g/100 g of dry leaf), in comparison to water solvent with 10.03 g/100 g, methanol solvent with 9.57 g/100 g and ethylacetate solvent with 3.72 g/100 g. Apparently, it was demonstrated that the yield of the extract increased as polarity of the solvent, and polarity solvents could extract much composition of persimmon leaf.

Antibacterial Activity

Eight species of common food spoilage and food borne pathogens

Table 1. Yields of the EPL (g /100 g dry persimmon leaf)

Solvent	Sample code	Yield[a]
Water	EPL-1	10.05±1.25A
75%Ethanol	EPL-2	14.98±1.81A
Methanol	EPL-3	9.57±1.30B
Ethylacetate	EPL-4	3.72±0.41C

[a] The values were means ± standard deviation of three replicate experiments. The mean values with different superscript letters were significantly different from one another ($p<0.05$) according to Duncan new multiple range test.

(four Gram-positive and four Gram negative) were used to evaluate the antibacterial effect of the EPL (Table 2). Inhibitory zones of 16~21 mm were considered as indicators for strong antibacterial effects, 11~15 mm for moderate sensitivities, and 10 mm for weak sensitivities. All of the EPL with selected solvent showed a wide antibacterial spectrum, which gave antibacterial properties against all test bacteria at selected concentrations. However, the EPL with 75% ethanol or water solvent showed higher inhibitory effect on the test bacteria than the one with methanol or ethylacetate solvent. *S. aureus, B. subtilis, B. cereus, E. coli, S. faecalis*, and *S. typhimurium* exhibited marked sensitivity to the EPL with 75% ethanol, but *P. mirabilis* and *P. aeruginosa* were more resistant than the other test bacteria to the same samples. In general, the antibacterial effect of the EPL on Gram-positive species was slightly strong than on Gram-negative ones.

Table 2. Inhibitory zone of the EPL on test bacteria

	Diameter of inhibitory zone (mm) [a]				
Strains	**EPL-1[b]**	**EPL-2[b]**	**EPL-3[b]**	**EPL-4[b]**	**Positive control[c]**
Gram-positive					
S. aureus	17 B	20 C	15 B	15 B	21 C
B. subtilis	16 B	21 C	15 B	16 B	20 C
B.cereus	15 B	20 C	14 B	15 B	20 C
S. faecalis	13 B	17 C	13 B	13 B	18 C
Gram-negative					
E. coli	15 B	19 C	13 B	15 B	19 C
S. typhimurium	14 B	16 B	12 B	12 B	19 C
P. mirabilis	12 B	13 B	10 B	10 B	17 C
P. aeruginosa	11 B	13 B	10 B	11 B	17 C

[a] The diameter (mm) was the mean of the three independent experiments (not including the 6mm diameter of the paper disk). The values with different capital letters in the same line were significantly different from the values for zone diameters with superscript a and from one another ($p<0.05$) according to Duncan new multiple range test.

Determination of MIC and MBC

All of the EPL in this work exhibited both bacteriostatic and bactericide activities. The MIC values and MBC values were listed in Table 3. The EPL with 75% ethanol exhibited the maximum inhibitory activity on most of the test strains with 0.4% (w/v) of MIC and/or MBC value except *P. mirabilis*, and *P. aeruginosa* with 0.79%. Generally, the inhibitory effect on test bacteria corresponded to the EPL with MIC and/ or MBC values ranging from 0.4~1.57%, and *S. aureus* gave the maximum sensitivity to all of the EPL. The weakest sensitivity to the EPL was observed on *P. mirabilis*, and *P. aeruginosa*. This is in agreement with the result of Table 2. In most case in Table 3, the MIC values were the same as the MBC values, and the samples at these levels showed inhibitory effect on bacterial growth by killing them. However, the EPL with methanol and ethylacetate solvent were the exception. They exhibited inhibitory effect on *P. mirabilis*, and *P. aeruginosa* respectively at 1.57%, but failed

Table 3. MIC and MBC of the EPL to test strains

	MIC and MBC [a] (%, w/v)				
Strain	**EPL-1[b]**	**EPL-2 [b]**	**EPL-3 [b]**	**EPL-4 [b]**	**Positive control**
Gram-positive					
S. aureus	0.4 A	0.4 A	0.79 B	0.4 A	0.2~0.4 A
B. subtilis	0.79 B	0.4 A	0.79 B	0.79 B	0.2~0.4 A
B.cereus	0.79 B	0.4 A	0.79 B	0.79 B	0.2~0.4 A
S. faecalis	1.57 C	0.4 A	1.57 C	0.79 B	0.2~0.4 A
Gram-negative					
E. coli	0.79 B	0.4 A	0.79 B	0.79 B	0.2~0.4 A
S. typhimurium	0.79 B	0.4 A	0.79 B	0.79 B	0.2~0.4 A
P. mirabilis	1.57 C	0.79 B	>1.57 C	>1.57 C	0.4 A
P. aeruginosa	1.57 C	0.79 B	>1.57 C	>1.57 C	0.4 A

[a] The values were the means of the average of three samples. The values with different capital letters in the same line were significantly different from one another ($p<0.05$) according to Duncan new multiple range test.
[b] EPL-1, EPL-2, EPL-3 and EPL-4 indicated the EPL with the solvent of Water, 75%Ethanol, Methanol and Ethylacetate respectively.

to kill the two bacteria at the same concentration. Therefore, the MBC values for these EPL on *P. mirabilis* and *P. aeruginosa* were greater than 1.57%.

Effects of EPL on the Survival of Test Bacteria

The antibacterial effects of the EPL on test bacteria were further measured by the growth curves obtained from cultures contained the EPL comparing to the ones obtained from the control without the EPL. Selected growth curves were shown in Figure 1.

Generally, the results were in agreement with those obtained in the agar diffusion experiments. The EPL resisted the growth of all of test bacteria in different degree. To *S. aureus, B. subtilis, B. cereus* and *E. coli*, strong inhibitory effects from the EPL were obtained at no less than moderate concentrations, and the EPL at the concentration of 0.4% (w/v) was more effective than potassium sorbate at 0.2% (w/v). To *S. faecalis* and *S. typhimurium*, the EPL at the concentration of 0.4% was less effective than potassium sorbate at 0.2%, but more effective than it at 0.1%. The EPL at the concentration of 0.8% also showed stronger activity than potassium sorbate at 0.4%. As to *P. mirabilis*, and *P. aeruginosa*, the EPL at the level of 0.8% was weaker than potassium sorbate at 0.4%. It must be emphasized that both of 0.8% EPL and 0.4% potassium sorbate showed complete inhibition to all of test strains except *P. mirabilis* and *P. aeruginosa.*

Application in Orange Juice

The market orange juice added the EPL was taken to be examined its application effect in food products. Figure 2 showed that the EPL at the level of 0.4% (w/v) exhibited efficacious antibacterial effect on combined bacteria responsible for juice spoilage. No bacterial growth was observed during 24h for the juice fortified with 0.4% or 0.8% EPL. After 48h, the control juice spoiled with odium smell and apparent microbial growth, but the one with 0.4% or 0.8% EPL showed extremely low bacterial growth. The juice with 0.4% EPL exhibited slow growth rate of bacteria with the same tendency of the curve by 0.2% potassium sorbate, whereas the one with 0.8% EPL kept lower growth rate of bacteria during storage time of the experiment. Therefore, it is obvious that the EPL can be added into orange juice to be used as a preservative.

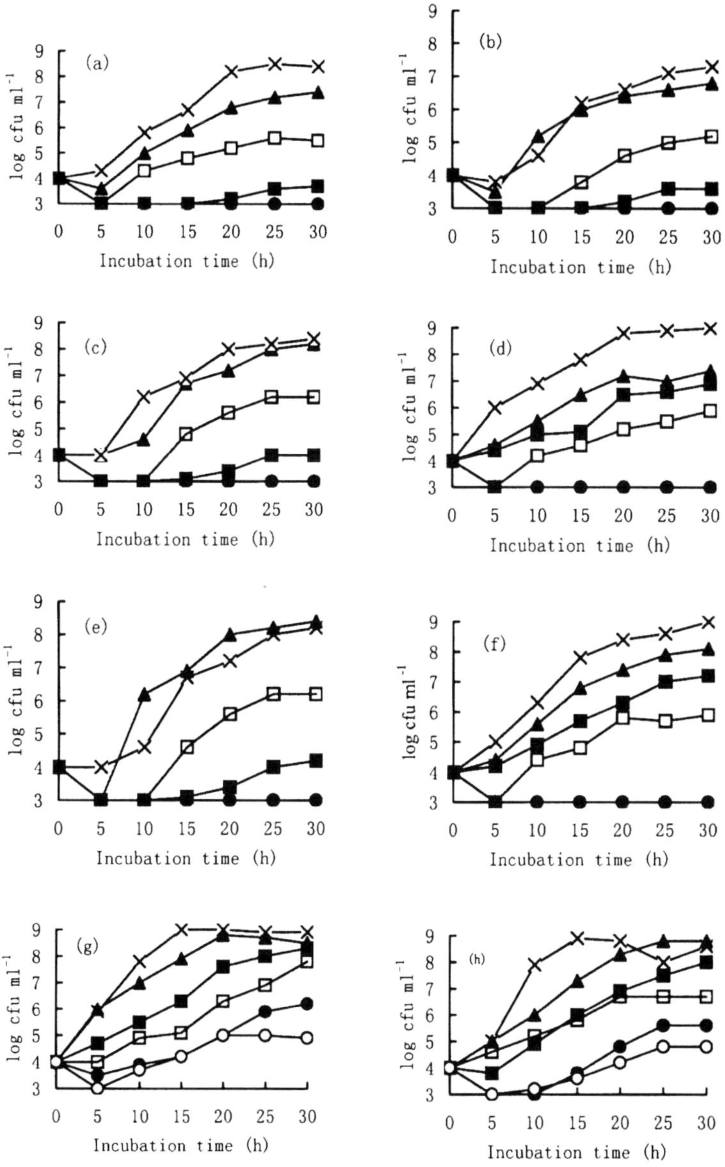

Figure 1. Growth curves of test bacteria. (a)*Staphylococcus aureus*, (b)*Bacillus subtilis*, (c)*Bacillus cereus*, (d) *Streptococcus faecalis*, (e)*Escherichia coli*, (f) *Salmonella typhimurium*, (g) *Proteus mirabilis*, (h)*Pseudomonas aeruginosa*: (×)The EPL 0.2%, (■)The EPL 0.4%, (●)The EPL 0.8%, (▲)Potassium sorbate 0.1%, (□)Potassium sorbate 0.2%, (○)Potassium sorbate 0.4%.

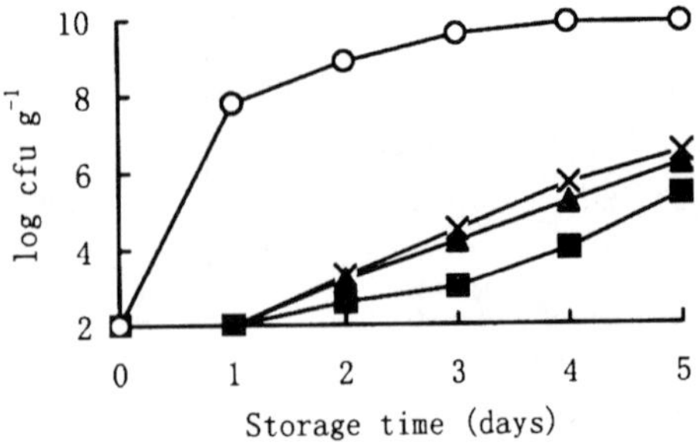

Figure 2. Influence of the EPL on spoilage bacteria of orange juice. (○) the control, (×)the EPL (0.4%), (■)the EPL (0.8%), (▲)potassium sorbate (0.2%).

DISCUSSION

The EPL opposes antibacterial activity against Gram-positive and Gram-negative species, and its bioactivity was stable to heat, acid and storage temperature. It has been reported that persimmon leaf contains much flavones and flavonoids, including astraglin, isquercitrin, kaempfetol-3-o-*β*-D-glucopyranoside, kaempferol, quercetin, rutin, kaempfetol 3-o-L-rhamnopyranoside, myricetin and so on. Those flavones and flavonoids are responsible for the bioactivity of the EPL, and the degree of hydroxylation of them might affect the antibacterial activity [6, 15, 16]. Quercetin can inhibit the growth of some Gram-positive bacteria, and luteolin is bacteriostatic against some Gram-positive bacteria. The number of hydroxyl groups in the B ring in flavones and flavonoids is associated with the antibacterial activity and the outer membrane of Gram-negative bacteria functions as a preventive barrier against hydrophobic compounds [17, 18]. In the present experiments, the EPL also showed more effective inhibition in general on Gram-positive bacteria than on Gram-negative ones.

It has also been reported that some organic acids are rich in persimmon leaf, including succinic acid, benzoic, salicylic, pyromucic acid, syringic acid and protocatechuic acid [6]. Those organic acids exhibited strong antibacterial activity on Gram-positive species, as well as Gram-negative ones. This could be used to explain that the susceptive difference of the EPL to Gram-positive bacteria and to Gram-negative ones was not very obvious in this study as only flavones and flavonoids led in previous reports.

In addition, triterpenes (such as betulic acid, oleanolic, ursolicacid and uvasol), tannins, volatile oil, coumarins and other compounds in EPL can be hypothesized to be partly responsible for the bioactivity of the EPL in different degree. Certainly, further experiments need to be made to prove it.

In conclusion, the EPL showed a wide spectrum and stable antibacterial activity on food spoilage and food-borne pathogens could be selected as an inhibitor to preserve food products where a naturally preservative is desire.

Persimmon leaf can be prepared to functional drinks and tea. Multi-bioactivity and functional effect of persimmon leaf made those food products healthy and therapeutic [7, 19]. Antioxidant compounds in Persimmon leaf could also prolong them storage time [8]. Therefore, the EPL will be potentially applied in functional food products development for health, therapy, prevention and food preservative purposed [20].

REFERENCES

[1] Matsuura shin and Jiunma Mumekazu. Components of the leaves of *Diospyrous kaki*. Ya kugaku Zasshi 1971; 90: 905

[2] Bosetto M. Dynamics of some elements in the fruits and foliage of *D. kaki* L. during masturation. Agric Ital 1986; (1-2): 7

[3] Nakamura Mitsuo. Separation of LAA from phenol in *Dispyros kaki* leaf by revesephase TLC. Gifu Daigaku Nogakubu Kenkyu Hokoku 1986;51: 23

[4] Lin LC. Constituents of *D. kaki* leaves ([22]). Chung-hua Yao Hsuth Tsa Chih 1988; 40: 195

[5] Middleton EJ and Kandaswami C. The impact of plant flavonoids on mammalian biology: implications for immunity, inflammation and cancer. In: The Flavonoids. Advances in Research Since 1986 (Eds). London; Chapman & Hall Harborne, 1994, pp. 619-52

[6] Guo M., Dong XP and Xu WP. The summary of study for persimmon leaves. Journal of Gansu College of TCM 2000; 17: 78-82

[7] Xu JW. Functional drink from persimmon leaves. Chinese Health and Nutrition 1993; 1: 36

[8] Tong C, Shcheng CH.and Yanzhen H. Study on Antioxidant Effect of the Persimmon Leaves. China Food Additives 1996; 4: 4-6

[9] Choi SW. Antioxidative activity of flavonoids in persimmon leaves. Food and Biotechnology 1996; 5: 119-23

[10] Hsieh PC. Antimicrobial effect of cinnamon extract. Journal of Agricultural

and Food Chemistry 2000; 38: 184-93

[11] Mall JL., Chen CP and Hsieh PC. Antimicrobial effect of extract from Chinese chive, cinnamon and corni fructus. Journal of Agricultural and Food Chemistry 2002; 49: 183-8

[12] Kim J, Wei CI.and Marshall MR. Antibacterial activity of some essential oil compounds against five foodborne pathogens. Journal of Agricultural and Food Chemistry 1995; 43: 2839-45

[13] José Manuel Cruz, José Manuel Domínguez and Herminia Domínguez. Antioxidant and antimicrobial effects of extracts from hydrolysates of lignocellulosic materials. Journal of Agricultural and Food Chemistry 2001; 49: 2459-64

[14] Sechi LA, Lezcano I and Nunez N. Antibacterial activity of ozonized sunflower oil. Journal of Applied Microbiology 2001; 90: 279-84

[15] Gorobets AV, Bandyukova VA. Diospyros flavonoids. Khim. Prir. Soedin1985; 5: 710

[16] Dorman HJD and Deans SG. Antimicrobial agents from plants: antibacterial activity of plant volatile oils. Journal of Applied Microbiology 2000; 88: 308-16

[17] Rauba JP, Remes S and Heinonen M. Antimicrobial effects of Finnish plant extracts containing flavonoids and other phenolic compounds. International journal of Food Microbiology 2000; 56: 3-12

[18] Puupponen-Pimiä R, Nohynek L and Meier C. Antimicrobial properties of phenolic compounds from berries. Journal of Applied Microbiology 2001; 90: 494-507

[19] Xie XL and Feng KS. Preparation for drink from persimmon leaves. Food Science 1989; 2: 26

[20] Gould GW. Industry perspectives on the use of natural antimicrobial and inhibitors for food applications. Journal of Food Protection 1996; Suppl: 82-6

Protective Effect of Crocin on Rat Heart Ischemia-Reperfusion Injury: Possible Mechanisms

Guang-Lin Xu* and Zhu-Nan Gong
*Center for New Drug Research & Development, College of Life Science, Nanjing Normal University, Nanjing 210097, China *Corresponding email: xudunlop@yahoo.com.cn*

INTRODUCTION

At present, plants have been a source of a number of effective drugs and a variety of plant extracts have been investigated for their use in treating cardiovascular diseases. Crocin (crocetin-digentibiose) was formerly an ethanol-extractable component of *Crocus Sativus* L. that has been demonstrated to prevent hypertension and atherosclerosis[1- 4]. Our laboratory has extracted crocin from the gardenia fruits successfully. In this study, we have found that administration of crocin improve injury induced by myocardial ischemia-reperfusion in animals. Although crocin has many cardiovascular effects, little is known about its mechanisms. Based on current and previous knowledge, we hypothesized that crocin might have a modulative effect on intracellular calcium. This study was designed to test this hypothesis by using cultured cardiomyocytes.

Cell cultures have been widely used to study the regulation of cell actions and neonatal cardiac cells in culture are generally used as a convenient in vitro model to specify the effects of regulating chemicals or factors. In this model, neonatal cardiac cells have been shown to exhibit similar characteristics as observed *in vivo* [5,6]. Myocardial injury induced by ischemia-reperfusion was closely related to the discrepancy in $[Ca^{2+}]i$ distribution, i.e. 'Ca^{2+} overload'. Studies showed that when hearts were reperfused after a long ischemia, the calcium in cytoplasm increased to 8-10 folds [7].

Since intracellular calcium is a major determinant of myocardial excitation-contraction coupling and uncontrolled changes in $[Ca^{2+}]i$ may result in pathological sequelae. We use fluo-3 as calcium indicator [8, 9] and KCl and AngII as stimulants in this study to investigate effects of crocin on intracellular calcium in primary-cultured myocardial cells of neonatal rats.

MATERIALS AND METHODS

Animals

Fifty Sprague-Dawley rats (220±18 g) and 2 to 3-day neonatal rats were provided by Experimental Animal Center of China Pharmaceutical University.

Materials

Fluo-3/AM was obtained from Molecular Probes Inc. AngII were purchased from Sigma. Crocin was extracted and purified from gardenia fruits as described previously [10]. Briefly, the gardenia fruits were extracted in the cold by triturating in 50% acetone and filtering under vacuum. The residues were applied onto a silica gel column, and crocin was eluted with 80% methanol. The chemical structure of purified crocin is illustrated in Figure 1. All other chemicals were commercial products of the highest grade available.

Surgical procedure and experimental design

The rats were anesthetized by intraperitoneal injection of 10% chloral hydrate (0.3ml /kg). A cannula was inserted into the trachea and the rats were ventilated with air using a respirator. All rats underwent thoracotomy at the fourth left intercostals space. The pericardium was opened and a silk suture was placed around the left anterior descending coronary artery approximately 2mm below its origin. Ischemia was induced by ligating the

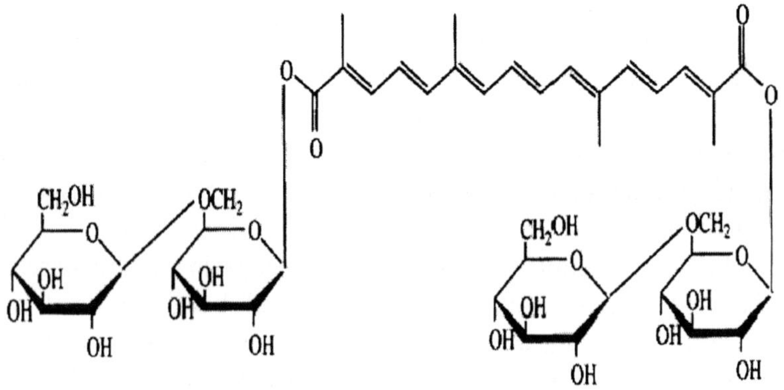

Figure 1. Chemical structure of crocin.

coronary. The chest was reopened and the ligature around the coronary artery was cut after thirty minutes to restore the blood flow. The duration of reperfusion lasted 30 minutes. Fifty rats were randomly assigned into 5 groups. (1) Group 1 (sham group, n=10). The coronary artery was surrounded by a silk but not ligated. (2) Group 2 (I/R group, n=10). This group consisted of rats undergoing ischemia 30 minutes and reperfusion 30 minutes. (3) Groups 3, 4 and 5 were administered i.v. crocin 5 mg/kg, 10 mg/kg, 20 mg/kg respectively before the onset of ischemia. The ventricular blood was collected for biochemical assays after 30 minutes of reperfusion.

Determination of creatine kinase (CK) activity (lactate dehydrogenase (LDH) efflux, superoxide dismutase (SOD) activity and malondialdehyde (MDA) content

CK and LDH are generally considered the markers of myocyte damage [11]. SOD activity and MDA content were evaluated as indices of oxygen free radical and lipid superoxide level. They were measured by using commercial kits (Jiancheng Bioengineering Institute, Nanjing, China).

Cell culture

Single-cell cultures were prepared according to previous description with minor modifications [12,13]. In summary, cells were obtained by brief, alternating cycles of room temperature gentle trypsinization and mechanical disaggregation. Briefly, hearts from 2- to 3-day old rats were excised and placed in filter-sterilized PBS solutions (NaCl 8.0 g/l, KCl 0.2 g/l, $Na_2HPO_4 \cdot 12H_2O$ 3.49 g/l, KH_2PO_4 0.2 g/l).The ventricles were isolated, cut into small pieces and incubated in PBS solution containing 0.06% trypsin (Sigma) at 37°C. The dissociated cells were pre-plated for 90 min in RPMI-1640 (Gibco, BRL) culture medium (pH 7.2) supplemented with 10% heat-inactivated newborn calf serum (Hang Zhou Si Ji Qing Biotech Co., China) and 1% antibiotics (penicillin G, 100 U/ml and streptomycin, 50U/ml).This pre-plating step removes most of the rapidly adhering non-myocyte cells. The suspended cells were subsequently collected, counted and diluted to a final concentration of 5×10^5cells/ml and then plated and incubated in 24-well trays with a volume of 1ml per well under humidified conditions (95% air/5% CO_2) at 37 °C. For determination of $[Ca^{2+}]i$, the cells were cultured on glass coverslip

(8mm×8mm) coated with 1% (W/V) gelatin.

Protocols for reagent applications

First, effects of various stimulants (KCl, AngII) on myocardiocyte $[Ca^{2+}]i$ were examined. In subsequent experiments, $[Ca^{2+}]i$ responses elicited by different stimulants were assessed under the condition of pretreatment with different concentrations of crocin. Cells were used for single application and repetitive determinations were not performed. KCl and AngII were dissolved in the physiological buffer.

Measurement of $[Ca^{2+}]i$ with fluo-3

The cells grown on glass coverslips were washed three times with 1ml Hank's buffered saline solution containing ($CaCl_2$ 0.14 g/l, KCl 0.4 g/l, KH_2PO_4 0.35 g/l, $MgCl_2 \cdot 6H_2O$ 0.10 g/l, $MgSO_4 \cdot 7H_2O$ 0.10 g/l, NaCl 8.0 g/l, $Na_2HPO_4 \cdot 12H_2O$ 0.12 g/l, D-glucose 1.0 g/l, Phenol Red 0.01, pH 7.2). The washed cells were loaded with fluo-3/AM (final concentration: 5 μM), which was dissolved in DMSO with 0.02% pluronic acid. The final DMSO concentration was less than 0.1% and had no effect on basal $[Ca^{2+}]i$. The cells were incubated for 20 minutes at 37°C on the stage of an inverted microscope (Axiovert S100, Zeiss).The loaded cells were then washed three times with warmed (37°C) buffer. The coverslip containing cells was placed in a self-made glass chamber and mounted on the stage of the inverted microscope. $[Ca^{2+}]i$ was measured in multiple cells simultaneously by fluorescent digital imaging with Axiovert S100 inverted microscope and Laser Sharp system (Issue 2.0) with the excitation wavelength set at 488nm and the emission recorded at 530nm.The fluorescent intensity (FI) changes of $[Ca^{2+}]i$ in the cytoplasm of each myocardiocyte was measured according to the fluorescence images. For comparison, the intensity of fluo-3 fluorescence for each region was divided by the resting fluorescence intensity before stimulation of the same region, after which the background fluorescence was subtracted. The resultant relative fluorescence intensity was used as an indicator of $[Ca^{2+}]i$ elevation. The systolic $[Ca^{2+}]i$, diastolic $[Ca^{2+}]i$ and spike $[Ca^{2+}]i$ frequency are presented according to previously described definitions [14].

Data analysis

All measurements were presented as the mean±SD. Tests for

significance were performed using Student's t-test. A p value <0.05 was considered significant.

RESULTS

Biological studies

In the sham hearts used as negative controls, the parameters CK, LDH, and MDA attained low levels, while SOD activity attained high level. In I/R group, these parameters (CK, LDH and MDA) were increased significantly (*vs* sham $p< 0.01$), and SOD activity was decreased (*vs* sham $p<0.01$), Compared with I/R group, the serum LDH concentrations in crocin 1 group were lowered significantly ($p< 0.05$). But the difference of CK concentrations, SOD activity and MDA production were not significant ($p>0.05$). Compared with I/R group, serum CK and LDH concentrations and MDA production in crocin 2 group were lowered significantly ($p< 0.01$), and SOD activity was increased significantly ($p<0.01$). The decease of CK, LDH, MDA in crocin 3 group was more remarkable than that in crocin 2 group.

Effects of crocin on resting [Ca^{2+}]i

Using a confocal imaging system, we monitored [Ca^{2+}]i changes in cultured neonatal cardiomyocytes loaded with fluo-3. In the resting state, cardiomyocytes have oscillatory [Ca^{2+}]i waves and diastolic, systolic, spike [Ca^{2+}]i frequency are regular and steady. The addition of crocin produced no significant changes in diastolic, systolic [Ca^{2+}]i and spike [Ca^{2+}]i frequency. Figure 2 shows a typical trace of resting [Ca^{2+}]i.

Effects of crocin on calcium responses to K^+ depolarization

Excitation-contraction coupling requires Ca^{2+} channels through voltage-dependent L-type Ca^{2+} channel and Ca^{2+}-induced calcium release from the SR (Sarcoplasmic Reticulum). To address whether crocin can affect calcium channels, myocardiocytes were depolarized with high K^+(60 mM). The results showed that depolarization with high K^+ led to an abrupt increase in [Ca^{2+}]i (FI : from 21.6 ± 9.4 to 56.7 ± 10.7, $p<0.01$, Figure 3). The addition of crocin, especially 10 μM, exhibited a biphasic curve after KCl treatment with an initial minor increase followed by a fast decline phase in which [Ca^{2+}]i returned to almost basal diastolic [Ca^{2+}]i level

(Figure 4,Table 1) . The spike [Ca^{2+}]i frequency was not clearly distinguished upon the addition of K^+ because of the tiny difference between the maximal and minimal [Ca^{2+}]i . Crocin did not return the [Ca^{2+}]i spike frequency to basal level. This suggests that crocin may reduce depolarization-induced Ca^{2+} influx.

Effects of crocin on AngII-induced responses in cardiomyocytes

Effects of AngII on cardiomyocyte [Ca^{2+}]i are presented in Figure 5. When the cells were exposed to AngII (10 nM), there was an initial sharp increase in peak [Ca^{2+}]i (FI: from 91.0±17.4 to 178.3±20.7, $p<0.01$) followed by a plateau phase in which [Ca^{2+}]i was still higher than basal state. Treating cells first with crocin followed by AngII showed that

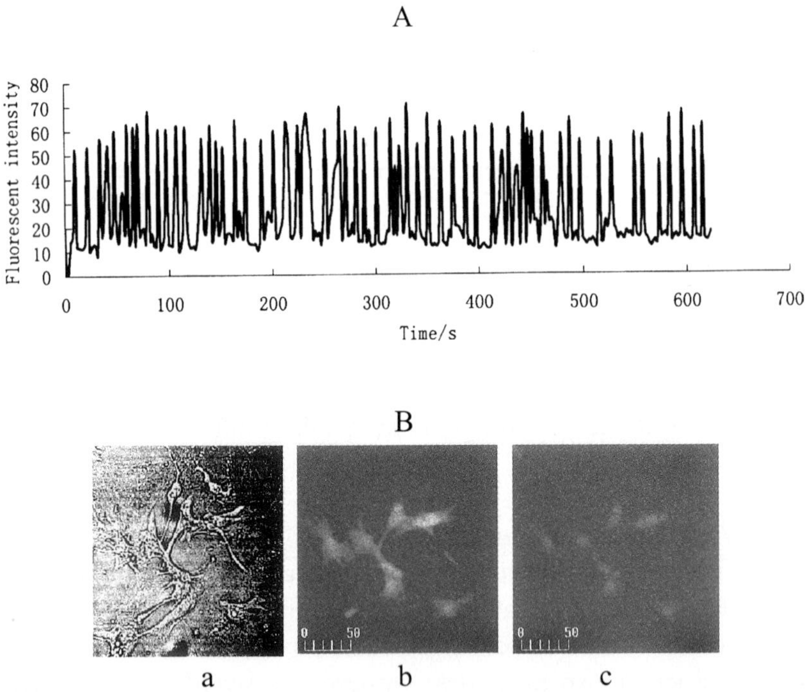

Figure 2. Resting [Ca^{2+}]$_i$ and oscillatory [Ca^{2+}]i waves in cardiomyocytes. A. Time course curve of static [Ca^{2+}]$_i$ expressed as fluorescent intensity. B. Photograph of cardiomyocytes in tested field of panel A (Magnification ×400, Bar =50 μm). a. Transmission microscopy graph. b. Fluorescent image corresponding to systolic [Ca^{2+}]$_i$ at 79.87s. c. Fluorescent image corresponding to diastolic [Ca^{2+}]$_i$ at 77.87s. All time course traces are from a representative single cell in tested field.

AngII -stimulated peak [Ca^{2+}]i elevation was not significantly reduced (Figure 6, Table 2). The spike [Ca^{2+}]i frequency could not be clearly identified upon the addition of AngII due to the tiny difference between the maximal and minimal [Ca^{2+}]i whether in the presence of crocin or not.

A

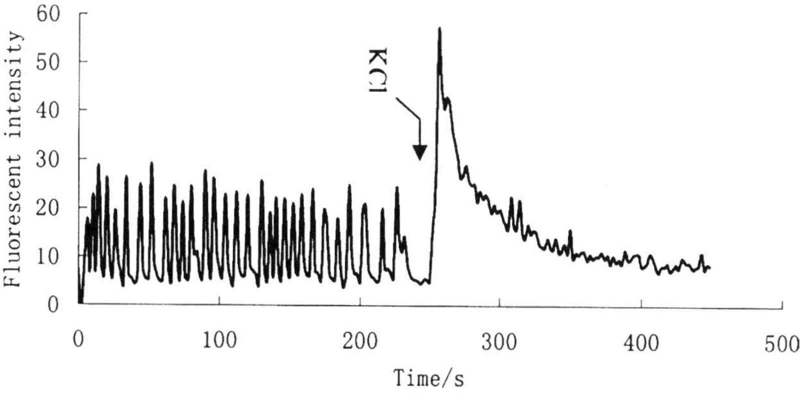

B

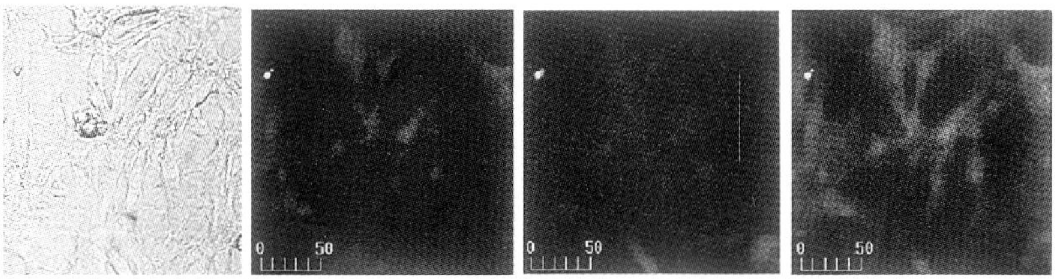

a b c d

Figure 3. Effect of KCl on [Ca^{2+}]$_i$ in cardiomyocytes. A. Time course of intracellular calcium induced by addition of KCl at 230.26s expressed as fluorescent intensity. B. Photograph of cardiomyocytes in tested field of panel A (Magnification ×400, Bar = 50μm). a. Transmission microscopy graph. b. Fluorescent image corresponding to basal systolic [Ca^{2+}]$_i$. c. Fluorescent image corresponding to basal diastolic [Ca^{2+}]$_i$. d. Fluorescent image at 258.26s after application of KCl. All time course traces are from a representative single cell in tested field.

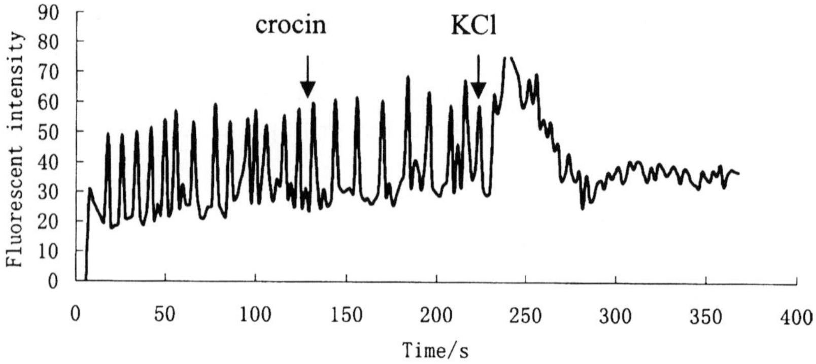

Figure 4. Effect of 10 μM crocin on KCl-induced $[Ca^{2+}]_i$ elevation in cardiomyocytes. Time course of $[Ca^{2+}]_i$ produced by successive application of crocin at 115.56s and KCl at 231.56s. All time course traces are from a representative single cell in tested field.

Table 1. Effect of crocin on serum CK0LDH0SOD and MDA in ischemia-reperfusion rats (mean ± SD, n=10)

	I/R	sham	Crocin1 (5mg/kg)	Crocin2 (10mg/kg)	Crocin3 (20mg/kg)
CK(U/l)	1793±291	629±89**	1607±198	1087±113**	907±146**
LDH(U/l)	517±41	225±28**	478±32*	347±37**	331±25**
SOD(U/l)	87.9±9.4	138.0±17.1**	78.1±5.1	118.0±9.7**	129.1±7.6**
MDA (nM)	10.01±1.27	4.07±0.41**	9. 8 0±0.76	6.24±0.53**	4.92±0.57**

*P<0.05; **P<0.01 compared with I/R group.

DISCUSSION

The results of *in vivo* model demonstrated that crocin exerts a beneficial effect on the ischemia-reperfusion rat hearts as crocin can:

1. Attenuate the damage of myocardial cell induced by I/R injury and decrease the release of CK, LDH from the damaged cells.
2. Improve the activity of SOD and lower the production of oxygen-derived free radicals to protect cardiomyocyte against damage caused by oxygen-derived free radicals.
3. Reduce MDA production, the end-product of peroxidation of cell membrane lipids, which indicates that crocin can maintain the mitochondrial integrity and function.

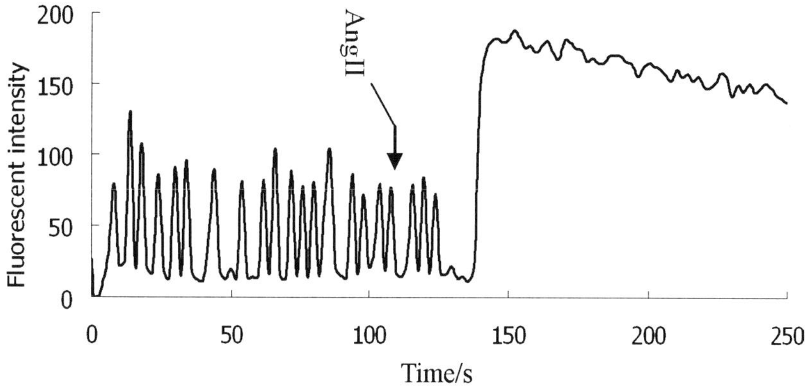

Figure 5. Effect of AngII on $[Ca^{2+}]_i$ in cardiomyocytes. Curve shows dynamic changes of intracellular calcium induced by addition of AngII at 140.62s expressed as fluorescent intensity. All time course traces are from a representative single cell in tested field.

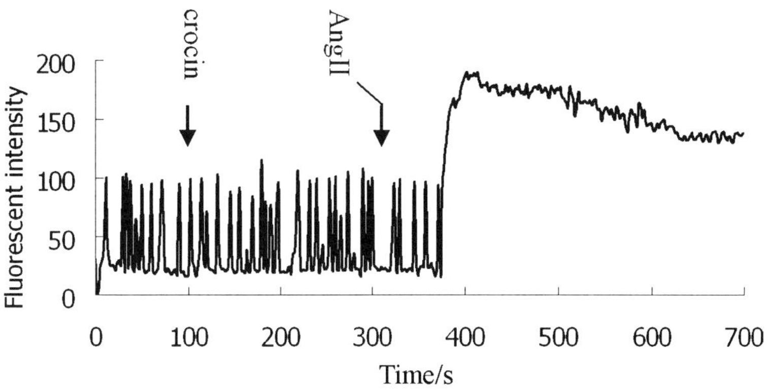

Figure 6. Effect of 10 μM crocin on AngII-induced $[Ca^{2+}]_i$ elevation in cardiomyocytes. Curve shows dynamic changes of $[Ca^{2+}]$i produced by successive treatment of crocin at 149.65s and AngII at 369.65s. All time course traces are from a representative single cell in tested field.

Intracellular calcium as a second messenger plays a key role in cell regulation and pathogenic processes. A pronounced disturbance in intracellular calcium homeostasis is a major subcellular pathophysiological feature of myocardial ischemia and reperfusion, resulting in an increase in intracellular calcium (i.e., Ca^{2+} overloading) with significant contribution to post-ischemic contractile dysfunction [15]. Several different lines of evidence suggest that in an ischemic insult, free cytosolic calcium rise from the beginning of ischemia, which may contribute to cell injury [16]. In a subsequent study, we utilized calcium-sensitive fluorescent indicator, Fluo-3/AM to measure the $[Ca^{2+}]_i$ in cultured single myocardial cell to study effects of crocin on AngII and KCl induced $[Ca^{2+}]_i$ elevation.

High–affinity AngII binding sites are found in neonatal cardiomyocytes and in the conduction system of rat hearts [17-19]. AngII mediated their intracellular effects through receptor-mediated stimulation of the breakdown of inositol phospholipids, resulting in the generation of inositol triphosphate and diacylglycerol [20,21] ,which led to significant $[Ca^{2+}]_i$ elevation. In the present study, the increase of $[Ca^{2+}]_i$ induced by AngII in myocardiocytes was observed. However, pretreatment with crocin appeared to exhibit no effect on this $[Ca^{2+}]_i$ increase suggesting that crocin may not affect AngII receptor-operated calcium channel.

It is reported that the major mechanism for $[Ca^{2+}]_i$ elevation in myocardiocytes is through voltage-dependent L-type Ca^{2+} channel. Depolarization with elevated K^+ in this study led to a significant L-Channel-mediated calcium influx, thus produced an abrupt increase with percentage of 170 %. The treatment of crocin before KCl almost completely inhibited the elevation of $[Ca^{2+}]_i$ induced by KCl. This suggests that crocin may reduce depolarization-induced calcium influx.

Table 2. Effect of crocin on intracellular calcium elevation induced by 10nM AngII and 60 mM KCl (mean ± SD, n=9)

Group	$[Ca^{2+}]_i$. elevation (%)
Control	------
AngII	96±10
0.1 μ M crocin + AngII	94±12
1.0 μM crocin + AngII	92±11
10 μ M crocin + AngII	96±9
KCl	170±19
0.1μ M crocin + KCl	147±20[#]
1.0μ M crocin + KCl	61±14[##]
10μ M crocin + KCl	10±6[##]

[#] $P<0.05$; [##] $P<0.01$ compared with KCl Group.

With these findings, we conclude that the regulation of crocin on $[Ca^{2+}]_i$ may relate to the blockage of calcium influx through L-type channel. As far as we know, this study demonstrates for the first time that crocin may have $[Ca^{2+}]_i$ modulation effects. Besides, crocin is a carotenoid in nature with unsaturated carbohydrate long chain, which is similar to membrane lipids. So crocin may occupy membrane calcium channels and block their opening.

Since elevation in intracellular calcium taking place after reperfusion plays a key role in the injury associated with ischemia-reperfusion, the results of this experiment may indicate that the protective effects of crocin on ischemia-reperfusion injury found in animals are associated with modulation of intracellular calcium. Despite the protective efficacy on injury in ischemia-reperfusion rats, the precise mechanism remains to be further explored.

REFERENCES

[1] Escribano J, Alonso G.L, Coca-Prados M, Fernandez JA. Crocin, safranal and picrocrocin from saffron (*Crocus sativus* L.) inhibit the growth of human cancer cells in vitro. Cancer Lett 1996; 100: 23-30

[2] Wang CJ, Chang MC, Lin JK. Inhibition of tumor promotion in benzo[±]pyrene-initiated CD-1 mouse skin by crocetin. Carcinogenesis 1995; 16: 187

[3] Gainer, JL. Method for treating hypertension.U.S.Pat.No.:4046880

[4] Gainer, Jr. Method for treating atherosclerosis. U.S.Pat.No.:3788468

[5] Guo W, Kamyia K, Cheng J, Toyama J. Changes in action potentials and ion currents in long-term cultured neonatal rat ventricular cells. Am J Physiol 1996; 271: C93-102

[6] Vornanen M. Contribution of sarcolemmal calcium current to total cellular calcium in postnatally developing rat heart. Cardiovasc Res 1996; 32: 400-10

[7] Silverman HS. Ionic basis of ischemic cardiac injury: insight from cellular studies. Cardiovasc Res 1994; 28: 581-8

[8] Kao JPY, Harootunian AT and Tsien RY. Photochemically generated cytosolic calcium pulses and their detection by fluo-3. J Biol Chem 1989; 264: 8179-84

[9] Trafford AW, Diaz ME, Eisner DA. Coordinated control of cell Ca^{2+} loading and triggered release from the sarcoplasmic reticulum underlies the rapid inotropic response to increased L-type Ca^{2+} current. Circ Res 2001; 88: 195-201

[10] Marie-Rose Van Calsteren Martine C, Bissonnette, Francois C, Christiane

D, Takahito Ichi, Yves LeBlanc JC, Daniel P et al. Spectroscopic characterization of crocetin derivatives from *Crocus sativus* and *Gardenia jasminoides* .J Agric Food Chem 1997; 45: 1055-61

[11] Handerson AR, Gerhardt W, Apple FS. The use of biochemical markers in ischemic heart diseases: summary of the roundtable and extrapolations. Clin Chim Acta 1998; 272: 93-100

[12] Gomez JP, Potreau D, Branka JE, Raymond G. Developmental changes in Ca^{2+} currents from newborn rat cardiomyocytes in primary culture. Pflugers Arch 1994; 428: 241-9

[13] Gomez JP, Potreau D. Effects of thapsigargin and cytopiazonic acid on intracellular calcium activity in newborn rat cardiomyocytes during their development in primary culture. J Cardiovas Pharmcol 1996; 27: 335-46

[14] Rhian MT, Jeannette F, Gaetan T, Ernesto LS. Intracellular Ca^{2+} modulation by angiotensin II and endothelin-1 in cardiomyocytes and fibroblasts from hypertrophied hearts of spontaneously hypertensive rats. Hypertension1996; 28: 797-805

[15] Bers DM. Excitation-contraction coupling and cardiac contractile force. Dordrecht. the Netherlands: Kluwer Academic Publishers;1991

[16] Wier WG, Balke CW. Ca^{2+} release mechanisms. Ca^{2+} sparks ,and local control of excitation-contraction coupling in normal heart muscle. Circ Res 1999; 85: 770-6

[17] Sechi LA, Grilfin CA, Grady EF, Kalinyak JE, Schambelan M. Characterization of angiotension II receptor subtupes in rat heart. Circ Res 1992; 71: 1482-9

[18] Baker KM, Singer MA. Identification and characterization of guinea pig angiotensin II ventricular and atrial receptors: coupling to inositol phosphate production. Circ Res 1988; 62: 896-904

[19] Touyz RM, Sventek P, Lariviere R, Thibault G, Farch J, Reudelhuber T, et al. Cytosolic calcium changes induced by angiotensin II in neonatal rat atrial and ventricular cardiomyocytes are mediated via angiotensin II subtype I receptors. Hypertension 1996; 27: 1090-6

[20] Rogers TB, Lokuta AJ. Angiotensin II signal transduction pathways in the cardiovascular system. Trends Cardiovasc Med 1994; 4: 110-6

[21] Marsden PA, Danthulur NR, Brenner BM, Ballermann BJ, Brock FA. Endothelin action on vascular smooth muscle involves inositol trisphosphate and calcium immobilization. Biochem Biophys Res Commun 1988; 158: 86-93

Protective Effects of *Herba Leonuri* in Ischemic Models

Xian Hu[a,b], Jian Chun Mao[c,*], Shan Hong Huang[b], Jian Sun[a], Ya Jun Wu[d], Wei Duan[b], Todd On[a] and Yi Zhun Zhu[a,e,*]

[a] *Department of Pharmacology, National University of Singapore, Singapore*

[b] *Department of Biochemistry, National University of Singapore, Singapore*

[c] *Department of Internal Medicine, Long Hua Hospital, Shanghai University of Traditional Chinese Medicine, Shanghai 200032.*

[d] *Department of Anatomy, National University of Singapore, Singapore*

[e] *Department of Pharmacology, School of Pharmacy, Fudan University, China*

* *Corresponding emails: maojianchunyichun@hotmail.com or phczhuyz@nus.edu.sg*

INTRODUCTION

Acute Myocardial Infarction (AMI) is considered as the leading cause of death worldwide [1].Therefore, prevention of AMI is of specially importance to reduce the mortality rate. AMI occurs as there is obstructed circulation in certain part of the heart. Most of cases are caused by thrombosis induced coronary arterial, typically left ventricular artery, occlusion [2]. A direct consequence is the reduction of blood supply to the affected region. Negative tissue remodeling and biochemical changes can be observed after AMI incidence [3].Thrombolytic medicines and mechanical reperfusion therapy are always applied to achieve rapid restoration of blood flow, which greatly aggrandize the survival rate after Myocardial Infarction (MI) challenge [2].

Several reports have pointed out that *Herba leonuri* (HL) extract could be used in the treatment of ischemic heart diseases [4-7]. HL, or Chinese Motherwort, is well-known for its cardiovascular effect. Pang (2001) discovered that the HL extract, when applied on isolated rate aorta, could inhibit the muscle relaxation induced by acetylcholine; and intravenous injection of HL extract to rat actually caused a transient increase in blood pressure[8]. In rats, intravenous injection of HL extract reduced the infarction size and protected subcellular structure of the myocardium against myocardial infarction induced by isoproperenol or pituitrin Wang 1983. In addition, HL can normalize ischemic EGF by decreasing heart rate in isoproterenol-treated rat model. Moreover, in ligation induced ischemic rabbit model, injection of HL significantly improve left ventricular systolic pressure as well as decrease the plasma level of creatine kinase, asparatate amino transferase and lactate dehydrogenase [9].

Although potential therapeutic applications of HL have been proposed, little was known about the mechanism beneath. Previous investigations in our lab using both *in vitro* and *in vivo* model have showed that HL extracts could alleviate AMI-caused damage on myocardium by eliminating oxidative stress [10]. An upregulated expression level of some apoptosis-related genes, as well as an increased capillary density, was unintentionally observed (unpublished data). In an attempt to verify these findings and to further explore the possible mechanism underlying, we aimed to examine the influence of HL extract on the expression of apoptosis- and angiogenesis-related genes both *in vitro* and *in vivo*. We hypothesized that the cardiac-protective impact of HL might be contributed by the attenuation of apoptosis and amplification of angiogenesis.

MATERIALS AND METHODS

Plant material

Sun dried whole *Herba leonuri* plant was originated from Sichuan Province, China. The raw material boiled in water for 45 minutes. The brownish aqueous extract was then concentrated in vacuum at 50°C (BUCHI Rotavapor R-144, waterbath B480). It was then condensated overnight at -80°C and the liquid was made into freeze-dried power. As previously described (10). Purified *Herba leonuri* (known as KardiGen™) was suppied by Herbatis Pte Ltd. (Singapore). Of 100g raw material, 8g powder was obtained. The powder was dissolved in distilled water (100mg/L).

Animal model

Healthy male Wistar rats weighing from 230g to 250g were obtained from the Laboratory Animal Centre, National University of Singapore. Eighteen rats were randomly allocated into three groups: 6 in sham group, and 12 in treatment group which were further divided into two subgroups:

water treated AMI-operated group (WT) and HL treated AMI-operated group (HLT). Rats in these two subgroups were given oral administration with water (4ml/kg/day: WT) and HL extract (400 mg/kg/day: HLT) for five weeks, i.e. one week before surgery and four week after surgery. At the end of five weeks, all rats were sacrificed to collect hearts. WT group was also used as control to show the change induced by AMI under normal circumstances. Detailed grouping can be found in Table 1.

Myocardial infarction (MI) was induced by ligation of the left anterior descending coronary artery of rats using a modified version of the technique described by Selye (1960)[11]. Successful occlusion was verified visually by the regional cyanosis of the myocardium distal to the ligation. Rats in the sham group were subjected to the same surgical procedure with the exception that the suture was passed under the coronary artery without ligation. After ligation, three and four sutures were used closed the thoracic cavity and muscle skin layer, respectively. With the surgical procedure used in this study, most of the rats developed an infarct size ranging from 40% to 60% of the left ventricular mass. Animals with infarct size less then 30% were excluded from further investigations.

Cell culture

Primary rat aortic VSMC were obtained from Ms. Kalaiselvi Kuppusamy (Department of Pharmacology, National University of Singapore). The cells were routinely grown in *D*ulbecco's *M*odified *E*agle's *M*edium (DEME) with 10% *F*etal *B*ovine *S*erum (FBS) and 2% gentamicin.

Hypoxia

Cells were distributed equally to standard six-well plate (1000 cell/well). HL extract was dissolved in distilled water at a concentration of 100mg/L and was further diluted using serum-free DEME to 100μg/ml when treated to cells. Before incubate the cells in the hypoxic chamber, cells were dosed as described in Table 1 and were incubated in serum-free DEME at 37°C for 1 hour in normoxic condition. A hypoxic condition was created by incubating the cells in an airtight Modular incubator chamber (Billups Rothenberg, California) with an atmosphere of 0.5% oxygen, 5% carbon dioxide and 94.5% nitrogen at 37°C for a period of 16 hours.

RNA isolation and qualification

RNA was isolated using TRIzol reagent (Invitrogen) according to the protocol provided, while RNA quantitation was determined by OD260 readings using spectrophotometer (BioSpec-1601, SHIMADZU). Quality

Table 1. List of *in vivo* animal groups and *in vitro* VSMC cell groups.

***In vivo* Groups**		Dosage	Number
Sham group	S group (water)	4mL/kg/day	6
Treatment Group	WT group (water)	4mL/kg/day	6
	HLT group(HL extract)	400mg/kg/day	6
***In vitro* Groups**		Dosage	Number of Wells
Hypoxic group	Water Hypoxia group (WH)	Distilled Water 30 μl per 3ml DEME	12
	HL Hypoxia group (HLH)	HL extract 100μg/ml	12
Normoxic group	Water normoxic Control group (WC)	Distilled Water 30 μl per 3ml DEME	12
	HL normoxic Control group (HLC)	HL extract 100μg/ml	12

Animal groups: Sham group, rats were treated with water 4 ml/kg/day orally without ligation of left anterior descending coronary artery; WT group, rats were treated with water 4 ml/kg/day orally with ligation of left anterior descending coronary artery; HLT group, rats were treated with HL extract 400 mg/kg/day orally with ligation of left anterior descending coronary artery.

Cell culture groups: WH group, VSMC cells were incubated with 3μl distilled water and 2970μl serum free DEME medium in normoxic condition for 1 hour before placed in hypoxia chamber for 16 hours; HLH group,VSMC cells were incubated with 3μl 100mg/L HL extract and 2970μl serum free DEME medium in normoxic condition for 1 hour before placed in hypoxia chamber for 16 hours; WC group, VSMC cells were incubated with 3μl distilled water and 2970μl serum free DEME medium for 17 hours under normoxic condition at; HLC group, VSMC cells were incubated with 3μl d100mg/L HL extract and 2970μl serum free DEME medium for 17 hours under normoxic condition. All incubation was carried out at 37°C.

of the RNA sample was assessed by the OD_{260}/OD_{280} ratio Only sample that had OD_{260}/OD_{280} values bigger than 1.6 were used.

Semi-quantitative reverse transcription-polymerase chain reaction

One microgram of total RNA was used for RT-PCR (Qiagen OneStep RT-PCR Kit) in order to check the expression of specific gene. The primers as well as the protocols used for RT-PCR are listed in Table 2.

Cell viability assay

Trypan blue dye method was used diluted in cell viability assay (Roche). Approximately 1000 cells were counted per well.

Statistical analysis

All data were expressed as mean ± SEM. Significance among groups were examined using one way ANOVA with the Bonferroi method. P values were calculated with 95% confidence interval (CI).

Table 2. List of Primers and Protocols used for Semi-Quantitative RT-PCR

Gene	Sense	Anti-Sense	RT-PCR Protocol
GAPDH	5'-CATGGTCTACATGTTCC AGT-3'	5'-GGCTAAGCAGTTGGTG GTGC-3'	94°C 6:00 mins 55°C 1:00 mins 72°C 8:00 mins (25 cycles)
Bcl-xl/xs	5'-TCACTTCCGACTGAAG AG TGA-3'	5'-AAAATGTCTCAGAGCA ACCGG-3'	94°C 6:00 mins 55°C 1:00 mins 72°C 8:00 mins (30 cycles)
P53	5'-ATGTTCCGAGAGCTGA ATGAGG-3'	5'-GGACTAGCATTGTCTTG TCAGC-3'	94°C 11:10 mins 57°C 1:10 mins 68°C 8:30 mins (35 cycles)
Fas	5'-AACATGAGAACATCCT GTGCC-3'	5'-TCCCTGCTCATGATGTC TACC-3'	94°C 6:00 mins 51°C 1:00 mins 72°C 8:00 mins (35 cycles)
VEGF	5'-CCATGAACTTTCTGCTC TTC-3'	5'-GGTGAGAGGTCTAGTT CCCGA-3'	94°C 10:00 mins 57°C 1:00 mins 68°C 8:30 mins (35 cycles)

RESULTS

In vivo gene expression of apoptosis and VEGF

The expression level of proapoptotic gene P53 and Fas were significantly lower in HL treated AMI-operated group than Water treated AMI-operated group ($p<0.05$). In contrast, this level of the anti-apoptotic gene Bcl-xl was statistically higher in HL treated AMI operated group. Meanwhile, no significant difference in the expression of Bcl-xs was observed ($p>0.05$), and no conclusion could be drawn due to the high S.D. Three out of four all tested genes expression were influenced by HL treatment toward apoptosis suppression direction, which suggested that HL had an overall potential anti-apoptotic effect *in vivo*. In contrast to AMI operated groups, expression of *VEGF* was undetectable in Sham operated group (Figure 1). This was in agreement with the idea that a burst in *VEGF* expression can be induced by conditions like tissue damage or hypoxia to trigger angiogenesis and hence met the need of increased oxygen demand. HL treated AMI-operated group had a much higher expression of *VEGF* than that of water treated AMI-operated group ($p<0.05$), which meant there might be a higher angiogenesis activity in the HL treated group.

In vivo coronary capillary measurement

Significantly lesser capillaries were observed in the left ventricles of the vehicle group (221.2±22.5) as compared to the HL-treated (316.8±16.5) groups ($p<0.05$). In the infarct area, myocardial necrosis was observed in the tissue sections obtained from all three groups. It was generally observed that there were virtually fewer vessels the vehicle group. HL group had a comparable distribution of vessels in the non-infarct area of left ventricles.

In vitro gene expression of apoptosis and VEGF

Similar to *in vivo* result, *P53*, *Fas*, *Bcl-xs* mRNA level had been dramatically decreased in HL treated hypoxia-operated group(HLT) compare to Water treated hypoxia-operated group(WT) ($p<0.05$)(Figure 2). Meanwhile, no statistically significantly change ($p>0.05$) in *Bcl-xl* expression level was observed between the two hypoxia groups. The possible reasons is that hypoxia induced apoptosis might be independent

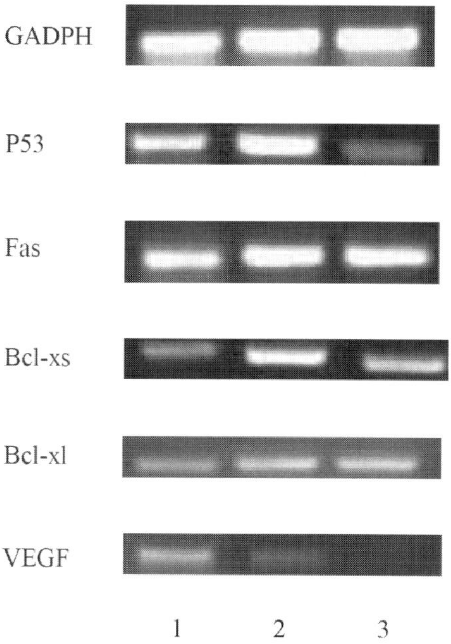

Figure 1. Apoptosis attenuating and angiogenesis promoting effect of HL on gene expression *in vivo*. A comparison among semi-quantified mRNA levels of *P53* (A), *Fas* (B), *Bcl-xl* (C), *Bcl-xs* (D), *VEGF* (E). *GAPDH* expression was used as an internal standard to normalize gene expression levels. Standard deviation was indicated. *$p<0.05$ compared to water treated AMI-operated group. $^{\#}p<0.05$ compared to sham-operated group.

Lane 1 = HL Treated AMI; Lane 2 = Water Treated AMI; Lane 3 = Sham

A

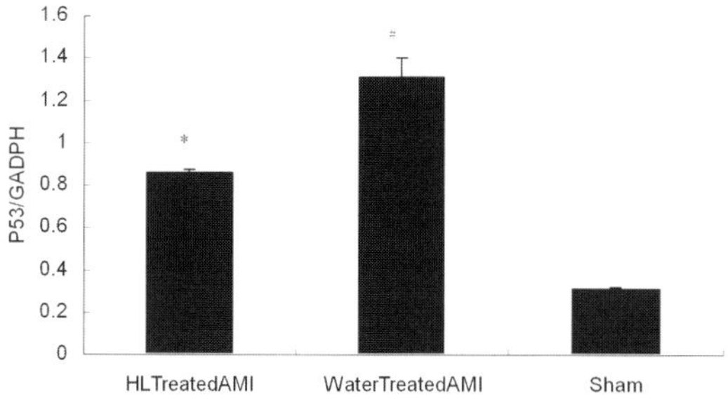

B

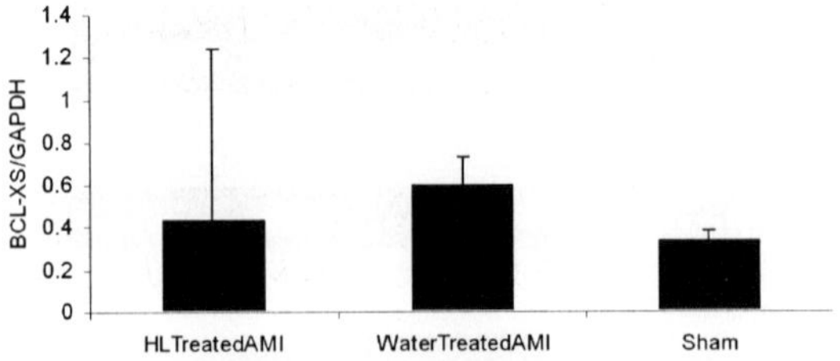
BCL-XS/GAPDH
1.4
1.2
1
0.8
0.6
0.4
0.2
0
HLTreatedAMI
WaterTreatedAMI
Sham

C

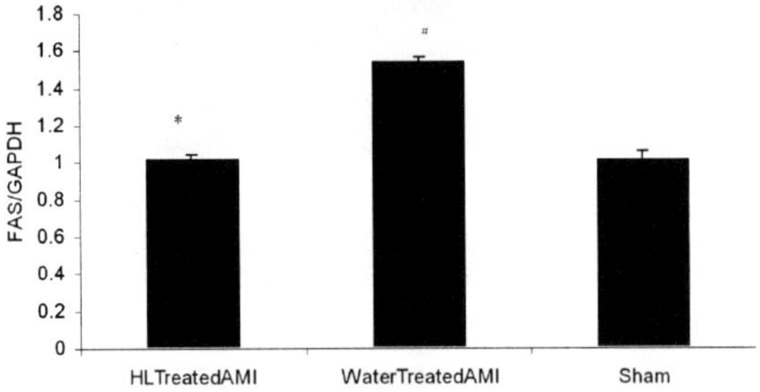
FAS/GAPDH
1.8
1.6
1.4
1.2
1
0.8
0.6
0.4
0.2
0
*
HLTreatedAMI
WaterTreatedAMI
Sham

D

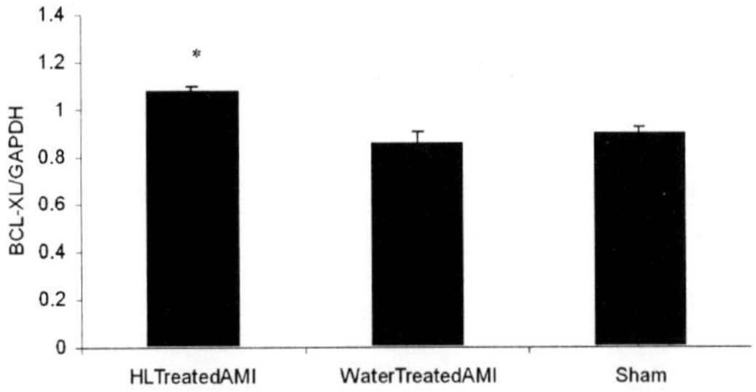
BCL-XL/GAPDH
1.4
1.2
1
0.8
0.6
0.4
0.2
0
*
HLTreatedAMI
WaterTreatedAMI
Sham

E

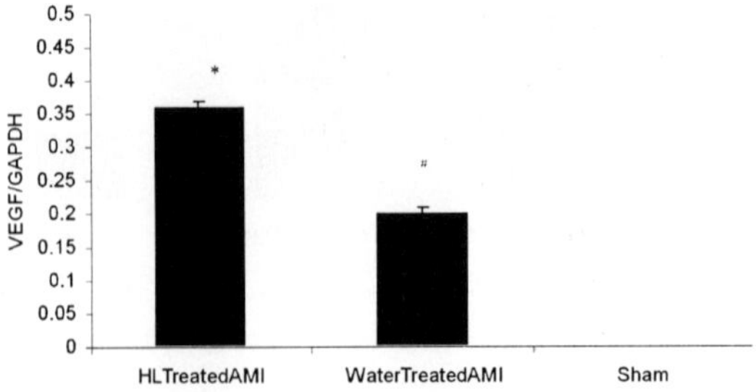
VEGF/GAPDH
0.5
0.45
0.4
0.35
0.3
0.25
0.2
0.15
0.1
0.05
0
*
HLTreatedAMI
WaterTreatedAMI
Sham

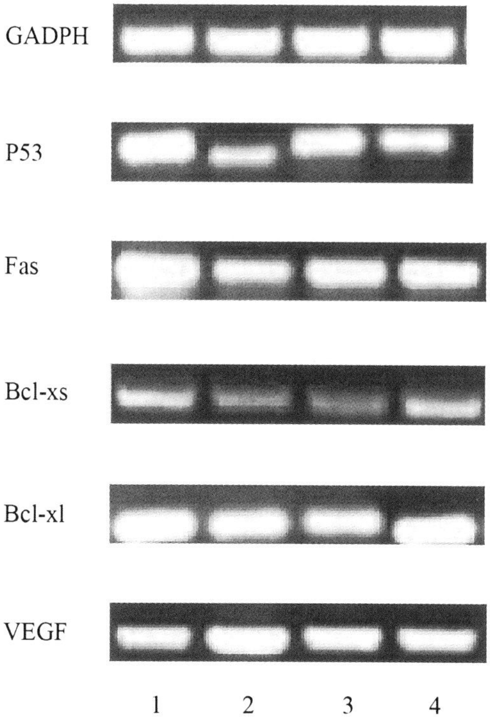

Figure 2. Apoptosis attenuating and angiogenesis promoting effect of HL on gene expression *in vitro*. A comparison among semi-quantified mRNA levels of *P53* (A), *Fas* (B), *Bcl-xl* (C), *Bcl-xs* (D), *VEGF* (E). *GAPDH* expression was used as an internal standard to normalize gene expression levels. Each group contains 6 wells of cells (standard six well culture plate) and each experiment was repeated twice. Standard deviation was indicated. $^*p<0.05$ compared to water-treated hypoxia group.

Lane 1 = Water Treated Hypoxia; Lane 2 = HL Treated Hypoxia; Lane 3 = Water Treated Control; Lane 4 = HL Treated Control

of *Bcl-xl* in VSMC. Besides, mRNA levels of the four genes examined HL treated normoxic group (HLC) and water treated normoxic group (WC) did not differ from each other statistically ($p>0.05$), which suggested that HL extract did not exert its action on neither apoptosis nor angiogenesis under normal condition in VSMC and this might somewhat explain the limited side-effect of HL. In consistent with our *in vivo* tests result, the *VEGF* level was statistically higher in HL treated hypoxia operated cell

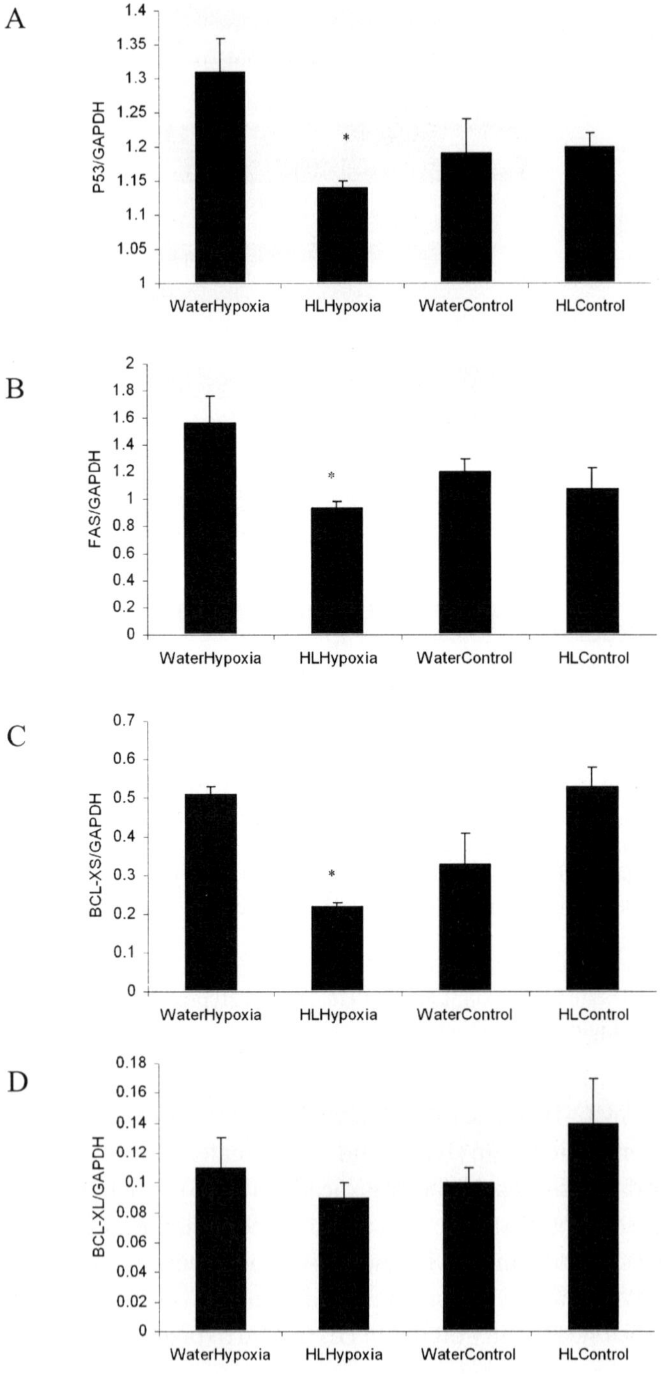
A
P53/GAPDH
1.4
1.35
1.3
1.25
1.2
1.15
1.1
1.05
1
*
WaterHypoxia
HLHypoxia
WaterControl
HLControl
B
FAS/GAPDH
2
1.8
1.6
1.4
1.2
1
0.8
0.6
0.4
0.2
0
*
WaterHypoxia
HLHypoxia
WaterControl
HLControl
C
BCL-XS/GAPDH
0.7
0.6
0.5
0.4
0.3
0.2
0.1
0
*
WaterHypoxia
HLHypoxia
WaterControl
HLControl
D
BCL-XL/GAPDH
0.18
0.16
0.14
0.12
0.1
0.08
0.06
0.04
0.02
0
WaterHypoxia
HLHypoxia
WaterControl
HLControl

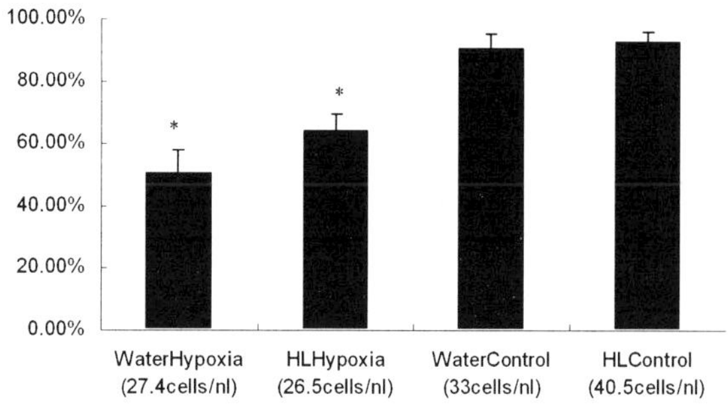

Figure 3. Comparison among cell viability trypan blue dye exclusion test result *in vitro* part. Each group contains 6 wells of cells (standard six well culture plates) and each experiment was repeated twice. Standard deviation and cell concentration were indicated. *$p<0.05$ compared to water-treated control group.

than water treated hypoxia operated cells ($p<0.05$) and no statistically difference was detected between the two normoxic group ($p>0.05$). Again, the result indicated that HL might act to promote capillary development by further up-regulating *VEGF* under hypoxic condition and this action was ceased under normoxic condition at current dosage.

The relative gene level was not normalized using cell number in our *in vitro* tests. The cells used for cell viability assay (trypan blue exclusion test) were treated by trypsin-EDTA as described in Material and Methods and hence not suitable for extracting RNA and further RNA level analysis. To avoid bias introduced by any difference between the cells that were subjected to RNA analyze and that were used for cell viability test, instead of normalizing the RNA level by cell count result, cell count result was used as a reference for interpreting RNA level.

***In vitro* cell viability**

Cell viability was tested using trypan blue exclusion test. HL treated hypoxia-operated group had a slightly higher viability than the water treated hypoxia-operated group ($p<0.05$) (Figure 3). And as expected, both the cell population and the cell viability of the hypoxia operated groups were significantly lower than that of the normoxic group ($p<0.05$). Again, no difference of cell viability between water treated normoxic group and HL treated normoxic group was detected indicating that HL extract do not harm the cell at this dosage ($p>0.05$). Interesting, although this difference was not statistically significant, the overall cell number in the HLC group was slightly higher then in the WC group ($p>0.05$).

DISCUSSION

HL may attenuate apoptosis

Depending on how severe the myocardial ischemia is, different level of cell death in the heart will occur. Many researchers used to believe that necrosis was the major mode of cardiomyocyte death and predominated during MI and other cardiovascular diseases [12]. However, there are ample evidences nowadays from *in vitro* and *in vivo* studies showing that terminally differentiated cardiomyocytes do undergo apoptosis [13]. Moreover, due to its sporadic occurrence and the prompt clearance of cells that undergo apoptosis by phagocytosis, apoptosis incidence in AMI could be grossly underestimated. In current study, our results confirmed that all alterations of tested apoptosis related genes were toward apoptosis promoting direction under AMI condition, suggesting that programmed cell death did occur and should be responsible for the tissue damage in AMI.

In MI, interruption of blood flow deprives myocardium of oxygen as well as other nutrients, leading to reduction in energy production, extremely altered redox states and hence severe oxidative stress. Until recently, the classical view is that oxidative stress induces exclusively necrosis. However, recent studies have shown that lower doses of free radicals can trigger apoptosis as well. It was reported that a increased exposure to hydrogen peroxide can cause apoptosis by up regulating *Fas-FasL* system, by releasing pro-apoptotic components from mitochondria, and by activating transcriptional factors such as *P53*[14].

As a transcription factor, *P53* is capable of regulating the expression of a range of downstream genes to exert certain protective or adaptive response, namely, apoptosis or growth arrest. These functions allow the cell to respond to a wide range of cellular stress, including hypoxia induced oxidative stress. In the animal model, AMI induced over expression of *P53* in both HL treated and water treated group suggesting the presence of apoptosis after AMI insult. And HL treatment in both experiment showed significant effects on down-regulating expression of *P53* gene compared to water treated groups.

HL treatment also reduced the expression of death receptor *Fas* compare to respective water-treated control group in both experiments indicating that HL extract might counteract with *Fas* up-regulation in AMI and hence might reduce the extent of cell loss. To induce apoptosis, prior binding of *Fas*L is required for *Fas*. *Fas*L expression is limited primarily to cells of hematopoietic origin or to the sites of immune privileges such as eyes and testes where they can trigger apoptosis of invading immune cells [15]. It remains unclear whether *FasL* is co-expressed with *Fas* in the acutely ischemic myocardium [16]

The *Bcl-2* family has unique influence on cardiomyocyte survival. One of the most interesting members of *Bcl-2* family, *Bcl-x*, has two isoforms, *Bcl-xl* and *Bcl-xs* [17,18]. *Bcl-xl*, which contains four BH domains and a transmembrane domain, has strong antiapoptotic effects. Administrations of insulin-like growth factors, hepatocyte growth factor, endothelin-1, or angiotensin-converting enzyme inhibitor can clearly attenuate cardiac I/R injury, and *Bcl-xl* induction by these signals has been suggested as a mechanism of cardioprotection [19-22]. However, so far there are no direct evidences of cardioprotection by *Bcl-xl* in MI injury. Previous studies showed that *Bcl-xs* had strong proapoptotic property and would inhibit anti-apoptosis function of *Bcl-xl* and *Bcl-2* [23]. Evidences from mRNA levels of *Bcl-xs* showed that it might participate in the apoptotic pathway in several systems, e.g., transient forebrain global ischemia, apoptosis in mammary epithelial cells and injured carotid artery [24,25].

The change of *Bcl-xl* expression was not observed after AMI operation in rat in our study, suggesting that it might not participate in the AMI-induced apoptosis. In contrast, expression of *Bcl-xs* was significantly increased after inducing AMI, indicating that *Bcl-xs* might be involved in ischemia-induced apoptosis. HL extract appeared to effectively increase the mRNA level of *Bcl-xl* and slightly *in vivo*, although not significantly reduce the expression of *Bcl-xs*. *In vitro,* HL treated hypoxia VSMC cell led to a significant reduction of *Bcl-xs* gene expression but no statistical change of *Bcl-xl* expression. The contradictory results from *in vivo* and *in vitro* might be explained by two reasons: 1) in our study only one cell type in the heart tissue was investigated *in vitro* while there might be different mechanisms when HL acted on VSMC alone; 2) high SD of the *Bcl-xs* data *in vivo* might be due to mis-operation, hence future verification was warranted. It has been reported that expressions of *Bcl-xl* and *Bcl-2* were modulated by two distinct mechanisms in the context of ischemia[26] and our observations do support this possibility.

The development of anti-ischemic drugs is still at a crossroad and it is hard to eliminate the negative incidence absolutely. Inhibition of cardiomyocyte apoptosis may represent a novel approach for treatment of cardiac disease. *In vivo* and *in vitro* studies have demonstrated that this complex network is somewhat controllable. Ideally, therapy should target on the earlier pathological events and prevent the occurrence of further complications. However, cell death is more than apoptosis in the heart tissue. It is not clear yet how much apoptosis contributes to the aetiology of heart disease. Some researchers criticize that apoptosis interference may cause more problem because inhibited apoptosis may convert to necrosis.

HL may promote angiogenesis

Neovascularization in the heart is not only critical for heart development, but also for tissue recovery after infarction. Angiogenesis is known to occur in response to mechanical, metabolic and inflammatory stimuli [27,28]. MI-induced hypoxia can trigger a number of metabolic processes to facilitate adaptation like angiogenesis that enable the heart to retain homeostasis. Previous study demonstrated that growth factor such as *VEGF* could induce vascularization in the ischemic area of the myocardium [29]. *VEGF* was effective in facilitating angiogenesis and improving collateral blood flow. These and subsequent studies raised much interest in the therapeutic angiogenesis in the form of growth factor manipulation such as gene therapy for patients with inoperable ischemic heart disease [30]. However, compared to gene therapy, researchers paid less attention to stimuli that triggered the synthesis of growth factors and the upregulation of their receptors. In our study, we found that HL extract effectively upregulated *VEGF* steady state mRNA in AMI or hypoxia condition both *in vivo* and *in vitro* compared to respective water-treated controls, indicating that HL treatment could be a novel potential therapeutic agents targeting at angiogenesis promoting aspect of MI management.

Generally speaking, the goal of therapeutic angiogenesis for the myocardium is to improve perfusion. Despite the merits of such treatments, many limitations and concerns concurrently exist, including a lack of long-term safety data, possible complicated delivery modalities, short half-life of growth factors and effects on organs other than heart [31]. However, HL extract does not have these issues. Chinese people have been orally consuming of HL extract for more than hundreds of years to cure various diseases. The long-term safety profile as well as the appropriate delivery modality is thus solved. Moreover, we found that HL extract could trigger further upregulation of endogenous VSMC only under hypoxic condition and under normoxic condition it did not mediate VSMC upregulation, suggesting that its action might be disease-specific. Nevertheless, currently only *in vitro* experiment supported this hypothesis. Further verification using animal model by mRNA analyzing and histological test is needed.

General discussion

Our previous study using the same animal model (only the sample size was different) showed that rats in HL treated AMI operated group had a higher survival rate (55.4%, or 31 out of 56) than that of rats in water-treated AMI operated group (40.5%, or 30 out of 74). The survival rate for the sham operated group was 100 % (16/16). This indicated that HL did reduce the mortality rate in some way and was overall beneficial to animals that suffered from AMI. Also, the results in this study demonstrated that the extract of *Herba leonuri* could attenuate the damage of myocardial infarction by means of alter the steady state mRNA level of *P53*, *Fas*, *Bcl-x* and *VEGF* to interfere with apoptosis and promote angiogenesis both *in vivo* and *in vitro*, suggesting two other possible molecular mechanisms of action for its effect on AMI.

Ischemic heart disease is associated with increased oxidative stress as well as impaired endogenous antioxidant systems. Previous research in our lab demonstrated that HL had an antioxidant effect that was equivalent, or even superior, to that of the standard antioxidants, namely ascorbic acid and Trolox. This effect is associated with the lower damage of myocardial infarction, especially in the acute phase. HL exerts its anti-oxidant effect through direct scavenging of free radicals such as HOCL and ONOOÉ, preserving activities of SOD and GPx as well as inhibiting lipid peroixdation by the depressed formation of MDA-like substances. Moreover, HL has the ability to scavenge superoxide as reported by Liu (2001)[7]. As such, the effect of HL of reducing tissue damaging effect might be a result of reduced oxidative stress, i.e. the alteration of apoptosis related genes studied could be a epiphenomenon of antioxidant effect of HL.

Meanwhile, it was found that *in vitro*, HL extract had no effect on studied genes under normoxic condition. This observation is in agreement with the principle of TCM. Unlike Western medicine, Chinese herbal medicine does not claim to cure illness or to kill pathogenic microbes. Instead, the purpose of TCM is the maintenance of healthy status. One of the main differences between TCM and Western medicine is that TCM treats the person by restoring harmony or balance in body instead of treating the particular diseases and Chinese herbalists consider all illness defined in Western medicine system as harmonic imbalances. As a result, in Chinese population, many healthy people also keep taking TCM to prevent illness and achieve perfect fitness.

Results in the present study showed that HL extract could effectively alleviate the pathological symptoms of AMI, mainly through reducing mRNA levels of pro-apoptotic genes and increasing expressions of anti-apoptotic and pro-angiogenic genes. Base on research in our lab, we believe that HL is a promising therapeutically agent for MI. Yet more details of the underlying molecular mechanisms, as well as the efficacy of long-term usage of HL, require future investigations.

REFERENCES

[1] Mochizuki S, Takeda N, Nagano M, Dhalla NS.Preface. In: the ischemic heart. Seibu Mochizuki, ed. Kluwer Academic Publishers. 1998

[2] Lange RA, Hillis LD. Reperfusion therapy in acute myocardial infarction. NEJM. 2002: 346,954-5

[3] Bing RJ. Some aspects of biochemistry of myocardial infarction. Cell Mol Life Sci. 2001:58, 351-5

[4] Chang CF, Li CZ. Experimental studies on the mechanism of anti-platelet aggregation action of motherwort. Zhong Xi Yi Jie He Za Zhi.1986: 6(1), 39-40

[5] Wang ZS, Li DW, Xia WJ, Qiu HQ, Zhu LY. The therapeutic effect of herba Leonuri in the treatment of coronary myocardial ischemia. J. Tradit. Chin. Med. 1988:8(2), 103-6

[6] Zhang CF, Jia YS, Wei HC, Zhu XM, Hui YM, Zhang CY, Mo QZ, Gong B. Studies on actions of extract of motherwort. J. Tradit. Chin. Med. 1982:2(4), 267-70

[7] Liu GS. Pharmacological research process of *Herba leonuri*(益母草的药理研究进展). Zhong Yao Cai 2001:25,71-2

[8] Pang S, Tsuchiya S, Horie S, Uchida M, Murayama T, Watanabe K. Enhancement of phenylephrine-induced contraction in the isolated rat aorta with endothelium by H2O-extract from an oriental medicinal plant Leonuri herba. Jpn. J. Pharmacol. 20*01: 86(2), 215-222.*

[9] Chen SR, Zheng HA, Chen HQ, Chen H, Dong RS, Zheng GW. Curative effect of herba Leonuri on myocardial ischemia and its mechanism. Zhongguo Yixue Luntan 2002:2(4)

[10] Sun J, Huang SH, Zhu YZ, Whiteman M, Wang MJ, Tan BKH. Anti-oxidative stress effects of *Herba leonuri* on ischemic rat hearts. Life Sci.2004: 76,3043-56

[11] Selye H,Bajusz E, Grasso S, Mendell P.Simple techniques for the surgical occlusion of coronary vessels in the rat. Angiology 1960:11, 398-407

[12] Kang PM, Haunstetter A, Aoki H, Usheva A, Izumo S. Morphological and molecular characterization of adult cardiomyocyte apoptosis during hypoxia and reoxygenation. Circulation Research 2000:87, 118-25

[13] Gill C, Mestril R, Samali A. Losing heart: the role of apoptosis in heart disease - a novel therapeutic target? Faseb. J. 2002:16: 135–46

[14] Chandra J, Samali A, Orrenius S. Triggering and modulation of apoptosis by oxidative stress. Free Radic. Biol. Med. 2000: 29, 323-33

[15] Mehmet H. Stroke treatment enters the *Fas* lane. Cell Death Differ.2001: 8,659-61

[16] Suda T, Takahashi T, Golstein P, Nagata S. Molecular cloning and expression of the *Fas* ligand, a novel member of the tumor necrosis factor family. Cell 1993: 75, 1169-78

[17] Boise LH, González-García M, Postema CE, Ding L, Lindsten T, Turka LA, Mao X, Nuñez G,Thompson CB. *Bcl-x*, a bcl-2-related gene that functions as a dominant regulator of apoptotic cell death. Cell 1993: 74, 597-608

[18] González-García M, Pérez-Ballestero R, Ding L, Duan L, Boise LH, Thompson CB, Nuñez G. Bcl-XL is the major bcl-x mRNA form expressed during murine development and its product localizes to mitochondria. Development 1994: 20, 3033-42

[19] Yamamura T, Otani H, Nakao Y, Hattori R, Osako M, Imamura H IGF-I differentially regulates Bcl-xL and Bax and confers myocardial protection in the rat heart. Am. J. Physiol. Heart Circ. Physiol. 2001:280, H1191–H1200

[20] Nakamura T, Mizuno S, Matsumoto K, Sawa Y, Matsuda H. Myocardial protection from ischemia/reperfusion injury by endogenous and exogenous HGF. J. Clin. Invest. 2000:106 ,1511–9

[21] Ogata Y, Takahashi M, Ueno S, Takeuchi K, Okada T, Mano H, Ookawara S, Ozawa K, Berk BC, Ikeda U, Shimada K, Kobayashi E. Antiapoptotic effect of endothelin-1 in rat cardiomyocytes *in vitro*. Hypertension 2003:41, 1156–63

[22] Kobara M, Tatsumi T, Kambayashi D, Mano A, Yamanaka S, Shiraishi J, Keira N, Matoba S, Asayama J, Fushiki S, Nakagawa M. Effects of ACE inhibition on myocardial apoptosis in an ischemia–reperfusion rat heart

model. J. Cardiovasc. Pharmacol. 2003: 41,880–9
[23] Minn AJ,Boise LH, Thompson CB. Bcl-x(S) anatagonizes the protective effects of Bcl-x(L). J Biol Chem 1996:271,6306-12
[24] Dixon EP, Stephenson DT, Clemens JA, Little SP. Bcl-X short is elevated following severe global ischemia in rat brains. Brain Res. 1997:776,222- 9
[25] Heermeier K, Benedict M, Li M, Furth P, Nuñez G, Hennighausen L. Bax and Bcl-xs are induced at the onset of apoptosis in involuting mammary epithelial cells. Mech. Dev. 1996:56, 197-207
[26] Igase M, Okura T, Kitami Y, Hiwada K. Apoptosis and Bcl-xs in the intimal thickening of balloon-injured carotid arteries. Clin. Sci. 1999:96, 605-12
[27] Hudlicka O, Brown M, Egginton S. Angiogenesis in skeletal and cardiac muscle. Physiol. Rev. 1992:72, 369–417
[28] Schaper W. New paradigms for collateral vessel growth. Basic Res. Cardiol. 1993:88,193–8
[29] Waltenberger J. Modulation of growth factor action: Implications forthe treatment of cardiovascular diseases. Circulation 1997:96,4083–94
[30] Sellke FW, Ruel M. Vascular growth factors and angiogenesis in cardiac surgery. Ann. Thorac. Surg.2003:75, S685–90
[31] Simons M, Bonow RO, Chronos NA, Cohen DJ, Giordano FJ, Hammond HK, Laham RJ, Li W, Pike Mm, Sellke FW, Stegmann TJ, Udelson JE, Rosengart TK. Clinical trials in coronary angiogenesis: Issues,problems, consensus: An expert panel summary. Circulation 2000:102,E73–E86

The Prophylactic Effects of Chinese Herbal Extract, 'Braintone®', on Stroked Wistar Rats

Lishan Low[a], Wanhui Wong[a], Wei Duan[b], Shufeng Zhou[c], Vincent Chou[d] and Yi Zhun Zhu[a,e]*

Department of [a]Pharmacology and [b]Biochemistry and [c]Pharmacy, National University of Singapore, Singapore 117597. [d]Hong Kong Health Care Association Ltd., Hong Kong. [e]Department of Pharmacology, School of Pharmacy, Fudan University, Shanghai 200032, China

**Corresponding email: zhuyz@shmu.edu.cn*

INTRODUCTION

Stroke is one of the most common diseases occurring in the world today, being the third leading killer in the industrialized world after heart disease and cancer [1]. Many experimental studies have shown that Losartan lowers the risk of vascular events, including stroke [2]. Losartan is a nonpeptide Angiotensin II AT1 receptor antagonist which relaxes peripheral blood vessels to lower blood pressure. Blockade of Angiotensin II formation with ACE inhibitors prevents and reverses alterations in cerebrovascular autoregulation, improves tolerance to hypotension, enhances neurological recovery and protects against focal cerebral ischemia [3-5].

The treatment of stroke has globally been focused on drugs like Losartan but natural herbs have been gaining attention as alternative medicines. Cells have two separate defense mechanisms – enzymatic and non-enzymatic - to protect them from oxidative damage. The generation of reactive oxygen species (ROS) after ischemic damage can offset the balance between the antioxidants and ROS, leading to the depletion of these antioxidants and accumulation of ROS. Many natural herbs are able to scavenge these free radicals. They do so by either neutralizing the effects of ROS through non-enzymatic reactions or by enhancing the activities of natural antioxidant enzymes found in the body like SOD and GST [6-8]. Braintone is an herbal supplement that consists of 4 main ingredients that are suspected to act as antioxidants. Both Braintone and Losartan are used to evaluate their effects on reduction of infarct volume through reduction in necrosis, as well as the improvement of neurological deficits resulting from stroke. The reduction in infarct volume or ischemic damage is associated with improvement in neurological outcome, although the degree of behavioural impairment is rarely correlated with the size of the infarct [9-10]. In this study in Wistar stroked rats, we evaluated the protective effects of Braintone on the mortality rate, behavioural deficits, infarct volume and capillary density in comparison with those of Losartan.

MATERIALS AND METHODS

Medicines

Braintone powder is supplied by the Hong Kong Health Care Association Ltd. Losartan is supplied by Pfizer, U.S.A..

Animal experiments

Male Wistar rats weighing (250g - 300g) were obtained from the Laboratory Animal Centre, National University of Singapore. 75 rats were randomly divided into four treatment groups and one sham group, as shown in Table 1. All the animals were housed under diurnal lighting conditions and allowed food and water *ad libitum*. The drugs were administered orally for 7 days daily before surgery to occlude the middle cerebral artery MCA). Stroke was induced by irreversible occlusion of the left MCA. Briefly, the rats were anaesthetized with chloral hydrate (350mg/kg, i.p). The left MCA was exposed by a subtemporal craniectomy and occluded using a cauterization pen from the point where it crossed the inferior cerebral vein to a point proximal to the origin of the lenticulo-striate branches.

Table 1. Medicines, dosage and the number of rats in the different treatment groups

Treatment Group (each n=15)	Dosage
Sham	–
Sham + Braintone	500mg/kg/day
Vehicle	20mg/kg/day
Braintone	500mg/kg/day
Losartan	1mg/kg/day

Rectal temperature was monitored and maintained at 37±°C by means of a heating blanket connected to the rectal probe during surgery [11-12]. After occlusion, the wound was closed with sutures. Sham rats were operated by omitting the occlusion only. Subsequently, the rats were sacrificed after the treatment period of 14 days by decapitation.

Infarct volume assessment

The area of infarct volume was assessed with 2,3,5-triphenyltetrazolium chloride (TTC) staining. The brains were removed without the cerebellum and cleared of overlying membranes. The cerebrum was sectioned into 8 pieces of 2mm thick coronal slices using a brain-sectioning block. The samples were stained with 0.1% TTC solution at 37°C for 30 minutes and then preserved in 4% formalin solution [13]. The fixed coronal slices of an animal were arranged as the sequence they were cut and scanned with 600dpi resolution. The infarct size was analyzed with an image analyzer system (Scion image for windows) and converted by integration (including correction for edema and atrophy) to the true infarct size of ischemic damage in the whole hemisphere using the equation indicated below:

$$S' = (S/Ti) * Tc$$

$$\text{Percent hemisphere true infarct size} = (\Sigma S'/\Sigma Tc) * 100$$

(S = Stroke area, Tc = Total area of contralateral hemisphere,
Ti = Total area of ipsilateral hemisphere)

Clinical evaluation

For the clinical evaluation, a scale of 0 to 5 was used to assess the behavioural and motor changes observed in the animal after the MCAO procedure using the neurological grading system developed by Bederson et al. [14]. A score of 0 is assigned when the animal, when suspended from the tail, exhibited normal behaviour that included extension of both forelimbs toward the floor. A score of 1 was assigned when the forelimb contralateral to the side of the MCAO was consistently flexed during the suspension and there was no other abnormality observed. The rats were then placed on the ground and allowed to move freely and were observed for circling behaviour. Rats that moved spontaneously in all directions but showed a monodirectional circling toward the paretic side when given a light jerk of the tail were assigned a score of 2. A score of 3 was assigned to rats that showed a consistent spontaneous contralateral circling. A score of 4 was assigned to rats that walked only when stimulated. A score of 5 was assigned to rats that were dead at the day of assessment. Rats that showed a higher clinical score portrayed all features of the lower grades as shown in Figure 1.

Evaluation of capillary density and morphology

The differentiation of coronary capillaries and evaluation of morphological changes after cerebral ischemia were performed using Hematoxylin and Eosin (H&E) Staining. At the end of treatment, the animal was perfused and fixed with 2% paraformaldehyde. The brain was then collected. Tissues collected were post-fixed in 2% paraformaldehyde for two hours. They were then transferred into 25% sucrose in phosphate

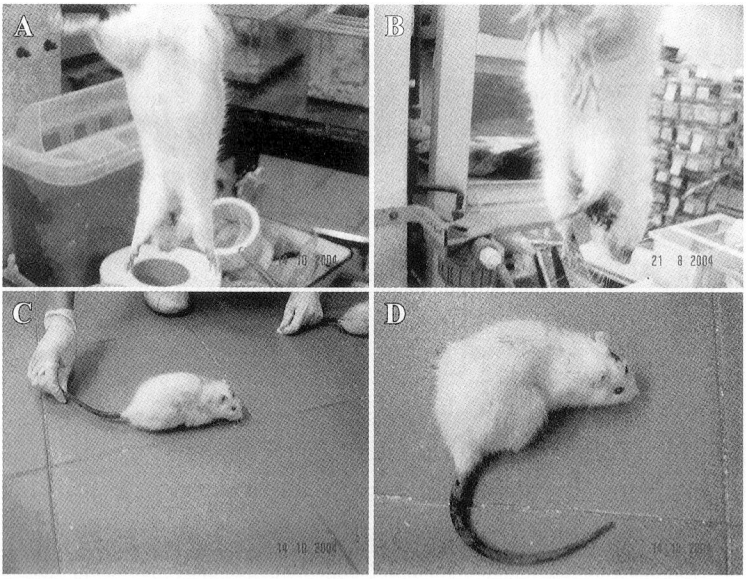

Figure 1. Neurological Grading System
(A) Rat extending both forelimbs towards the floor (Score 0)
(B) Contralateral forelimb is flexed during suspension after MCA occlusion (Score 1)
(C) When a slight jerk was done to the tail of the rat, the rat showed monodirectional circling towards the paretic side (Score 2)
(D) Rat showed consistent spontaneous contralateral circling (Score 3)

buffer to allow the sample dehydration. Brain tissues containing infarct area were then selected out and dipped in M-1 Embedding Matrix (Tissue Tek) for frozen sectioning, frozen in liquid nitrogen and stored at -80ºC. The cerebral sections were then cut from the middle part of the frozen brain at 20μm using a LEICA CM1800 Cryostat. 4 – 6 sections will be obtained from each brain. H&E staining was carried out and the sections were examined and photographed using a fluorescent Leica® microscope (Leica Microsystems, IL, USA).

Statistical analysis

All results obtained were analysed using Microsoft Excel. The standard deviations and p-values were obtained using two-tailed student's t-test to test the significance of differences between groups.

RESULTS

Mortality rate after stroke

The mortality rates for the period of study are shown in Figure 2. Mortality rates were compiled according to the number of rats dead at day 13 as a percentage of the total number of rats in the group. Rats that died before the induction of stroke on day 7 are discounted, as well as rats that died less than 24 hours after the induction of stroke. The mortality rate of all the treatment groups at the end of the treatment period were 0% (0/15) for Sham, 0 % (0/15) for Sham + Braintone-treated, 0% (0/15) for Braintone-treated and 0% for Losartan-treated group except for the Vehicle group at 6.7% (1/15).

Infarct volume assessment

As shown in Figures 3 and 4, the Vehicle group has the highest infarct volume followed by the Losartan-treated group with the Braintone-treated group having the lowest value. However, no significant difference was found in infarct volume between the Vehicle, Braintone- and Losartan-treated groups due to big variations in the infarct volume. However Braintone-treated group had the smallest infarct volume.

Clinical evaluation

The neurological scores for the vehicle-treated group were significantly higher than for the sham-operated group on day 10 and day

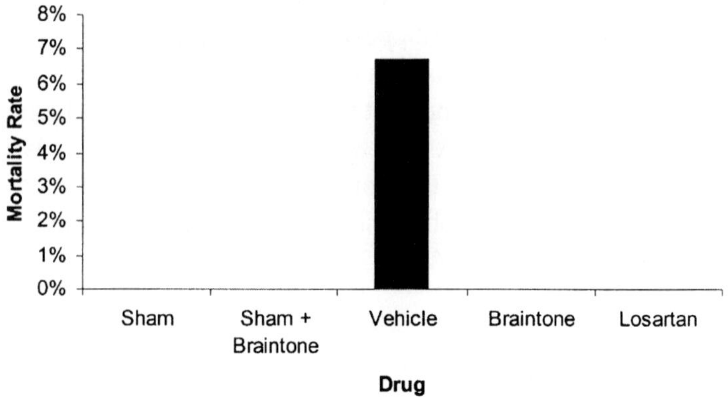

Figure 2. Mean mortality rate of the different treatment groups after stroke

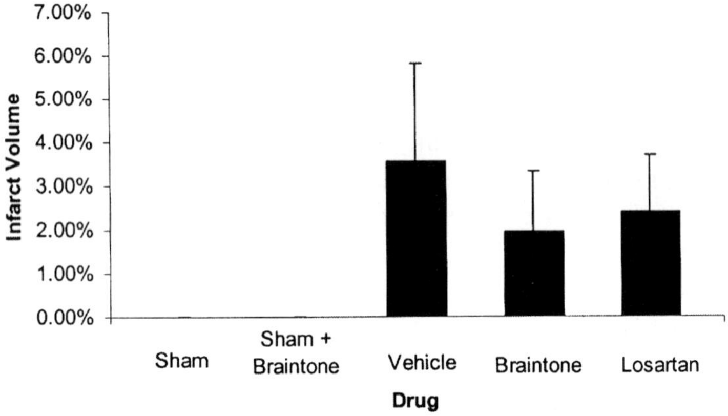

Figure 3. Infarct volumes of the different treatment groups after stroke. Data is presented as mean ± SD; each group, n=3.

13 ($p<0.01$). For the former, the score on day 8 was significantly higher than that for the Braintone-treated group; no significant difference was found between vehicle- and Losartan-treated groups. On day 10, the vehicle-treated group had a higher score than both Braintone- and Losartan-treated groups, though the differences were not significant. On day 13, the score for the vehicle-treated group was significantly higher than that for the Braintone-treated group ($p<0.01$) and the Losartan-treated group ($p<0.05$) [Table 2; Figure 5].

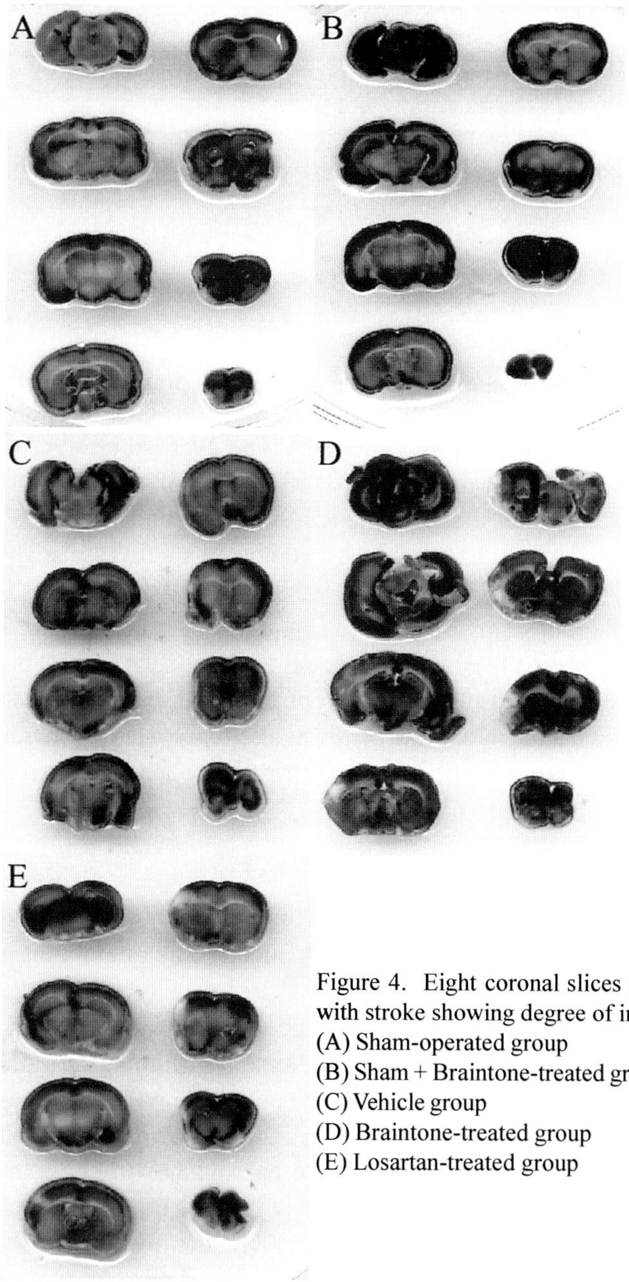

Figure 4. Eight coronal slices of brains of rats with stroke showing degree of infarct volume.
(A) Sham-operated group
(B) Sham + Braintone-treated group
(C) Vehicle group
(D) Braintone-treated group
(E) Losartan-treated group

Table 2. Neurological Scores over a 14-days course following stroke

Treatment Groups	Day 8	Day 10	Day 13
Sham	0	0	0
Sham + Braintone	0	0	0
Vehicle	1.5±1.13	1.2±1.32††	1.6±1.45††
Braintone	0.6±0.72*	0.8±0.83	0.4±0.51**
Losartan	0.8±0.74	0.4±0.64	0.6±0.51*

Evaluation of capillary density and morphology

Blood vessels of the brains of rats with stroke were counted under a microscope by amplifying by 100 times. The average number of microvessels of each group was calculated in 3 sections of each brain and 3 samples for each group. As shown in Figures 6 and 7, both Braintone- and losartan-treated groups had the highest number of microvessels. The capillary density of the brains with stroke was significantly increased in the Losartan-treated group compared to the vehicle-treated group ($p<0.05$). The capillary density in the Braintone-treated group was the highest but this difference was not statistically significant due to the large SD values. The extent of fragmented nuclei and cells was also observed to be higher for the vehicle-treated group than the other treated groups, as shown in Figure 7.

DISCUSSION

This study showed that Braintone produced similar outcomes in terms of mortality rate, infarct volume and capillary density compared to Losartan. However, Braintone was also able to lower the extent of necrosis of the cells which resulted in a lower infarct volume and neurological score. This will also lead to a lower mortality rate occurring with lesser cell damage. Braintone helps to significantly improve neurological behaviours after induction of stroke compared to vehicle and was shown to be better than Losartan in improving neurological deficits. Losartan does not possess antioxidant properties, with its main effect being the lowering blood pressure. Thus, it was not found to lower the infarct volume as well as Braintone.

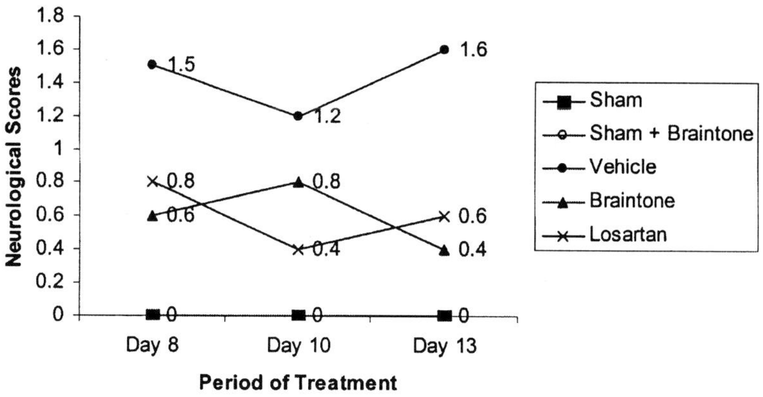

Figure 5. Neurological Scores after stroke over a 14-day course of drug treatment. Data is presented as mean ± SD; each group, n=15.

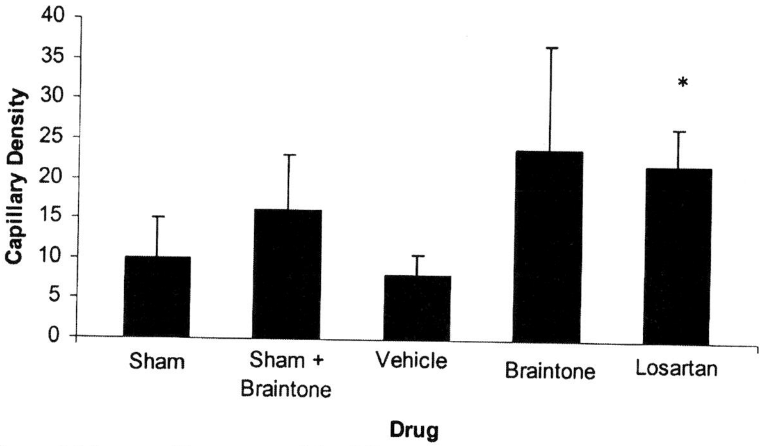

Figure 6. Mean capillary count of the different treatment groups after stroke Data is presented as mean ± SD; each group, n=3.

H&E staining showed the extent of cell death and vascularization in coronal sections of the brain. Necrotic cells generally appear more fragmented and less compact compared to healthy cells [15]. Due to cell lysis and DNA damage, the cells fragment and cause fuzzy edges around the nuclei. In Braintone- and Losartan-treated groups, the degree of brain injury was less compared to the vehicle-treated group.

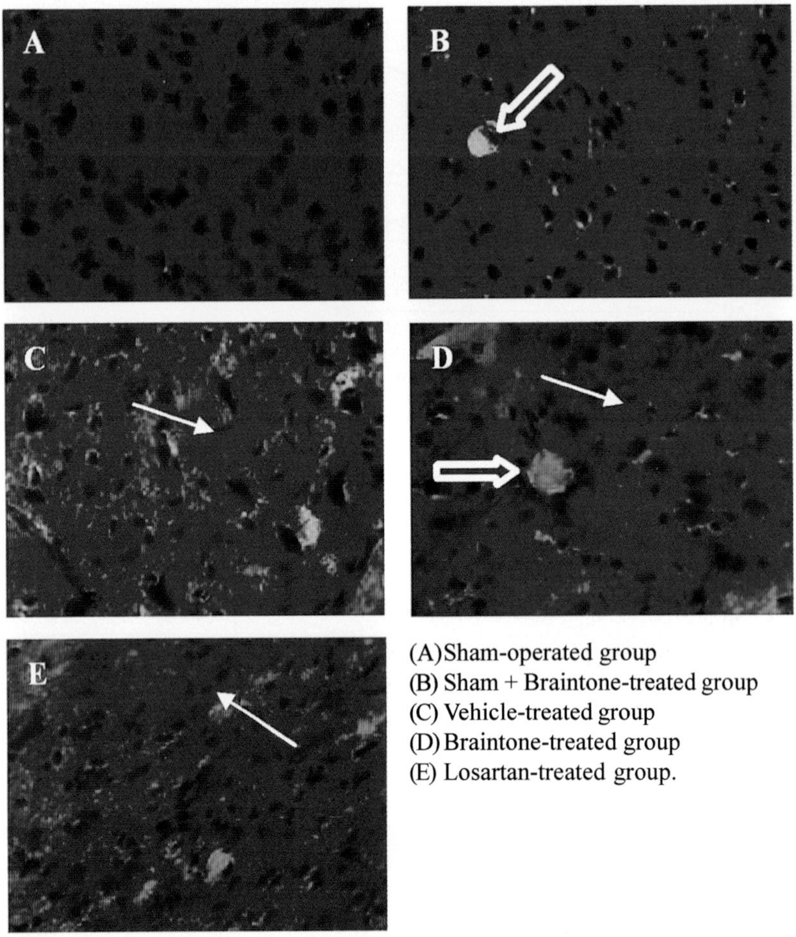

(A) Sham-operated group
(B) Sham + Braintone-treated group
(C) Vehicle-treated group
(D) Braintone-treated group
(E) Losartan-treated group.

Figure 7. Light photomicrographs (40 x magnification) of brain sections with Hematoxylin and Eosin staining, indicating the degree of morphological changes.

Neurons close to normal-appearing nuclei but slightly shrunken indicate slight damage. They are represented by white arrowheads while boxed arrows represent capillaries. In the Sham-operated and Sham + Braintone-treated groups, the cells are not fragmented and the nuclei are compacted together. The Braintone- and Losartan-treated groups have cells that are fragmented but to a lesser extent than those seen for the Vehicle group.

The number of capillaries found in the brain sections stained with H&E was the highest in the Braintone-treated rats compared with the other groups, suggesting that Braintone has an effect on angiogenesis. Even without stroke induction, Braintone was able to promote angiogenesis as seen in the high capillary density of the sham + Braintone-treated group. With the increase in capillary density, more blood can be delivered via alternative routes other than the damaged MCA branches. Newly formed vessels could increase blood flow and hence provide a compensatory response to correct the imbalance between the perfusion capacity of vessels and the need for oxygen and nutrients by the brain, especially at the penumbra region [16].

Braintone appears to be able to produce beneficial effects in the brains of rats with stroke similar to Losartan. However, the potential of using Braintone globally for stroke still needs to be evaluated by clinical trials. The mechanisms involved between the two drugs differ and can be further determined by measuring the blood flow, blood pressure, apoptotic gene expression and effect on antioxidants enzymes through antioxidant assays. Further evaluation will need to be conducted on the individual purified components as well as combinations of two or more of the ingredients present in Braintone. Braintone has opened a new avenue for using Chinese herbal extracts to prevent ischemic stroke.

ACKNOWLEDGEMENT

The authors would like to acknowledge research grant from National Medicine Research Council, Singapore (R-184-000-037-213 and R-184-000-109-213) and partial support from the Hong Kong Health Care Association Ltd., Hong Kong.

REFERENCES

[1] Miller L.P. Stroke Therapy: basic, preclinical, and clinical directions; 1999. p. 3-423

[2] Dahlöf B, Devereux RB, Kjeldsen SE, Julius S, Beevers G, et al. Cardiovascular morbidity and mortality in the Losartan Intervention for Endpoint Reduction in Hypertension Study (LIFE): a randomised trial against atenolol. Lancet 2002; 359: 995–1003

[3] Yasuaki N, Takeshi I, Juan MS, and Kathryn S. Angiotensin II AT_1 Blockade Normalizes Cerebrovascular Autoregulation and Reduces Cerebral Ischemia in Spontaneously Hypertensive Rats. Stroke 2000; 31:2478-86

[4] Torup M, Waldemar G, Paulson OB. Ceranapril and cerebral blood flow autoregulation. J Hypertens 1993; 1: 399–405

[5] Dai WJ, Alexandra F, Thomas H, Thomas U, Juraj C, and William MA. Blockade of Central Angiotensin AT1 Receptors Improves Neurological Outcome and Reduces Expression of AP-1 Transcription Factors After Focal Brain Ischemia in Rats. Stroke 1991; 30: 2391-9

[6] Ji XY, Tan B KH, Zhu YC, Linz W, Zhu YZ. Comparison of cardioprotective effects using ramipril and DanShen for the treatment of acute myocardial infarction in rats. Life Sci 2003; 73:1413-26

[7] Sun J, Tan B KH., Huang SH., Whiteman M, Zhu YZ. Effects of natural products on ischemic heart diseases and cardiovascular system. Acta Pharmacol Sin 2002; 23: 1-7

[8] Zhu YZ, Huang SH, Whiteman M, Tan B KH, Zhu YC. Antioxidants in Chinese herbal medicines: a biochemical perspective. Nat Prod Rep 2004; 21:478-89

[9] Menzies SA, Hoff JT, Betz AL. Middle Cerebral Artery Occlusion in Rats: A Neurological and Pathological Evaluation of a Reproducible Model. Neurosurgery 1992; 31(1): 100-7

[10] Rogers DC, Campbell CA, Stretton JL, Mackay KB. Correlation between motor impairment and infarct volume after permanent and transient middle cerebral artery occlusion in the rat. Stroke 1997; 47: 469-74

[11] Lu Q, Zhu YZ, Wong P TH. Neuroprotective effects of candesartan against cerebral ischemia in spontaneously hypertensive rats. Neuroreport 2005; 16:1963-7

[12] Zhu YZ, Chimon GN, Zhu YC, Lu Q, Li B, Hu HA, Yap EH, Lee HS, Wong P TH. Expression of angiotensin II AT2 receptor in the acute phase of stroke in rats. Neuroreport 2000; 11: 1191-4

[13] Bederson JB, Pitts LH, Germano SM, Nishimura MC, Davis RL, Bartkowski HM. Evaluation of 2,3,5-triphenyltetrazolium chloride as a stain for detection and quantification of experimental cerebral infarction in rats. Stroke 1986; 17: 1304–8

[14] Bederson JB, Pitts LH, Miles Tsuji BS, Nishimura MC, Davis RL, Bartkowski H. Rat middle cerebral artery occlusion: evaluation of the model and development of a neurologic examination. Stroke 1986; 17: 472–6

[15] Levitan and Kaczmarek. The *Neuron: Cell and Molecular Biology*, Oxford University Press. 2002

[16] Sun YJ, Jin KL, Lin X, Childs J, Mao XO, Logvinova A and Greenberg DA. VEGF-induced neuroprotection, neurogenesis, and angiogenesis after focal cerebral ischemia. J. Clin Inves 2003; 111: 1843-51

Therapeutic Applications of Ceylon Tea: Potential and Trends

Tissa Amarakoon[a,]*, Shang Hong Huang[b] and Ranil De Silva[b].
[a]Biochemistry Division, Tea Research Institute, Talawakelle, Sri Lanka,
[b]Dept. of Anatomy, Faculty of Medical Sciences, University of Sri Jayewardenepura, Nugegoda, Sri Lanka
**Corresponding email: tissa61@yahoo.co.uk*

INTRODUCTION

Health benefits have been attributed to tea since the beginning of consumption in 27th century B.C. In the recent past tea and its components have been studied very intensively in relation to their beneficial effects on humans. Out of the many phytochemicals found in tea, polyphenolic flavonoids have been identified as the main bioactive components which provide beneficial effects after consumption. The radical scavenging and mineral complexing antioxidant activity of the flavonoids are the predominant mechanisms by which the beneficial effects are imparted. Many other mechanisms such as altering the endothelial function, direct binding with carcinogens, inducing Phase I and II enzymes, apoptosis, inhibition of transcription factors and anti-angiogenesis are also exhibited by tea components. All these may be synergistically acting to impart the beneficial effects.

Research on bioactive compounds of tea carried out so far had mainly focused on effects of tea on cardiovascular disease, cancer and oral health. This paper discusses the chemical components in different types of tea, their bioavailability, and biological activities and how these activities could act to reduce the risk of certain diseases.

CHEMICAL COMPOSITION OF TENDER SHOOTS OF TEA

There are two main types of tea, ‘green tea’ and ‘black tea’, produced in the world. Both are processed products of tender shoots of *Camellia sinensis* and the differences are due to the method of processing (Table 1).

Table 1. Chemical composition of tender shoots of tea (*Camellia sinensis*)

Cold water-soluble		
Flavanols:	Epigallocatechin gallate (EGCG)	9 - 13
	Epigallocatechin (EGC)	3 - 6
	Epicatechin gallate (ECG)	3 - 6
	Epicatechin (EC)	1 - 3
	Gallocatechin (GC)	1 - 2
	Catechin (C)	1 – 2
Flavonols and their glycosides		3 – 4
Leucoanthocyanins		2 – 4
Phenolic acids		4
Total polyphenols		**27 - 40**
Caffeine		3 – 4
Amino acids:	Theanine	2
	Others	2
Carbohydrates		4
Organic acids		0.5
Volatile compounds		0.01
Partially hot water-soluble		
Polysaccharides:	Starch	2 – 5
	Other	12
Protein		15
Ash (inorganic material)		5
Water-insoluble		
Cellulose		7
Lignin		6
Lipids		3

Major water soluble chemical components in the tender shoots are the polyphenolic flavonoids. Flavanols (flavan-3-ols) are the predominant flavonoids in fresh tea leaves (Figure 1). These are commonly called as catechins.

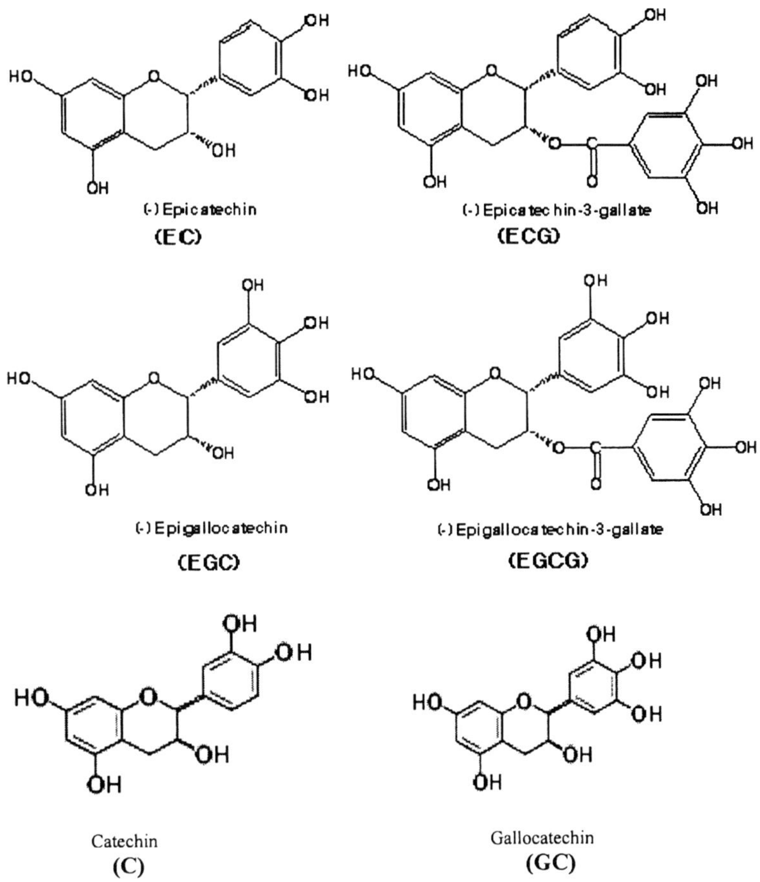

Figure 1. Major flavan-3-ols (catechins) in fresh shoots of tea.

In addition, flavonols such as quercitin, myricitin and kaempferol are also found in fresh leaves of tea (Figure 2). These are mostly present as their glycosides.

BLACK TEA PROCESSING

In black tea processing the plucked shoots are withered to reduce the moisture content by approximately 50%. Then the shoots are rolled by mechanical rollers to macerate and break them. This process of breaking up the leaves facilitate the contact of certain enzymes and substrates, which

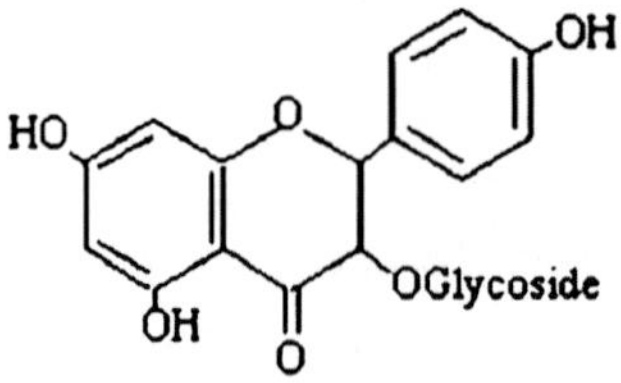

Kaempferol Glycoside

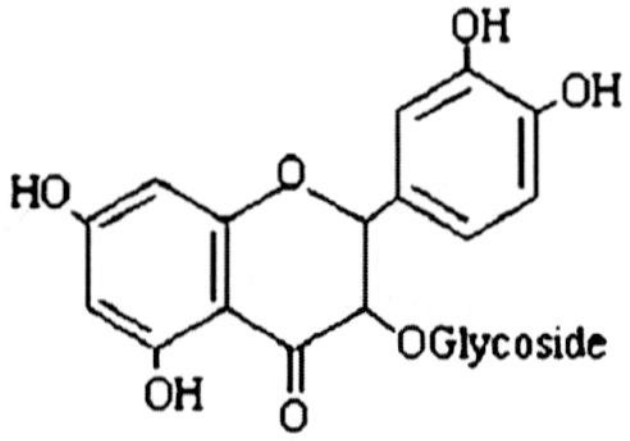

Quercitin Glycoside

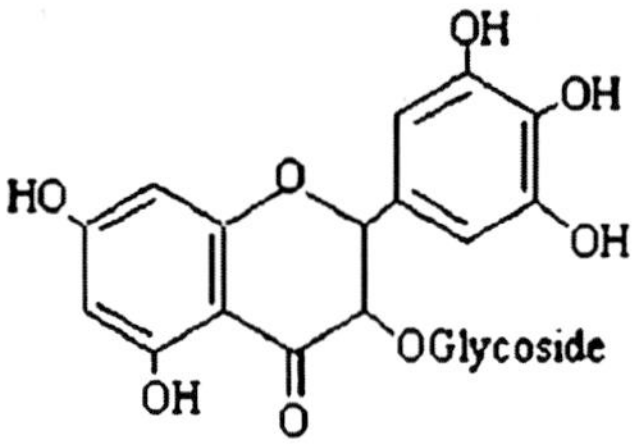

Myricitin Glycoside

Figure 2. Major flavonol glycosides in fresh shoots of tea

are compartmentalized in the intact shoots. Therefore, rolling starts a series of enzymatic reactions. Most important reaction is the conversion of flavan-3-ols (commonly called as catechins), the predominant component in the fresh shoots, to what is known as theaflavins and thearubigins, dimeric and polymeric compounds. Then the tea is dried at high temperature to reduce the moisture content to about 3% to improve the keeping qualities. High temperatures in dryers denature the enzymes and halt the enzymatic reactions [1].

GREEN TEA PROCESSING

Major difference in green tea processing is that after plucking, shoots are subjected to heat by steaming or pan firing (keeping on a heated pan for a short time) (Figure 3). This inactivates the enzymes in the shoots and therefore, after rolling, enzymatic reactions do not take place. Thus the chemical composition of green tea is similar to that of fresh tender shoots.

CHEMICAL COMPOSITION OF TEA BREW

The important components in tea, as far as health is concerned, are the water soluble components that come into the cup on brewing (Table 2). A typical brew of tea contains 0.35% tea solids in water. Therefore, tea (without milk and sugar) does not contain significant amounts of major

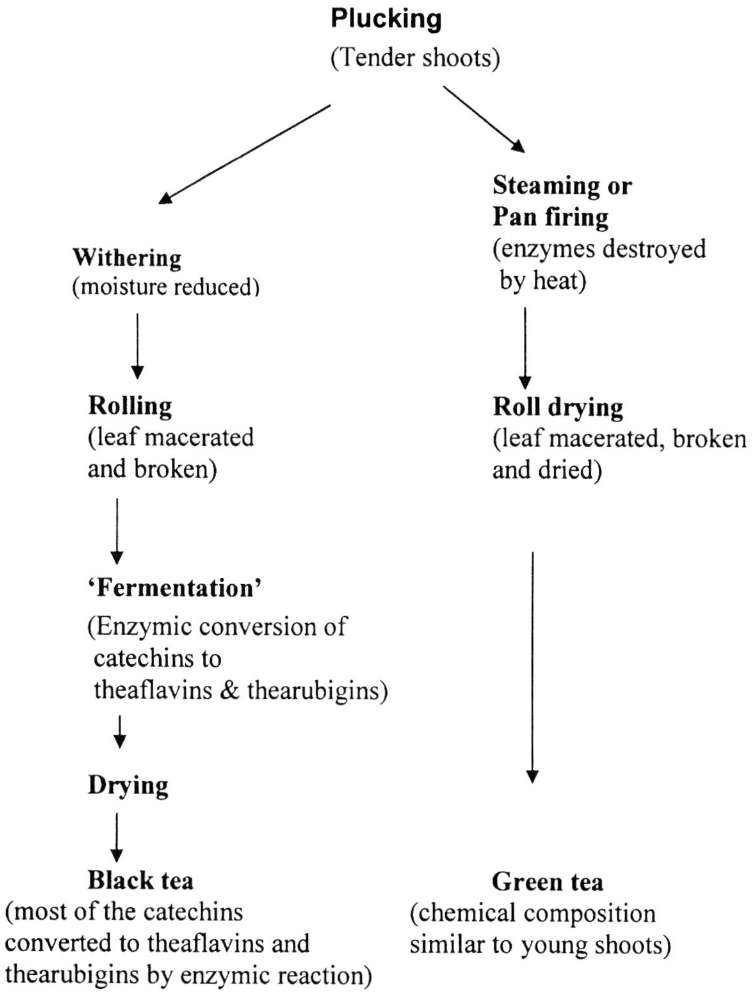

Figure 3. Steps in black and green tea processing.

nutrients (carbohydrates, protein and lipids). The calorific value of tea is almost zero.

Table 2. Chemical composition of black and green tea brew
As % of extracted solids

Black tea	Green tea	
Catechins	9%	30%
Theaflavins	4%	-
Simple polyphenols	3%	2%
Flavonols	1%	2%
Other polyphenols (including thearubigin)	23%	6%
Total polyphenols	***40%***	***40%***
Amino acids: Theanine	3%	3%
Other Amino Acids	3%	3%
Protein/Peptides	6%	6%
Organic acids	2%	2%
Carbohydrates: Sugars	7%	7%
Other Carbohydrates	4%	4%
Lipids	3%	3%
Methyl Xanthins: Caffeine	3%	3%
Other Methyl Xanthins	<1%	<1%
Minerals: Potassium	5%	5%
Sodium	<0.05%	<0.05%
Other minerals	5%	5%
Volatile (aroma) compounds	Trace	Trace

However, two non-nutrient components in tea, caffeine and polyphenols are important as these have *in vivo* biological activities. Caffeine, the well known stimulant, acts on the central nervous system, increases alertness and reduces feelings of drowsiness and fatigue. The polyphenols are mainly responsible for the most beneficial effects reported for tea. Antioxidant activities of these polyphenols appear to be the mode of action for most of the beneficial effects [2].

BIOAVAILABILITY OF TEA COMPONENTS

Bioavailability and pharmacokinetics of tea components in different tissues of the body are less intensively studied than their biological activities. However, it is known that polyphenols undergo degradation by colonic bacteria and enter the enterohepatic circulation. It is likely that, during their transport and distribution in the body, polyphenols are bound to proteins.

Differences in polyphenol bioavailability at different sites, under varying conditions, may be one reason for the variable results obtained in studies investigating biological activities of tea. It was known for a long time that different types of tea polyphenols show different degrees of bioavailability [3]. Recent advances in biochemical techniques have given some information on the fate of tea components following their ingestion. In some studies, the chemical constituents of tea were measured in tissues directly. In other studies the changes of parameters, such as the antioxidant activity in tissues, have been employed to estimate the tea components in tissues, both in animal models and in humans.

Consumption of three cups of strong tea per day (2 g of tea per cup), for two weeks, increases total polyphenol levels in human blood by 25 per cent [4]. Although it has been suggested that milk added to tea binds polyphenols and reduces their bioavailability, Hollman *et al.* have shown that the addition of milk does not alter the bioavailability of tea flavonols [5].

In a study by Nakagawa and Miyazawa [6] with rats, 60 minutes after a single oral administration of epigallocatechin-3-gallate (EGCG), at 500 mg/kg body weight, the levels of EGCG were: plasma 12.3 nmol/ml, liver 48.4 nmol/g, brain 0.5 nmol/g, small intestinal mucosa 565 nmol/g and colonic mucosa 68.6 nmol/g. These values indicate that EGCG is absorbed from the digestive tract, particularly through the intestinal mucosa.

He and Kies [7] investigated the impact of green and black tea consumption on polyphenol concentration in blood, urine and faeces. Ten healthy adults received a laboratory-controlled constant diet comprising of normal food. A 56-day study period was divided randomly into four 14-day periods in which no tea, green tea, black tea or decaffeinated black tea were given three times a day. Green tea consumption resulted in the highest urinary and faecal elimination, and the highest retention of polyphenols in the blood, followed by black tea and decaffeinated black tea.

The bioavailability of prominent black tea catechins were assessed in humans drinking tea throughout the day. After 5 days of consuming a low-flavonoid diet, subjects drank a black tea preparation containing 15.48, 36.54, 16.74, and 31.14 mg of epigallocatechin (EGC), epicatechin (EC), epigallocatechin gallate (EGCG) and epicatechin gallate (ECG), respectively, at four time points (0, 2, 4 and 6 h). Blood, urine and faecal specimens were collected over a 24- to 72-h period and catechins were quantified by HPLC. Plasma concentrations of EGC, EC and EGCG increased significantly relative to the baseline concentrations. Plasma EGC, EC and EGCG peaked at 5 h, and ECG peaked at 24 h. Urinary excretion of EGC and EC, which peaked at 5 h, was raised relative to baseline amounts, and faecal elimination of all four catechins was increased relative to baseline. Approximately 1.68 per cent of ingested catechins were present in the plasma, urine and faeces, and the bioavailability of the gallated catechins was lower than that of the non-gallated forms [8].

A study was conducted by Shahrzad [9] to find the bioavilability of gallic acid (GA) in humans. Black-tea brew was found to contain 93 per cent of its GA in free form. After the administration of a single oral dose of acidum gallicum tablets or tea (each containing 0.3 mmol GA) to 10 volunteers, plasma and urine samples were collected over various time intervals. Concentrations of GA and its metabolite, 4-O-methylgallic acid (4-O-MGA), were determined. GA from both tablets and tea was rapidly absorbed and eliminated with mean half-lives of 1.19 ± 0.07 and 1.06 ± 0.06 h, and mean maximum concentrations in the plasma of 1.83 ± 0.16 and 2.09 ± 0.22 μmol/l, respectively. After oral administration of tablets and black tea, 36.4 ± 4.5 and 39.6 ± 5.1 per cent of the GA dose were extracted in urine as GA and 4-O-MGA, respectively. The relative bioavailability of GA from tea, compared with that from the tablets, was 1.06 ± 0.26, showing that GA is as available from drinking tea as it is from taking GA tablets.

Plasma and urinary concentration of the flavonols, quercetin and kaempferol, were measured after intake from onions and black tea. Fifteen human subjects were given black tea, 1600 ml/day, or fried onions, 129 g/day. Tea provided 49 mg of quercetin and 27 mg of kaempferol/day, while onion provided quercetin only, at 13 mg/day. The concentration of quercetin after tea increased four-fold from baseline levels and three-fold after onion. The concentration of kaempferol after tea increased about six-fold [10]. Absorption of EGCG in rats had been studied using radio-labelled EGCG. After oral administration of [4-(3)H] EGCG, the radioactivity in blood, the major tissues, urine, and faeces was measured at periods of time. The radioactivity in blood and most tissues was low at 4 h, began to increase after 8 h, peaked at 24 h, and then decreased. Major urinary elimination of radioactivity occurred in the 8-24 h period, and the cumulative radioactivity excreted by 72 h was 32.1% of the dose [11].

In a similar study radioactive [^{3}H] EGCG solution was directly administered into the stomachs of CD-1 male and female mice, and the radioactivity in the digestive tract, various organs, blood, urine and faeces was measured after 1, 6 and 24 h. Radioactivity was detected after one hour in all organs. Radioactivity was detected in the digestive tract, liver, lung, pancreas, mammary glands, skin, brain, uterus, ovary and testes. Within 24 hours, 6.4 per cent (males) and 6.6 per cent (females) of the total administered radioactivity were excreted in the urine, and 33.1 per cent (males) and 37.7 per cent (females) were eliminated in the faeces. A second administration after 6 h increased the radioactivity in blood, brain, liver, pancreas, bladder and bone by four to six times [12].

Following the oral administration of EC, EGC, ECG and EGCG to rats, the presence of all these compounds was detected in the blood of the portal vein and measured using HPLC and MS [13].

Similarly, the absorption of EGCG into the circulatory system of rats was studied after oral administration of 50-mg doses. The concentrations in the plasma peaked about half an hour after administration and then decreased quickly [14].

Lee *et al.* [15] studied the human plasma levels of EGCG, EGC, ECG and EC after consumption of 1.2 g of green tea in warm water. One hour later, the levels were EGCG 46-268 ng/ml, EGC 82-206 ng/ml, and EC 48-80 ng/ml. ECG was not detected.

In a study by Nakagawa *et al.* [16], healthy human subjects were given 3, 5, or 7 capsules of green-tea extract orally (corresponding to 225, 375, and 525 mg of EGCG and 7.5, 12.5, and 17.5 mg of EGC, respectively). The plasma EGCG and EGC concentrations before administration were both below the detection limit (< 2 pmol/ml), but 90 min after administration the concentrations increased to 657, 4300, and 4410 pmol of EGCG/ml, and 35, 144, and 255 pmol of EGC/ml, in the subjects who received 3, 5, and 7 capsules, respectively. These concentrations of EGCG and EGC in the plasma were statistically different, and dependent on the dose of green-tea extract given.

Nakagawa *et al.* [17] investigated the effect of green-tea catechin supplementation on the antioxidant capacity of human plasma. Eighteen healthy male volunteers, who orally ingested green-tea extract (254 mg of total catechins/subject), had 267 pmol of EGCG/ml in the plasma 60 min after administration. The levels of a plasma oxidant, phosphatidylcholine hydroperoxide (PCOOH), attenuated from 73.7 pmol/ml in the control, to 44.6 pmol/ml in the catechin-treated subjects, being correlated inversely with the increase in plasma EGCG level. Of the total catechins consumed by humans, a large percentage passes out unchanged in the faeces [18] and, of that absorbed into the body, approximately 60 per cent are excreted in the urine, and the rest in the bile [19]. Catechins are absorbed into the blood in the portal vein of humans and rats [20], and are thereby conveyed from the gastrointestinal tract to the liver.

Relatively less number of studies was carried out on the bioavailability of theaflavins from black tea probably due to the difficulties of detection in tissues as they are present as glycosides or sulphates. Therefore, instead of direct measurements, biological activities such as antioxidant activity in different body tissues have been used to estimate the tissue distribution of TF. However, improved methods to detect TF in tissues by digestion of the plasma, urine, or tissue samples with beta-d-glucuronidase and sulfatase, followed by extraction with ethyl acetate and subsequent separation by reversed-phase high-performance liquid chromatography (HPLC) has been reported [21].

In a similar study HPLC-MS procedure has been developed for the analysis of theaflavin in human plasma and urine. Levels were measured after enzymatic deconjugation, extraction into ethyl acetate, and separation by HPLC, using tandem mass spectrometry as a detecting system. Two healthy volunteers consumed 700 mg theaflavins, equivalent to about 30 cups of black tea. The maximum concentration detected in blood plasma was 1.0 microg l^{-1} in a sample collected after 2 h. The concentration in urine also peaked after 2 h at 4.2 microg l^{-1} [22].

Tea polyphenols undergo extensive metabolism, and within eight hours of ingestion over 90 per cent of the catechins are excreted. After 24 hours, they are not detectable in the blood. Tea polyphenols, unlike many other compounds, have the ability to penetrate to the tissues of the brain, overcoming the so-called blood-brain barrier. This, together with their antioxidant and iron-chelating capacity, may lead to their becoming useful in the treatment of neurodegenerative diseases resulting from oxidative stress [23].

ANTIOXIDANT ACTIVITY OF TEA

Tea polyphenols are efficient scavengers of free radicals both *in vitro* and *in vivo*. Because of the catechol and gallate moieties (adjacent OH groups) in these molecules they could chelate metal ions (Figure 4). Thus these molecules could reduce the metal mediated LDL oxidation and production of metal ion mediated very reactive free radicals in cellular fluids (Figure 5). In addition, due to their low redox potentials they could neutralize free radicals by providing H atoms.

Hydroxyl groups at positions 5 and 7, and oxygen at position 1, make the carbons at positions 6 and 8 of the flavonoid molecule strongly nucleophilic. Thus, the carbons at positions 6 and 8 can form C-O or C-C bonds with reactive species and neutralise them, causing the flavonoid molecules to undergo oxidative polymerisation. These properties are also important in direct binding to chemical carcinogens which results in their neutralisation.

$$H_2O_2 + Fe^{2+} \longrightarrow Fe^{3+} + OH^{\bullet} + OH^{-}$$

$$H_2O_2 + Cu^{2+} \longrightarrow Cu^{3+} + OH^{\bullet} + OH^{-}$$

Figure 4: Free metal ion mediated radical formation. Eg. Fenton reactions

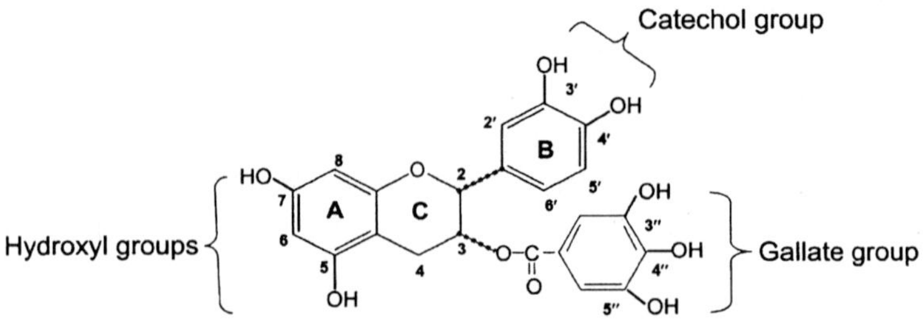

Figure 5: Important functional groups for antioxidant activity

In early research on tea and health, many studies were carried out and the antioxidant activity (or free radical scavenging capacity) of *C. sinensis* extracts and individual tea flavonoids *in vitro* were established. However, as Du Toit et *al* [24] caution, *in vitro* results cannot always be extrapolated to conditions in the human body, owing to variations in bioavailability of tea components. In subsequent *in vivo* studies, the antioxidant activity of green, black and oolong teas, and of tea components, was established.

In a recent *in vitro* anti-oxidant tests, the Ceylon tea extract (Dilmah) showed similar strong anti-oxidant properties (Figure 6) as the popular well-known Chinese green tea (Qin Dao and Bi Xue Chun) in the market by TEAC (trolox equivalent anti-oxidant capacity) value. Though the scavenging effects of Ceylon and green teas on HOCl as determined by Pyrogallol red bleaching assay were lower than that for ascorbic acid (Figure 7) which is used as a positive control, both Ceylon and green teas were still ranking as top anti-oxidant plants among the natural products.

In an *in vitro* study by Zhao *et al.* [25], the scavenging ability of tea polyphenols on superoxide anions were found to be comparable with that of vitamins C and E. Superoxide anions were generated by irradiation of a system containing riboflavin and 5, 5-dimethyl-1-pyrroline-1-oxide, and their formation was measured by electron spin resonance. A green-tea polyphenol extract, vitamin C and vitamin E were added separately, and their scavenging rates on the generated superoxide anions measured. The rates were 72 per cent, 96 per cent and 23 per cent for tea polyphenols, vitamin C, and vitamin E, respectively.

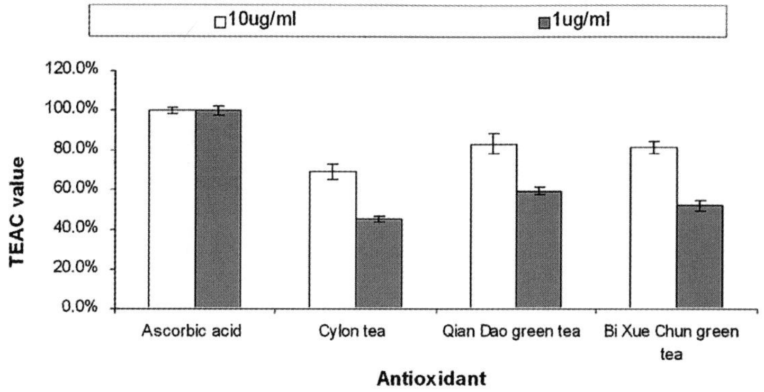

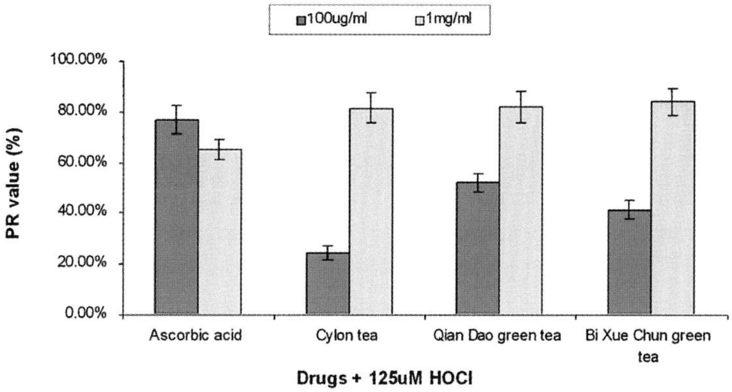

Figure 7: The scavenging effects of Ceylon and green teas on HOCl as determined by Pyrogallol red bleaching assay

Food from plant sources in general is rich in antioxidants. However, the antioxidant capacity of black and green tea is much higher than that of many vegetables as tea contain uniquely high amounts of flavonoids [26].

Du Toit *et al.* [27] used the stable free radical of polyaromatic hydrocarbon, 2, 2-diphenyl-1-picrylhydrazyl (DPPH), to measure and compare the radical scavenging capacities of fruits, vegetables, black, green and oolong teas, and herbal teas. The results are given in vitamin C equivalents. There are no statistically significant differences in the scavenging capacities of black, green and oolong teas. One or two cups of tea are equal in antioxidant capacity to five portions of fruits or vegetables, or 400 mg of vitamin C.

Paganga *et al.* [28] compared the antioxidant capacities of a serving (100g) of fresh fruit and vegetables with previous reports on 150 ml servings of different beverages (500 ml in the case of beer), and expressed the results in Trolox equivalents. The following were found to be equal in their antioxidant activities: one glass of red wine, 12 glasses of white wine, two cups of tea, four apples, five portions of onion, 5.5 portions of egg plant, 3.5 glasses of blackcurrant juice, 3.5 glasses of beer, seven glasses of orange juice, 20 glases of apple juice.

Yen and Chen [29] also tested oolong, green and black tea extracts for their scavenging activity on FRs *in vitro*. Activity against peroxidation was determined by incubating tea samples with linoleic acid, and measuring oxidation in the reaction mixtures by spectrometry, after colour generation with ferric chloride and thiocyanate. Oolong tea showed 73 per cent inhibition of linoleic acid peroxidation, and both green and black tea showed an inhibition of 40 per cent. Superoxide anions were generated by a non-enzymic system, and their oxidative activity in reaction mixtures containing tea samples were determined spectrophotometrically by measuring the reduction of nitro blue tetrazolium. Oolong, green and black teas gave 75 per cent, 58 per cent and 52 per cent inhibition of oxidative activity, respectively. Hydroxyl radicals were generated by the reaction of hydrogen peroxide with ferrous ions and trapped in 5, 5-dimethyl-pyrroline-N-oxide (DMPO). The DMPO-OH formed, a measure of the oxidative activity of hydroxyl, was determined by electron paramagnetic resonance (EPR) spectrometry. Oolong, green and black teas reduced oxidative activity, as shown by the EPR signal, by 92 per cent, 91 per cent and 77 per cent, respectively. The carcinogenic effect of polyaromatic hydrocarbons may be due to the *in situ* generation and oxidative activity of their free radicals. Therefore, the scavenging effect of the tea extracts on the polyaromatic hydrocarbon, DPPH, was also measured. Oolong, green and black teas reduced the oxidative activity of DPPH radicals by 54.6 per cent, 59.4 per cent and 49.0 per cent, respectively.

Scott *et al.* [30] found that catechin and epicatechin have a scavenging action on hydroxyl, superoxide, trichloromethyl peroxyl radicals, and hypochlorous acid.

In another study, catechin was found to be the most effective of the tea components in quenching singlet oxygen [31].

Rice-Evans *et al.* [32] measured the antioxidative capacity of several plant-derived polyphenols, including quercetin, myricetin, epicatechin and catechin which are found in tea. The total antioxidant activity against aqueous phase radicals were measured and compared with the activity of the water-soluble analogue of tocopherol, Trolox. Results are given as the Trolox equivalent antioxidant capacity (TEAC). The TEAC values for quercetin, myricetin, epicatechin and catechin were 4.7, 3.1, 2.5 and 2.2, respectively. These results indicate that the tea components have a greater antioxidant potential than Trolox.

Theaflavins in black tea were shown to be better scavengers of superoxide than gallocatechins [33].

Another approach to demonstrate antioxidant activity of tea is the sparing of endogenous antioxidants and antioxidant enzymes when tea is introduced in the diet or by other means.

In hypercholesterolemic rabbit model, both green and black tea in drinking water resulted in a 1.2 fold increase in Vitamin E after 8 weeks [34]. Addition of green tea catechins to plasma resulted in sparing of endogenous α-tocopherol [35]. In a similar experiment addition of green tea catechins to LDL resulted in a dose dependent sparing of α-tocopherol at concentrations ranging from 2 to 20 microM [36].

The decrease in tissue glutathione (GSH) under oxidative stress conditions is also reduced by tea components. In one such study, providing green tea extract in drinking water prevented liver GSH decrease induced by ethanol administration in rats [37]. In another study, liver glutathione decrease in rats induced by bacterial lipopolysaccharide was synergistically reduced by Vitamin E and black tea extracts [38].

ANTI-INFLAMMATORY ACTIVITY

Inflammation is a key element in the response of the innate immune system to a variety of challenges, including those provided by bacterial and viral infection as well as by damaged or dying host cells. Free radical production, resulting in oxidative stress, is also a part of the inflammatory response. Prolonged inflammation could damage the host cells itself and therefore, inflammations play a major role in the pathogenesis of cardiovascular diseases and other diseases such as rheumatoid arthritis and cataract. Studies have shown that tea components could alter the adverse effects of inflammatory response.

In a cross-sectional study with 1031 healthy men, relationship between tea consumption and inflammatory markers in blood were measured. Blood samples were analysed for C-reactive protein (CRP), serum amyloid A (SAA), serum haptoglobin and plasma fibrinogen. It was found that CRP, SAA and haptoglobin were significantly associated with tea consumption indicating that tea drinking acts to attenuate systemic inflammation [39].

Huang and his co-workers had elegantly demonstrated the antiinflammatory activity of tea using a CD-1 mice model. A single topical application of TPA to ears of CD-1 mice induced a time- and dose-dependent increase in edema as well as formation of proinflammatory cytokines interleukin-1beta (IL-1beta) and interleukin-6 (IL-6) in mouse ears. A single topical application of equimolar of black tea constituents (TF, theaflavin-3-gallate, theaflavin-3'-gallate, and theaflavin-3,3'-digallate) strongly inhibited TPA-induced edema of mouse ears. Application of TFs mixture to mouse ears 20 min prior to each TPA application once a day for 4 days inhibited TPA-induced persistent inflammation, as well as TPA-induced increase in IL-1beta and IL-6 levels. The decreased amounts of AA metabolites prostaglandin E(2) (PGE(2)) and leukotriene B(4) (LTB(4)) levels indicate that TF had also inhibited arachidonic acid (AA) metabolism via both cyclooxygenase (COX) and lipoxygenase pathways. Oral administration of TFs or the hot water extract of black tea leaves also significantly inhibited TPA-induced edema in mouse ears. Authors conclude that black tea constituents, TF and its derivatives, had strong anti-inflammatory activity *in vivo* which may be due to their ability to inhibit AA metabolism via lipoxygenase and COX pathways. Inhibitory effects of black tea theaflavin derivatives on 12-O-tetradecanoylphorbol-13-acetate-induced inflammation and arachidonic acid metabolism in mouse ears [40].

A study using SKH-1 hairless mouse model had found that green tea could alter UV radiation exposure induced mitogen-activated protein kinases (MAPK) and NF-kappaB signalling pathways. This demonstrates that green tea could reduce the UV radiation induced inflammation. [41].

In a cell culture study it was found that EGCG from green tea inhibited TNF-alpha-mediated IL-8 gene expression dose dependently. EGCG inhibited TNF-alpha-mediated activation of IkappaB kinase and subsequent activation of the IkappaB alpha/NF-kappaB pathway [42]. Later the same group found that EGCG markedly inhibited IL-1 beta-mediated IL-1 beta receptor-associated kinase (IRAK) degradation and the signalling events downstream from IRAK degradation, namely IKK activation, Ikappa B alpha degradation, and NF-kappa B activation. In addition, EGCG inhibited phosphorylation of the p65 subunit of NF-kappa B [43].

It has also been shown that theaflavin-3,3'-digallate from black tea blocked phosphorylation of IkappaB from cytosolic fraction and reduced lipopolysacchride-induced nuclear accumulation of transcription factor NF-kappaB p65 and p50 subunits. This resulted in inhibiting the induction of inducible NO synthase transcription [44].

A polyphenol rich extract from black tea decreased bacterial lipopolysaccharide (endotoxin) induced IL-6 production and antioxidant acute phase protein (caeruloplasmin) concentration while increasing liver glutathione concentration in rats. Antioxidant activity of tea polyphenols may have contributed to decrease in IL-6 concentrations [45].

The anti-inflammatory action of tea components may be due to multiple activities of tea flavonoids. Free radicals increase the cascade of events in the inflammatory process. Therefore, antioxidant activity may play a role in attenuation of inflammation. In addition, modulation of transcription factors and NO synthase may also contribute in the anti-inflammatory activity.

TEA AND CARDIOVASCULAR DISEASE

Coronary heart disease (CHD) and cerebrovascular disease are the main types of cardiovascular (CVD) diseases. CHD is the number one cause of death and disability in adults worldwide. The CHD progresses in stages by forming atherosclerotic plaques in one or more branches of coronary arteries. The initiation occurs in sub endothelium of the arteries with lipid deposition and lipid oxidation. Followed by impaired vasodilation and inflammation. The narrowing of artery could result in platelet derived thrombosis.

Oxidation of lipids is thought to play a significant role in atherogenesis. High low density lipoprotein (LDL) cholesterol in blood results in accumulation of LDL in the intima of arteries. LDL itself is not considered atherogenic and it contributes to lesion formation after oxidation [46].

The Dilmah Ceylon tea was further shown its protective effects in a hypoxic (similar as ischemic) condition of vascular smooth muscle cells (VSMCs). The hypoxic VSMCs treated with Ceylon tea group showed a significantly lower ($p<0.005$) quantity for the oxidized DNA bases 8-OH Adenine and FAPy-Guanine when compared to Vehicle. 8-OH Guanine, as a DNA damage biomarker, showed that its level in Ceylon tea treated group is lower than in vehicle group (Table 3).

LDL OXIDATION

Many studies have shown the ability of tea brews and its polyphenolic components to inhibit lipid oxidation both *in vitro* and *in vivo.* Suppression of oxidation by oolong tea had been demonstrated in an oxidation model using human LDL. The oxidation of LDL induced by 2-2'-azobis 4-methoxy-2, 4-dimethyvaleronitrile was suppressed by oolong tea in a dose dependent manner [47].

The antioxidative effects of green tea catechins against cholesterol oxidation were examined in an *in vitro* lipoprotein oxidation system. The antioxidative activity against copper catalyzed LDL oxidation was in the decreasing order (-)-epigalocatechin gallate (EGCG)=(-)-epicatechin gallate (ECG)>(-)-epicatechin (EC)=(+)-catechin (C)>(-)-epigallocatechin (EGC) [48].

Inhibitory effect of green tea flavonoids on LDL oxidation had been demonstrated in Osteogenic Disorder Shionogi (ODS) rats. ODS rats can not synthesize ascorbic acid and therefore, subjected to increased oxidative stress. LDL oxidation lag time was significantly longer in rats fed green tea flavonoids [49].

The acute effects of ingestion of black and green tea on *ex vivo* Cu(2+)-induced lipoprotein oxidation was investigated using 20 healthy men. Compared with the water control, there was a greater lag time for

Table 3. Oxidized DNA base products were analyzed and quantified by GC/MS and Vehicle group was set as positive control. Results are expressed as mean±S.D. of 3 determinations performed in triplicate and expressed as nmol/mg DNA. *significantly difference v.s. Vehicle group. # significantly difference v.s. Vehicle and hypoxia treated with Ceylon tea groups.

	DNA base product	Non-hypoxia	Hypoxia (Vehicle)	Hypoxia + Ceylon Tea
1	5-OH, Me Hydantoin	0.12±0.16	0.13±0.19	0.13±0.06
2	5-Formyl Uracil	0.17±0.05#	0.98±0.01	0.89±0.65
3	5-OH Uracil	0.03±0.00	0.05±0.00	0.05±0.03
4	5-(OH, Me) Uracil	0.04±0.01	0.08±0.02	0.05±0.03
5	5-OH Cytosine	0.04±0.01#	0.24±0.01	0.14±0.02
6	FAPy Adenine	0.08±0.07	0.16±0.05	0.12±0.04
7	8-OH Adenine	0.06±0.02	0.49±0.02	0.09±0.02*
8	2-OH Adenine	0.04±0.02	0.06±0.01	0.02±0.02
9	FAPy Guanine	0.03±0.02	0.08±0.01	0.02±0.02*
10	8-OH Guanine	0.07±0.06	0.23±0.04	0.12±0.03

black tea and a similar trend for green tea indicating a mild acute effect on ex vivo lipoprotein oxidation in human serum. Authors suggest that absence of effects in some previous studies may be due to the isolation of LDL particles from polyphenolic compounds that are present in the aqueous phase of serum. Therefore, in this study acute effects of ingestion of black and green tea on *ex vivo* Cu(2+)-induced lipoprotein oxidation was studied without prior isolation of lipoproteins from serum [50].

The ability of a range of dietary flavonoids to inhibit low-density lipoprotein (LDL) oxidation was tested in an *in vitro* study. Quercetin which is present in both green and black tea was found to be the most effective inhibitor of oxidative damage to LDL *in vitro* [51].

In vitro LDL lipoprotein oxidation model was used to assess the relative antioxidant activity of the green and black tea on a cup-serving basis. The antioxidant activity as determined by the lag time was found to be 186-338 min for green tea and 67-277 min for black tea indicating slightly higher activity for green tea on per serving basis [52]. In a similar study catechins in green tea and theaflavins in black tea were found to have similar activity in an *ex vivo* human LDL oxidation model [53].

These *in vitro* and *ex vivo* studies show that both green and black tea inhibit LDL oxidation. However, whether tea consumption increases the resistance of LDL oxidation *in vivo* depend on tea flavonoid absorption from the digestive tract and bioavailability in required concentration. A limited number of human *in vivo* studies have provided mixed results. In some studies protection of LDL oxidation by tea components were not observed [54, 55]. While in some studies mild protective effects were exhibited.

In a cross over study with 22 healthy male non-smokers it was found that green tea consumption reduced the malondialdehyde modified LDL concentration. They drank 7 cups/day of water for 2 weeks and drank 7 cups/day of green tea for the next 2 weeks. Of the 22 subjects, 20 had been in the habit of drinking green tea before the study. Plasma catechins concentrations significantly decreased at the end of the water period and then increased at the end of the green tea period. The ratio of MDA-LDL/LDL-cholesterol (0.74 +/- 0.21 vs. 0.65 +/- 0.20, $p < 0.02$) significantly decreased at the end of the green tea period suggesting that green tea consumption may inhibit LDL oxidation *in vivo* [56].

In another *in vivo* study, 14 healthy volunteers consumed 750 mL black tea/d for 4 wk. After the subjects had consumed tea for 4 wk, the lag time before LDL oxidation was significantly ($P < 0.01$) prolonged from 54 to 62 min. This minor prolongation indicates a mild LDL protection action of black tea [57]. In a similar study, 12 healthy male volunteers aged 28—42 years drank 600 mL of green tea daily for four weeks. Oxidized low-density lipoprotein (ox-LDL) profiles were measured at baseline and after two and four week ingestion of green tea. The levels of ox-LDL significantly decreased after four weeks of green tea ingestion [58].

Evidence from these studies suggests that tea consumption may have a mild effect on LDL oxidation in humans.

EFFECT ON PLASMA LIPID LEVELS

The relationship between tea consumption and plasma lipid levels has been assessed in several epidemiological studies.

Effect of the addition of two cups of GT (containing approximately 250 mg of total catechins) to a controlled diet in a group of healthy volunteers was compared with a group following the same controlled diet but not consuming GT. After 42 days, consumption of GT caused a statistically significant but moderate decrease in LDL cholesterol (from 119.9 to 106.6 mg/dL, $P<.05$) with respect to control [59].

Black tea consumption on lipid and lipoprotein concentrations in mildly hypercholesterolemic adults was studied in a blinded randomized crossover study (7 men and 8 women, consuming a controlled diet for 3 wk/treatment). Five servings/d of tea were compared with a placebo beverage. Five servings/d of tea reduced total cholesterol by 6.5% and LDL cholesterol by 11.1% [60].

Princen [61] and others found that black tea and green tea had no effect on plasma lipid levels in a randomized, placebo-controlled study. Healthy male and female smokers (aged 34+/-12 years, 13 to 16 per group) consumed during a 4-week period 6 cups (900 mL) of black or green tea or water per day, or received 3.6 grams of green tea polyphenols per day as a supplement (equivalent to the consumption of 18 cups of green tea per day). Consumption of black or green tea had no effect on plasma cholesterol and triglycerides, HDL and LDL cholesterol.

Loktionov and co workers [62] found that Apolipoprotein E (ApoE) genotype modulates the effect of black tea on blood lipids. In a 4-week randomized crossover trial to compare the effect of six mugs of black tea per day v. placebo on blood lipids, it was found that out of four ApoE genotype variants only two genotypes show a reduction of blood lipids with tea consumption. In the other two genotypes there was no difference.

Thus, the evidence from *in vitro* studies, animal model studies and epidemiological studies suggest that inclusion of tea in the diet may reduce total and LDL cholesterol and their oxidation and therefore, may reduce the risk of coronary heart disease. Effects on lipids are not the only mechanism by which tea may influence CHD.

ENDOTHELIAL FUNCTION

Endothelial vasomotor dysfunction is related to oxidative stress and is an early phase of atherosclerosis; it occurs in those with atherosclerosis and other risk factors for coronary disease such as hypertension and hypercholesterolemia [63].

Duffy *et al.* [64] quantified endothelial function as 'flow mediated dilatation' (FMD) of the brachial artery in the human arm. Sixty-six patients with coronary artery disease were given black tea and water in a cross-over design. Short-term effects were examined two hours after the consumption of 450 ml of tea or water, and long-term effects after the consumption of 900 ml of tea or water every day for four weeks. Plasma flavonoids increased after short- and long-term tea consumption. Vasomotor function of the brachial artery was examined initially and following each intervention, with vascular ultrasound. Both short-term and long-term tea consumption raised the FMD to about the same extent. Consumption of water had no effect. Acute tea consumption, following on chronic tea consumption, gave an additional increase in FMD over that from the two types of consumption separately. Equivalent doses of caffeine had no short-term effect on FMD. That tea acts directly on the endothelium is shown by the fact that its consumption has no effect on the response to nitroglycerin (which is an endothelium–independent vasodilator), on the size of the resting artery, or on blood pressure or heart rate.
A similar Australian study [65] showed that five cups of black tea every day for four weeks, compared with hot water consumption, gave significant endothelium–dependent dilatation and endothelium–independent dilatation in the brachial artery.

A randomised crossover study had shown that Green tea reverses endothelial dysfunction in healthy smokers. Twenty healthy male smokers were randomised to consume green tea or hot water in a crossover design. At baseline and two hours after consumption of 400 ml green tea or hot water the response of forearm blood flow (FBF) to reactive hyperaemia, an index of endothelium dependent vasodilatation, and to sublingual administration of glyceryl trinitrate (GTN), an index of endothelium independent vasodilatation were measured. Consumption of green tea significantly increased FBF during reactive hyperaemia in smokers. In contrast, green tea had no effect on FBF at rest or GTN induced vasodilatation. These results suggest that green tea consumption reverses endothelial dysfunction in healthy smokers through endothelium dependent vasodilatation [66].

Results of the above studies indicate that one way in which black tea may reduce cardiovascular risk is by improved vasodilation of conduit arteries.

PLATELET AGGREGATION AND BLOOD CLOTTING

Tea flavonoids could inhibit the platelet aggregation to a certain extend. Some studies indicate that tea components could alter the concentrations of molecules which are involved in cell adhesion.
Tea polyphenols have been found to inhibit platelet aggregation and blood coagulation, strengthen the walls of blood vessels, and interact with catecholamines in anti-inflammatory activity [67].

In a cross-over epidemiological study in Australia with 22 men, consumption of five cups of tea per day was compared with hot water consumption for five weeks, and the plasma levels of molecules, called p-selectin, which promote cell adhesion were measured. It was found that tea drinking lowers p-selectin levels, indicating that tea acts as an anti platelet-aggregator and so reduces thrombus formation [68].

In a similar placebo control, cross-over study in the U.S., six cups of black tea (3.6 g tea solids, 1.2 g tea flavonoids) per day for three weeks reduced platelet function induced by collagen [69].

Mitane *et al.* [70] studied the inhibitory effects of tea catechins on rabbit platelet aggregation *in vitro*. It was found that 0.025, 0.05, 0.1 and 0.2 mg/ml of epigallocatechin gallate (EGCG) isolated from green tea inhibited platelet aggregation in a dose-dependent manner. At the 0.2 mg/ml level, aggregation was completely inhibited

These studies indicate that regular tea consumption could decrease the possibility of thrombus formation to a certain degree.

EPIDEMIOLOGICAL STUDIES ON TEA AND CARDIOVASCULAR DISEASES

The results from epidemiological studies strongly suggest a protective role for tea and TEA flavonols with respect to cardiovascular diseases. Some studies have found that tea consumption has a protective effect against the risk of coronary heart disease (CHD), possibly caused by flavonoids acting as antioxidants [71].

The first indication of the protective effect of a high consumption of black tea on ischaemic heart disease was given by the Boston Collaborative Surveillance Programme in 1972 [72]. A non-significant risk reduction for a myocardial infarction of 34 per cent was found for drinkers of more than six cups of tea, as against non-tea drinkers.

In a 25-year cross-cultural correlation study of 16 cohorts (12,763 men; 40-59 years) in seven European countries, flavonol and flavone intake in the diet was found to be inversely related to mortality from CHD. They concluded that average flavonoid intake may partly contribute to differences in coronary heart disease mortality across populations [73]

One of the cohorts of this study, a Dutch cohort of 805 men (65-84 years), was in a 5-year study, the Zutphen Elderly Study. It was found here that the risk of dying from CHD was significantly lower in men with a high daily intake of tea flavonol and flavone: a reduction in mortality risk of more than 50 per cent at an average flavonol intake of 42 mg/day. The daily flavonoid intake had an inverse relationship to the incidence of myocardial infarction. Sixty-one per cent of the dietary flavonoids came from black tea, as against 13 per cent from onions and 10 per cent from apples, the other major sources. The incidence of myocardial infarction for three levels of flavonoid intake, low (0-19.0 mg/day), medium (19.1-29.9 mg/day) and high (more than 29.9 mg/day) were, per 1000 persons, 16.2, 13.8 and 7.6, respectively [74]. This effect was independent of the major established risk factors. A follow-up after 10 years confirmed the earlier findings, this time a dose-response relationship between flavonol intake and CHD mortality being established [75].

In a 20-year study of a Finnish cohort of 5,133 men and women (30–69 years old), death from CHD was inversely related to flavonol and flavone intake, but not strongly [76]. The relative risks between the highest intake (over 5 mg/day) and the lowest intake (less than 2.5 mg/day) were 0.73 and 0.67 for men and women, respectively.

Another study, with 34,789 male health professionals (40-75 years) in the USA, found a weak inverse association between intake of flavonol (7–40 mg/day) and flavone, and CHD mortality, but only for those with a history of CHD [77].

An apparently contradictory result to other epidemiological studies (where there was a negative or zero relationship) was given by the Caerphilly Study in Wales on a population of men: flavonol intake (14-43 mg/day) was found to be *positively* related to ischemic heart disease [78].

According to the authors, the reasons for the contradiction were that heavy tea drinkers in Wales were from a social class characterised by an unhealthy life style (smoking and high fat-intake). In more sophisticated groups, tea drinking is commonly associated with a healthy lifestyle.

A study in Finland using 25,372 smokers has found that intake of flavonols show an inverse relation to myocardial infarction (but only a weak inverse relation to CHD mortality) [79].

The association of tea and flavonoid intake with myocardial infarction was examined in the general Dutch population, using the data from the Rotterdam Study of 4,807 men and women, aged 55 years and over, who had no history of myocardial infarction [80]. The study lasted from 1990 to 1997. Data were adjusted for age, gender, body mass, smoking, education level, and daily intake of alcohol, coffee, polyunsaturated and saturated fats, fibre, vitamin E and total energy. During 5.6 years of follow-up, 146 first myocardial infarctions occurred, 30 of which were fatal. The relative risk of myocardial infarction was lower in tea-drinkers, with an intake of over 375 ml per day, than in non-tea-drinkers. The inverse relationship with tea drinking was stronger for fatal than for non-fatal events. The authors conclude that an increased intake of tea and flavonoids may contribute to the primary prevention of ischemic heart disease.

In a cross-sectional study with 1,371 men in Yoshimi, Japan, increased green tea consumption was associated with decreased concentrations in the serum of total cholesterol and triglyceride, an increase in HDL cholesterol and a decrease in LDL cholesterol. Three levels of tea were consumed: 0-3, 4-9, and > 9, cups. Total cholesterol for these levels were, respectively, 4.85, 4.76 and 4.58 mmol/l; triglyceride 1.65, 1.60 and 1.45 nmol/l; HDL cholesterol 36.4, 36.5 and 37.4 per cent; and LDL cholesterol 62.5, 62.6 and 61.7 per cent. Although the values are statistically significant and the conclusion is made, from these results, that green tea may protect from cardiovascular disease, it may be argued also that the relatively small changes in the serum values can have no clinical significance [81].

In a similar study with men in Northern Kyushu, Japan, green tea consumption was found to be inversely related to serum cholesterol levels. Four levels of daily green tea were consumed: 0-2, 3-5, 6-8 and > 9, cups. Total cholesterol for these levels was 193, 190, 187 and 185 mg/dl, respectively. However no association was found with triglyceride or HDL cholesterol [82].

A study in Norway, with 9,856 men and 10,233 women, showed that black tea consumption was inversely related to total serum cholesterol and systolic blood pressure [83]. Four levels of daily black tea were consumed: 0-1, 1-2, 3-4 and > 5, cups. In the men, total serum cholesterol for these levels were 6.24, 6.20, 5.96 and 6.19 mmol/l, respectively, and blood pressure were 136.2, 135.0, 135.9 and 133.1 mm, respectively. In the women, total serum cholesterol were 6.11, 5.96, 5.89 and 5.92 mmol/l, respectively, and blood pressure was 131.7, 130.2, 127.9 and 127.2 mm, respectively.

A study in Taiwan using in 1507 subjects (711 men and 796 women) with newly diagnosed hypertension it was found that habitual tea drinking reduce the risk of developing hypertension in the Chinese population. Six hundred subjects (39.8%) were habitual tea drinkers, defined by tea consumption of 120 mL/d or more for at least for one year. Compared with non-habitual tea drinkers, the risk of developing hypertension decreased by 46% for those who drank 120 to 599 mL/d and was further reduced by 65% for those who drank 600 mL/d or more after adjusting for age, sex, socioeconomic status, family history of hypertension, body mass index, waist-hip ratio, lifestyle factors (total physical activity, high sodium intake, cigarette smoking, alcohol consumption, and coffee drinking), and dietary factors (vegetable, fruit, unrefined grain, fish, milk, visible-fat food, and deep fried food intake) [84].

However, a study in Israel, with 3,858 men and 1,511 women, showed no significant association between black tea consumption and serum cholesterol levels (Green and Harari, 1992). However in this study, only a small percentage of subjects consumed tea at the higher levels. The percentiles of men consuming the five levels of tea daily (0, 1-2, 3-4, 5 and > 5 cups) were 23.4, 61.1, 11.9, 1.5 and 1.9, respectively. The corresponding percentiles of women were 27.4, 58.9, 11.3, 1.3 and 0.9, respectively [85].

Also, the Scottish Heart Health Study (with 10,359 people in 22 Scottish districts consuming a mean of 4 cups of black tea per day) found no relationship between tea consumption and diagnosis of CHD or serum cholesterol levels [86].

A meta-analysis of the available data indicates that three cups of black tea per day results in a 11 per cent reduction in risk of myocardial infarction [87].

Data from *in vitro,* animal model and epidemiological studies indicate that tea drinking may reduce the risk of CHD. This may be due to multiple mechanisms; antioxidant, anti-inflammatory, anti-coagulant, antihypertensive and improvement of endothelial function.

TEA AND CANCER

Anticarcinogenic potential of tea, specially its flavonoids, has been found in numerous *in vitro* and *in vivo* studies. The main mechanisms of anticarcinogenecity appears to be through antimutagenecity of tea flavonoids due to antioxidant activity and ability of these molecules to bind directly with chemical carcinogen molecules in biological systems. Tea flavonoids are easily oxidised to the corresponding quinones in a reversible reaction. Flavonoids can therefore act as both hydrogen acceptors and donors. This makes it possible for them to interact with the active form of most chemical carcinogens and convert them into inactive forms.

Inhibition of heterocyclic amine formation and inducing Phase II enzymes also play a major role in anticarcinogenic activity of tea. Molecular mechanisms such as flavonoid mediated apoptosis and inhibition of transcription factors such as NFêB and AP-1 have also been identified as potential mechanisms in preventive pathways. Antiangiogenesis also has been identified as a potential mechanism of inhibition.

Identification of tea flavonoids to inhibit carcinogenesis in rodent models [88] and subsequently in epidemiological studies with human cohorts [89] in 1980s resulted in numerous investigations on tea and cancer.

Research, both *in vivo* and *in vitro*, had indicated that tea and tea polyphenols interfere with or inhibit carcinogensis at initiation or during the further stages, by different mechanisms. Further, tea consumption decreases the growth rate of tumour cells and prevents metastasis and the formation of large tumours [90].

FRs and ROSs are implicated in the initial damage to DNA (mutagenesis) and in the further stages of carcinogensis. It is now well known that the antioxidant properties of tea flavonoids play a major role in cancer prevention [91].

Bunkova and others [92] found that two common kinds of green tea (Chinese Gunpowder and Japanese Sencha) exhibited high antimutagenic activity in the Ames test (24.7+/-3.7% and 34.1+/-2, 1% of inhibition without metabolic activation; 74.9+/-1.7% and 62.7+/-4.3% of inhibition with metabolic activation, respectively) as well as in *S. cerevisiae* D7 test (Gunpowder: 62.7+/-5.7% of Trp convertants inhibition and 52.6+/-5.3% of Ilv revertants inhibition; Sencha: 45.6+/-4.2% of Trp convertants inhibition, 50.0+/-4.8% of Ilv revertants inhibition). In the cytogenetic analysis of peripheral blood lymphocytes (CAPL) method reduced number of aberrant cells as well as decreased number of chromosome breaks was observed using both green tea extracts.

Antimutagenic activity of black tea polyphenols theaflavins (TF) and thearubigins (TR) were demonstrated in a study where a significant decrease was observed in mutagenicity against a known carcinogen, benzo[a]pyrene (B[a]P). Different concentrations of TF and TR decreased chromosomal aberrations (CA) and sister chromatid exchange (SCE) *in vitro* and *in vivo* Salmonella assay using bone marrow cells of mice [93].

The protective effects of three tea extracts (green tea, GTE; oolong tea, OTE; and black tea, BTE) and five tea polyphenols (epicatechin, EC; epicatechin gallate, ECG; epigallocatechin, EGC; epigallocatechin gallate, EGCG; and theaflavins, TFs) on B[a]P-induced DNA damage in Chang liver cells had been demonstrated using the comet assay [94]. Many other similar studies have demonstrated the antimutagenic activity of tea extracts and tea components.

Tea flavonoids could alter the activity of cytochrome P450 (CYP) enzymes involved in the activation of procarcinogens and phase II enzymes, largely responsible for the detoxification of carcinogens. Activation of phase II detoxifying enzymes, such as UDP-glucuronyl transferase, glutathione S-transferase, and quinone reductase by flavonoids results in the detoxification of carcinogens and represents one mechanism of their anticarcinogenic effects. A number of flavonoids are potent non-competitive inhibitors of sulfotransferase 1A1 (or P-PST); this may represent an important mechanism for the chemoprevention of sulfation-induced carcinogenesis [95].

Exposure of rats to green tea (2.5%, w/v), as the sole drinking fluid, for 4 wks significantly increased the UDP glucuronosyl transferase activity [96] Another study in rats had found that black tea (2.5% w/v) as sole source of drinking water induced hepatic CYP1A2 activity and expression. But cytosolic glutathione S-transferase activity and glucuronosyl activity was only modestly induced in the group of animals receiving tea [97].

Recent research has shown that, in addition, tea components interfere with the other mechanisms of cancer promotion. Thus, they alter the expression of certain genes. Down-regulation of four genes, NF-kappaB-inducing kinase, death-associated protein kinase 1, rhoB, and tyrosine-protein kinase, and up-regulation of one gene, retinoic acid receptor alpha 1, were effected by tea components [98]. Other studies also have shown attenuation of the activation of the gene, kappaB, and resulting production of polypeptides called cytokines certain of which act as growth promoters in some tumours [99, 100, 101].

Cell-to-cell signal transduction is also important in the development of cancer. Tea components are involved in the inhibition of these signal transduction pathways [102, 103].

For the growth of cancer cells, one important requirement is an adequate blood supply. Therefore, generation of new blood vessels (angiogenesis) to the cancer tissue is essential. In the treatment of cancer, some drugs are targeted at preventing of angiogenesis (angioprevention). Tea components also show angioprevention activity [104, 105].

Receptors for vascular endothelial growth factor (VEGF) play a major part in angiogenesis. Physiological concentrations of tea catechins, EGCG, CG and ECG, induce rapid blocking of these receptors, suggesting that the anticancer properties of tea extracts may be due to their inhibition of VEGF-dependent angiogenesis [106].

Induction of programmed cell death, or apoptosis, could help to reduce the progression of cancer. Recent studies have shown that tea can induce apoptosis [107, 108].

Potential carcinogens in food, specially in cooked meat, are heterocyclic aromatic amines (HAA). Tea components reduce the mutagenicity of HAA [109].

Tea components are also effective against radiation injury in general. They have been found to give protection against DNA scission induced by β-rays [110] and UV radiation [111, 112, 113].

Tea components are able to reduce carcinogenicity from smoking. Of the numerous carcinogens in cigarette smoke, the most potent appears to be 4-(methylnitrosoamino) -1-(3-pyridyl)-1-butanone (NNK). Many studies have shown that tea components could inhibit tumour formation by NNK [114].

There are several reports indicating that both green and black tea extracts give protection against the development of cancers and have anticarcinogenic effects at different sites in animal models. These findings are especially significant because the positive effects in animals are produced by tea at concentrations usually consumed by humans [115].

A large number of experiments with animal models has been conducted to find the effect of both black and green tea on cancer in different body organs, such as the stomach, colon, lung, liver and skin. Both black and green tea extracts inhibit carcinogenesis induced by a wide range of chemical agents and radiation, such as UV radiation.

Tumours in male Sprague-Dawley rats, normally induced by subcutaneous injection with N-nitrosomethylbenzylamine (NMBzA), were inhibited by green tea and black tea extracts given orally [116]. Sixty-five per cent of rats treated with NMBzA for five weeks developed oesophageal tumours, 39 weeks after the treatment commenced. Either decaffeinated green tea or decaffeinated black tea, provided as the only fluid for drinking, during the five weeks of treatment with NMBzA, reduced tumour incidence by 70 per cent. When the tea preparations were given after the NMBzA treatment period, the tumour incidence was reduced by 50 per cent. This study shows that the reduction in tumour formation is not by tea components reacting directly on the carcinogen, since tea and NMBzA were administered by two different routes, orally and subcutaneously, respectively. It is suggested that the tea inhibits carcinogen activation, or prevents oxidative damage to DNA by ROSs produced during carcinogen metabolism.

Ohishi *et al.* [117] found that epigalocatechin gallate (EGCG) could reduce colon carcinogenesis induced in rats by the carcinogen, asoxymethane.

Epidemiological studies on the effect of tea on human cancers have given mixed results [118]. There have been a range of such studies: on cancers of the mouth, pharynx, oesophagus, stomach, colon, rectum, pancreas, lung, mammary glands, bladder and the kidney.

Although dose-related effects have not been demonstrated, several studies suggest a lowered risk of digestive tract cancers among tea drinkers, especially those consuming green tea [119, 120].

A study in the Netherlands using 3,692 people has shown that black tea consumption is inversely related to the incidence of bladder cancer [121]. A population-based study in the US, with 1452 bladder-cancer subjects, 406 kidney-cancer subjects and 2434 control subjects, also found an inverse relationship between tea consumption and bladder cancer [122]. However, there was no association with kidney cancer.

Another study of a population in the south-western United States has shown that hot black tea consumption is associated with a lower incidence of squamous cell carcinoma in the skin [123].

In a Chinese population, it was found that green tea consumption reduced lung cancer in non-smokers, but not in smokers [124].

With 1,160 hospital patients in Japan, followed from 1988 to 1999, it was found that drinking green tea lowered the risk of recurrence of breast cancer [125].

However, in some other epidemiological studies on breast cancer, no association was found. Thus, in a large population-based prospective cohort study in Sweden (the Mammography Screening Cohort), comprising 59,036 women, aged 40-76 years, no association were found between tea consumption and breast cancer [126]. Similarly, no association was found between green tea-drinking and cancer of the stomach, colon, rectum, liver, gall bladder, pancreas, lung, breast and bladder, in a prospective study in Hiroshima, Japan, where 38,540 people were followed from 1979 to 1994 [127].

As with the coronary heart diseases *in vitro*, animal model and epidemiology studies indicate that tea drinking could reduce the risk of carcinogenesis. Tea components act in different stages of carcinogenesis through different mechanisms.

CONCLUSION

Tea had been consumed as a beverage for centuries. Today, global consumption of tea is second only to water. Safety of tea as a beverage is demonstrated by the absence of reported adverse effects although it has been habitually consumed by large populations for a long time.

Chemistry of the tea brew shows that it does not contain significant amounts of nutrients. Therefore, it could be regularly consumed with out any additions to the daily calorie intake.

Direct measurements and indirect measurements on their activities in body tissues indicate that a small proportion of tea flavonoids are absorbed from the digestive tract and reach different tissues in the body. Tea flavonoids are quickly metabolised and generally exhausted in 24 h.

In vitro and *in vivo* studies demonstrate that both green tea and black tea flavonoids are strong antioxidants. This is the main mechanism by which tea reduce the risk of oxidant mediated pathogenesis of degenerative diseases.

Myriad of *in vitro*, animal model and epidemiological studies indicate that regular consumption of black and green tea could reduce the risk of cardiovascular disease and carcinogenesis. Altering the endothelial function, direct binding with carcinogens, inducing Phase I and II enzymes, apoptosis, inhibition of transcription factors, anti-angiogenetic activity and anti-hypertension activity of tea flavonoids also help in reducing the risk of these diseases. It could also improve the oral health and increase the populations of beneficial micro-organisms in the digestive tract.

Careful scrutiny of the results on tea and health show that beneficial effects of tea components are mild effects, as could be expected from an item in the diet. However, results from epidemiological studies indicate that regular consumption of tea could actually reduce the risk of these diseases.

Therefore, regular consumption of tea could be considered as a part of a modern day healthy diet.

REFERENCES

[1] Samaraweera DSA. Technology of tea processing. In: Handbook on Tea. (Sivapalan, P., Kulasegaram, S., Kathiravetpillai ,A., eds) 1989 Tea Research Institute of Sri Lanka, Talawakelle, Sri Lanka. pp. 158-207

[2] Modder WWD, Amarakoon AMT. Tea and Health, Tea Research Institute of Sri Lanka, Talawakelle, Sri Lanka. 2002

[3] Das NP, Griffiths LA, Studies on flavonoid metabolism. Biochem J 1969; 115:831-36

[4] He YK, Kies C. Grccn and black tea consumption by humans: impact on polyphenol concentrations in faeces, blood and urine. Plant Foods and Human Nutrition 1994; 46: 221-29

[5] Hollman PC, Van het Hof KH, Tijburg LB, Katan MB. Addition of milk does not affect the absorption of flavonols from tea in man. Free Radical Research 2001; 34: 297-300

[6] Nakagawa K, Miyasawa T. Absorption and distribution of tea catechin (-) epigallocatechin-3-gallate in the rat. Journal of Nutritional Science and Vitaminology 1997; 43: 679-83

[7] He YK, Kies C. Green and black tea consumption by humans: impact on polyphenol concentrations in faeces, blood and urine. Plant Foods and Human Nutrition 1994; 46: 221-29

[8] Warden BA, Smith LS, Beecher GR, Balentine DA, Clevidence BA. Catechins are bioavailable in men and women drinking black tea throughout the day. Journal of Nutrition 2001; 131: 1731-37

[9] Shahrzad S, Aoyagi K, Winter A, Koyama A, Bitsch I. Pharmacokinetics of gallic acid and its relative bioavailability from tea in healthy humans. Journal of Nutrition 2001; 131: 1207-10

[10] De Vries JHM, Hollman PCH, Meyboom S. Plasma concentrations and urinary excretion of antioxidant flavonols quercetin and kaempferol as biomarkers for dietary intake. American Journal of Clinical Nutrition 1998; 68: 60-65

[11] Kohri T, Matsumoto N, Yamakawa M, Suzuki M, Nanjo S, Hara Y, *et al.* Metabolic fate of (-)-[4-(3)H]epigallocatechin gallate in rats after oral administration. Journal of Agricultural and Food Chemistry 2001 ; 49: 4102-12

[12] Suganuma M, Okabe S, Oniyama, M. Wide distribution of [^{3}H] (-)-epigallocatechin gallate, a cancer preventive tea polyphenol, in mouse tissue. Carcinogenesis 1998; 19: 1771-76

[13] Okushio K, Matsumoto N, Kohri T. Absorption of tea catechins into rat portal vein. Biological and Pharmaceutical Bulletin 1996; 19: 326-29

[14] Unno T, Takeo T. Absorption of (-) epigallocatehcin gallate into the circulation system of rats. Bioscience, Biotechnology and Biochemistry 1995; 59: 1558-59

[15] Lee KJ, Wang ZY, Li H. Analysis of plasma and urinary tea polyphenols in human subjects. Cancer Epidemiology, Biomarkers and Prevention 1995; 4: 393-99

[16] Nakagawa K, Okuda S, Miyazawa T. Dose-dependent incorporation of tea catechins, (-)-epigallocatechin-3-gallate and (-)-epigallocatechin, into human plasma. Bioscience Biotechnology and Biochemistry 1997; 61:1981-85

[17] Nakagawa K, Ninomiya M, Okubo T, Aoi N, Juneja LR, Kim M. *et al.* Tea catechin supplementation increases antioxidant capacity and prevents phospholipid hydroperoxidation in plasma of humans. Journal of Agricultural and Food Chemistry 1999; 47: 3967-73

[18] Hara Y. Influence of tea catechins on the digestive tract. J Cell Biochemistry 1997; Suppl 27: 52-58

[19] Brown S, Griffiths, LA. New metabolites of the naturally occurring mutagen, quercetin, the pro-mutagen, rutin and taxifolin. Experimentia 1981; 39: 198-200

[20] Okushio K, Matsumoto N, Kohri T. Absorption of tea catechins into rat portal vein. Biological and Pharmaceutical Bulletin 1996; 19: 326-29

[21] Lee MJ, Prabhu S, Meng X, Li C, Yang CS. An improved method for the determination of green and black tea polyphenols in biomatrices by high-performance liquid chromatography with coulometric array detection. Anal Biochem. 2000; 279: 164-69

[22] Mulder TP, van Platerink CJ, Wijnand Schuyl PJ, van Amelsvoort JM. Analysis of theaflavins in biological fluids using liquid chromatography-electrospray mass spectrometry. J Chromatogr B Biomed Sci Appl. 2001; 760: 271-79

[23] Levites Y, Weinreb O, Maor G, Youdim MB, Mandel S. Green tea polyphenol (-)-epigallocatechin-3-gallate prevents N-methyl-4-phenyl-1,2,3,6-tetrahydropyridine-induced dopaminergic neuro- degeneration. Journal Neurochem 2001; 78: 1073-82

[24] Du Toit R., Volsteedt Y. and Apostolides Z. Comparison of the antioxidant content of fruits, vegetables and teas measured as vitamin C equivalents. Toxicology 2001;166, 63-69

[25] Zhao B, Li X, He R. Scavenging effect of extracts of green tea and natural antioxidants on active oxygen radical. Cell Biophysics 1989; 14: 175-85

[26] Cao G, Sofic E, Prior R.L. Antioxidant capacity of tea and common vegetables. Journal of Agricultural and Food Chemistry 1996; 44: 3426-31

[27] Du Toit R, Volsteedt Y, Apostolides Z. Comparison of the antioxidant content of fruits, vegetables and teas measured as vitamin C equivalents. Toxicology 2001; 166: 63-69

[28] Paganga G, Miller N, Rice-Evans C.A. The polyphenolic content of fruit and vegetables and their antioxidant activities. What does a serving constitute? Free Radic Res 1999; 30: 153-62

[29] Yen GC, Chen HY. Antioxidant activity of various tea extracts in relation to their antimutagenecity. Journal of Agricultural and Food Chemistry 1995; 43: 27-32

[30] Scott BC, Butler J, Halliwell B, Aruoma OJ. Evaluation of the antioxidant action of ferrulic acid and catechins. Free Radical Research Communications 1993; 19: 241-53

[31] Tournaire C, Croux S, Maurette MT, Beck I, Hocquaux M, Braun AM *et al.* Antioxidant activity of flavonoids: efficiency of singlet oxygen quenching. Journal of Photochemistry and Photobiology B. 1993 ; 19: 205-15

[32] Rice-Evans CA, Miller NJ, Bolwell PG, Bramley PM, Pridham JB. The relative antioxidant activity of plant-derived polyphenolic flavonoids. Free Radical Research 1995; 22: 375-83

[33] Jovanovic SV, Simic MG. Antioxidants in Nutrition. Annals of the New York Academy of Sciences. 2000; 899: 326-34

[34] Tijburg LB, Wiseman SA, Meijer GW, Weststrate JA. Effects of green tea, black tea and dietary lipophilic antioxidants on LDL oxidizability and atherosclerosis in hypercholesterolaemic rabbits., Atherosclerosis. 1997;135, 37-47

[35] Lotito SB, Fraga CG. Ascorbate protects (+)-catechin from oxidation both in a pure chemical system and human plasma., Biol Res. 2000; 33,151-57

[36] Zhu QY, Huang Y, Tsang D, Chen ZY. Regeneration of alpha-tocopherol in human low-density lipoprotein by green tea catechin. J Agric Food Chem. 1999;47:2020-25

[37] Skrzydlewska E, Ostrowska J, Farbiszewski R, Michalak K. Protective effect of green tea against lipid peroxidation in the rat liver, blood serum and the brain. Phytomedicine 2002; 9, 232-38

[38] Amarakoon AMT, Tappia PS, Grimble RF. Endotoxin induced production of interleukin-6 is enhanced by vitamin E deficiency and reduced by black tea extract. 1995; Inflamm Res 44:301-05

[39] De Bacquer D, Clays E, Delanghe J, De Backer G. Epidemiological evidence for an association between habitual tea consumption and markers of chronic inflammation. Atherosclerosis. 2006; [Epub ahead of print]

[40] Huang MT, Liu Y, Ramji D, Lo CY, Ghai G, Dushenkov S, Ho CT. Inhibitory effects of black tea theaflavin derivatives on 12-O-tetradecanoylphorbol-13-acetate-induced inflammation and arachidonic acid metabolism in mouse ears.Mol Nutr Food Res. 2006; 50: 115-22

[41] Afaq F, Ahmad N, Mukhtar H. Suppression of UVB-induced phosphorylation of mitogen-activated protein kinases and nuclear factor kappa B by green tea polyphenol in SKH-1 hairless mice. Oncogene. 2003; 18;22(58):9254-64

[42] Chen PC, Wheeler DS, Malhotra V, Odoms K, Denenberg AG, Wong HR. A green tea-derived polyphenol, epigallocatechin-3-gallate, inhibits IkappaB kinase activation and IL-8 gene expression in respiratory epithelium. Inflammation. 2002; 26: 233-41

[43] Wheeler DS, Catravas JD, Odoms K, Denenberg A, Malhotra V, Wong HR. Epigallocatechin-3-gallate, a green tea-derived polyphenol, inhibits IL-1 beta-dependent proinflammatory signal transduction in cultured respiratory epithelial cells. J Nutr. 2004; 134: 1039-44

[44] Lin YL, Tsai SH, Lin-Shiau SY, Ho CT, Lin JK. Theaflavin-3,3'-digallate from black tea blocks the nitric oxide synthase by down-regulating the activation of NF-kappaB in macrophages. Eur J Pharmacol. 1999; 367: 379-88

[45] Amarakoon AMT, Tappia PS, Grimble RF. Endotoxin induced production of interleukin-6 is enhanced by vitamin E deficiency and reduced by black tea extract. Inflamm Res. 1995; 44: 301-05

[46] Shaw PX. Rethinking oxidized low-density lipoprotein, its role in atherogenesis and the immune responses associated with it. Arch Immunol Ther Exp (Warsz). 2004;52:225-39

[47] Kurihara H, Fukami H, Toyoda Y, Kageyama N, Tsuruoka N, Shibata H, *et al* Inhibitory effect of oolong tea on the oxidative state of low density lipoprotein (LDL). Biol Pharm Bull. 2003 May;26(5):739-42

[48] Osada K, Takahashi M, Hoshina S, Nakamura M, Nakamura S, Sugano M. Tea catechins inhibit cholesterol oxidation accompanying oxidation of low density lipoprotein in vitro. Comp Biochem Physiol C Toxicol Pharmacol. 2001; 128: 153-64

[49] Kasaoka S, Hase K, Morita T, Kiriyama S. Green tea flavonoids inhibit the LDL oxidation in osteogenic disordered rats fed a marginal ascorbic acid in diet. J Nutr Biochem. 2002 Feb;13(2):96-102

[50] Hodgson JM, Puddey IB, Croft KD, Burke V, Mori TA, Caccetta RA *et al.* Acute effects of ingestion of black and green tea on lipoprotein oxidation. Am J Clin Nutr. 2000 May;71(5):1103-7

[51] O'Reilly JD, Sanders TA, Wiseman H. Flavonoids protect against oxidative damage to LDL in vitro: use in selection of a flavonoid rich diet and relevance to LDL oxidation resistance ex vivo? Free Radic Res. 2000 Oct;33(4):419-26

[52] Richelle M, Tavazzi I, Offord E. Comparison of the antioxidant activity of commonly consumed polyphenolic beverages (coffee, cocoa, and tea) prepared per cup serving. J Agric Food Chem. 2001 Jul;49(7):3438-42

[53] Leung LK, Su Y, Chen R, Zhang Z, Huang Y, Chen ZY. Theaflavins in black tea and catechins in green tea are equally effective antioxidants. J Nutr. 2001 Sep;131(9):2248-51

[54] McAnlis GT, McEneny J, Pearce J, Young IS. Black tea consumption does not protect low density lipoprotein from oxidative modification. Eur J Clin Nutr. 1998 Mar;52(3):202-6

[55] van het Hof KH, de Boer HS, Wiseman SA, Lien N, Westrate JA, Tijburg LB. Consumption of green or black tea does not increase resistance of low-density lipoprotein to oxidation in humans. Am J Clin Nutr.1997 Nov;66(5):1125-32

[56] Hirano-Ohmori R, Takahashi R, Momiyama Y, Taniguchi H, Yonemura A, Tamai SGreen tea consumption and serum malondialdehyde-modified LDL concentrations in healthy subjects. J Am Coll Nutr.2005 Oct;24(5):342-6

[57] Ishikawa T, Suzukawa M, Ito T, Yoshida H, Ayaori M, Nishiwaki M. Effect of tea flavonoid supplementation on the susceptibility of low-density lipoprotein to oxidative modification. Am J Clin Nutr. 1997 Aug;66(2):261-6

[58] Sung H, Min WK, Lee W, Chun S, Park H, Lee YW, Jang S, Lee DH The effects of green tea ingestion over four weeks on atherosclerotic markers. Ann Clin Biochem. 2005 Jul;42(Pt 4):292-7

[59] Erba D, Riso P, Bordoni A, Foti P, Biagi PL, Testolin G Effectiveness of moderate green tea consumption on antioxidative status and plasma lipid profile in humans. J Nutr Biochem. 2005 Mar;16(3):144-9

[60] Davies MJ, Judd JT, Baer DJ, Clevidence BA, Paul DR, Edwards AJ, Wiseman SA, Muesing RA, Chen SC Black tea consumption reduces total and LDL

cholesterol in mildly hypercholesterolemic adults. J Nutr. 2003 Oct;133(10):3298S-3302S

[61] Princen HM, van Duyvenvoorde W, Buytenhek R, Blonk C, Tijburg LB, Langius JA, Meinders AE, Pijl H. No effect of consumption of green and black tea on plasma lipid and antioxidant levels and on LDL oxidation in smokers. Arterioscler Thromb Vasc Biol. 1998 May;18(5):833-41

[62] Loktionov A, Bingham SA, Vorster H, Jerling JC, Runswick SA, Cummings JH Apolipoprotein E genotype modulates the effect of black tea drinking on blood lipids and blood coagulation factors: a pilot study. Br J Nutr. 1998 Feb;79(2):133-9

[63] Balentine DA. The role of tea flavonoids in cardiovascular health. Proceedings of 2001 International Conference on O-Cha (tea) Culture and Science, Session III, 2001. Shizuoka, Japan; pp. 84-89

[64] Duffy SJ, Keaney JF.Jr., Holbrook M, Gokce N, Swerdloff PL, Frei B, Vita JA. Short- and long-term tea consumption reverses endothelial dysfunction in patients with coronary artery disease. Circulation 2001; 104, 151-156

[65] Hodgson JM, Puddey IB, Burke V, Watts GF, Beilin LJ. Regular ingestion of black tea improves brachial artery vasodilator function. Clinical Science 2002; 102, 195-201

[66] Nagaya N, Yamamoto H, Uematsu M, Itoh T, Nakagawa K, Miyazawa T et al. Green tea reverses endothelial dysfunction in healthy smokers. Heart 2004;90:1485-1486

[67] Tijburg LBM, Mattern T, Folts JD, Weisgerber UM, Katan MB. Tea flavonoids and cardiovascular disease: a review. Critical Reviews in Food Science and Nutrition 1997; 37: 771-785

[68] Hodgson JM, Puddey IB, Mori TA, Burke V, Baker RI, Beilin L.J. Effects of regular ingestion of black tea on haemostasis and cell adhesion molecules in humans. European Journal of Clinical Nutrition . 2001;55, 881-886

[69] Balentine, DA. The role of tea flavonoids in cardiovascular health. Proceedings of 2001 International Conference on O-Cha (tea) Culture and Science, Session III, Shizuoka, Japan; 2001;pp. 84-89

[70] Mitane YS, Miwa M, Okada S. Platelet aggregation inhibitors in hot water extracts of green tea. Chemical and Pharmaceutical Bulletin 1990; 38: 790-793

[71] Thelle DS. Coffee, tea and coronary heart disease. Curr Opin Lipidol.1995; 6: 25-27

[72] Boston Collaborative Surveillance Program. Coffee drinking and acute myocardial infarction. Lancet ii, 1972; 1278-81

[73] Hertog MGL, Kromhout D, Aravanis C, Blackburn H, Buzina R, Fidanza F. *et al* Flavonoid intake and long-term risk of coronary heart disease and cancer in the Seven Countries Study. Archives of Internal Medicine 1995; 155: 381-386

[74] Hertog MGL, Feskens EJM, Hollman PCH, Katan MB, Kromhout D. Dietary antioxidant flavonoids and the risk of coronary heart disease: the Zutphen Elderly Study. Lancet 1993; 342: 1007-11
[75] Hertog, MGL, Feskens, EJM, Kromhout, D. Antioxidant flavonols and coronary heart disease risk. Lancet 1997; 349: 699
[76] Knekt, P, Jervinen, R, Reunanen, A, Maatela, J. Flavonoid intake and coronary mortality in Finland: A cohort study. British Medical Journal 1996; 312: 478-81
[77] Rimm EB, Katan, MB, Ascherio, A, Stampfer, MJ, Willett, WC. Relation between intake of flavonoids and risk of coronary heart disease in male health professionals. Annals of Internal Medicine 1996; 125: 383-89
[78] Hertog, MGL, Sweetnam, PM, Fehily, AM, Elwood, PC, Kromhout, D. Antioxidant flavonols and ischemic heart disease in a Welsh population of men: The Caerphilly Study. American Journal of Clinical Nutrition 1997 ; 65: 1489-94
[79] Hirvonen T, Pietinen P, Virtanen M, Ovaskainen ML, Hakkinen S, Albanes D, Virtamo J. Intake of flavonols and flavones and risk of coronary heart disease in male smokers. Epidemiology 2001; 12: 62-67
[80] Geleijnse JM, Launer LJ, Van Der Kuip DA, Hofman A, Witteman JC. Inverse association of tea and flavonoid intakes with incident myocardial infarction: the Rotterdam Study. American Journal of Clinical Nutrition 2002; 75: 880-86
[81] Imai K, Nakachi K. Cross sectional study of effects of green tea on cardiovascular and liver diseases. British Medical Journal. 1995; 310: 693-96
[82] Kono S, Shinchi K, Ikeda N, Yanai F, Imanishi K. Green tea consumption and serum lipid profiles: a cross-sectional cohort study in Northern Kyushu, Japan. Preventive Medicine 1992; 21: 526-31
[83] Stensvold I, Tverdal A, Solvoll K, Foss OP. Tea consumption, relationship to cholesterol, blood pressure and coronary total mortality. Preventive Medicine. 1992; 21: 546-53
[84] Yang YC, Lu FH, Wu JS, Wu CH, Chang CJ. The protective effect of habitual tea consumption on hypertension. Arch Intern Med. 2004;164: 1534-40
[85] Green MS, Harari G. Association of serum lipoproteins and health related habits with coffee and tea consumption in free living subjects examined in the Israeli CORDIS study. Preventive Medicine, 1992; 21: 532-45
[86] Brown CA, Bolton-Smith C, Woodward M, Tunstall-Pedoe H. Coffee and tea consumption and the prevalence of coronary heart disease in men and women: results from the Scottish Heart Health Study. Journal of Epidemiology and Community Health 1993; 47: 171-175
[87] Balentine, D.A. The role of tea flavonoids in cardiovascular health. Proceedings of 2001 International Conference on O-Cha (tea) Culture and Science, Session III, Shizuoka, Japan; 2001. pp. 84-89

[88] Fujiki H, Komari A, Suganuma M, Chemoprevention of cancer. In: Bowden GT, Fischer SM, Eds. Comprehensive Toxicology. London: Pergamon Elsevier Science 1997; 12: 453-71

[89] Imai K, Suga K, Nakachi K. Cancer-preventive effects of drinking green tea among a Japanese population. Preventive Medicine 1997; 26: 769-75

[90] Weisburger, JH. Antimutagenesis and anticarcinogenesis, from the past to the future. Mutation Research 2001; 480-481: 23-25

[91] Weisburger, JH. Tea and prevention of heart disease and cancer. Proceedings of 2001 International Conference on O-Cha (tea) Culture and Science, Session III, 2001 Shizuoka, Japan; pp. 90-94

[92] Bunkova R, Marova I, Nemec M. Antimutagenic properties of green tea. Plant Foods Hum Nutr. 2005 Mar;60(1):25-9

[93] Halder B, Pramanick S, Mukhopadhyay S, Giri AK. Inhibition of benzo[a]pyrene induced mutagenicity and genotoxicity by black tea polyphenols theaflavins and thearubigins in multiple test systems. Food Chem Toxicol. 2005 Apr;43(4):591-7

[94] Yen GC, Ju JW, Wu CH. Modulation of tea and tea polyphenols on benzo(a)pyrene-induced DNA damage in Chang liver cells. Free Radic Res. 2004 Feb;38(2):193-200

[95] Moon YJ, Wang X, Morris ME. Dietary flavonoids: Effects on xenobiotic and carcinogen metabolism. Toxicol In Vitro. 2006 Mar;20(2):187-210. Epub 2005 Nov 11

[96] Bu-Abbas A, Clifford MN, Ioannides C, Walker R. Stimulation of rat hepatic UDP-glucuronosyl transferase activity following treatment with green tea. Food Chem Toxicol. 1995 Jan;33(1):27-30

[97] Yoxall VR, Parker DA, Kentish PA, Ioannides C. Short-term black tea intake modulates the excretion of urinary mutagens in rats treated with 2-amino-3-methylimidazo-[4,5-f]quinoline (IQ): role of CYP1A2 upregulation. Arch Toxicol. 2004 Aug;78(8):477-82

[98] Okabe S, Fujimoto N, Sueoka N, Suganuma M, Fujiki H. Modulation of gene expression by (-)epigallocatechin gallate in PC-9 cells using a cDNA expression array. Biological and Pharmaceutical Bulletin 2001; 24: 883-86

[99] Levites Y, Youdim MB, Maor G, Mandel S. Attenuation of 6-hydroxydopamine (6-OHDA)-induced nuclear factor-kappa B (NF-kappa B) activation and cell death by tea extracts in neuronal cultures. Biochemical Pharmacology 2002; 63: 21-29

[100] Surh YJ, Chun KS, Cha HH, Han SS, Keum YS, Park KK, *et al.* Molecular mechanisms underlying chemopreventive actions of anti-inflammatory phytochemicals: down regulation of COX-2 and iNOS through suppression of NF kappa-B activation. Mutation Research 2001; 480-481: 243-68

[101] Amarakoon AMT, Tappia PS, Grimble RF. Endotoxin induced production of interleukin-6 is enhanced by vitamin E deficiency and reduced by black tea extract. Inflammation Research 1995; 44: 301-05

[102] Yang CS, Maliakal P, Meng X. Inhibition of carcinogenesis by tea. Annual Review of Pharmacology and Toxicology 2002; 42: 25-54

[103] Wiseman S, Mulder T, Rietveld A. Tea flavonoids: bioavailability *in vivo* and effects of cell signalling pathways *in vitro*. Antioxidants & Redox Signalling 2001; 3: 1009-21

[104] Tosetti F, Ferrari N, De Flaora S, Albini A. Angioprevention: angiogenesis is a common and key target for cancer chemopreventive agents. FASEB Journal 2002; 16: 2-14

[105] Bertolini F, Fusetti L, Rabascio C, Cinieri S, Martinelli G, Pruneri G. Inhibition of angiogenesis and induction of endothelial and tumour cell apoptosis by green tea in animal models of human high-grade non-Hodgkin's lymphoma. Leukaemia 2000 ; 14: 1477-82

[106] Lamy S, Gingras D, Beliveau R. Green tea catechins inhibit vascular endothelial growth factor receptor phosphorylation. Cancer Research 2002; 62: 381-85

[107] Zhang G, Miura Y, Yagasaki K. Induction of apoptosis and cell cycle arrest in cancer cells by *in vivo* metabolites of teas. Nutrition and Cancer 2000; 38: 265-73

[108] Hayakawa S, Kimura T, Saeki K, Koyana Y, Aoyagi Y, Noro T. *et al.* Apoptosis inducing activity of high molecular weight fractions of tea extracts. Bioscience, Biotechnology, and Biochemistry 2001 ; 65: 459-62

[109] Stavric B, Matula TI, Klassen R, Downie RH. The effect of teas on the *in vitro* mutagenic potential of heterocyclic aromatic amines. Food and Chemical Toxicology 1996; 34: 515-23

[110] Yoshioka H, Kurosaki H, Yoshinaga K, Saito K, Yoshioka H. Beta-ray induced scission of DNA in tritiated water and protection by a green tea percolate and (-) Epigallocatechin gallate. Bioscience Biotechnology and Biochemistry 1997; 61: 1560-63

[111] Record IR, Dreosti IE. Protection by tea against UV A+B induced skin cancer in hairless mice. Nutrition and Cancer 1998; 32: 71-75

[112] Lou YR, Lu YP, Xie JG, Huang MT, Conney AH. Effect of oral administration of tea, decaffeinated tea and caffeine on the formation and growth of tumours in high risk SKH-1 mice previously treated with ultraviolet B light. Nutrition and Cancer 1999; 33: 146-53

[113] Nomura M, Ma WY, Huang C, Yang CS, Bowden GT, Miyamoto K, *et al.* Inhibition of ultraviolet B-induced AP-1 activation by theaflavins from black tea. Molecular Carcinogenesis 2000; 28: 148-55

[114] Chung FL. The prevention of lung cancer induced by a tobacco-specific nitrosamine-induced lung tumorigenesis by compounds derived from cruciferous vegetables and green tea. Annals of the New York Academy of Sciences 1999; 686: 186-201

[115] Dreosti IE, Wargovich MJ, Yang CS. Inhibition of carcinogenesis by tea: The evidence from experimental studies. Critical Reviews in Food Science and Nutrition 1997; 37: 761-70

[116] Wang ZY, Wang LD, Lee MJ, Ho CT. Inhibition of N-nitrosomethylbenzlamine induced oesophageal tumorigenesis in rats by green tea and black tea. Carcinogenesis. 1995; 16: 2143-48

[117] Ohishi T, Kishimoto Y, Miura S, Shiota G, Kohri T, Hara Y, *et al.* Synergistic effects of (-)epigallocatechin gallate with Sulindac against colon carcinogenesis of rats treated with azoxymethane. Cancer Letters 2002; 177: 49-56

[118] Blot WJ, McLaughlin JK, Chow W. Cancer rates among drinkers of black tea. Critical Reviews in Food Science and Nutrition 1997; 37: 739-60

[119] Blot WJ, Chow W-H, McLaughlin JK. Tea and cancer: a review of the epidemiological evidence. European Journal of Cancer Prevention 1996; 5: 425-438

[120] Blot WJ, McLaughlin JK, Chow W. Cancer rates among drinkers of black tea. Critical Reviews in Food Science and Nutrition 1997; 37: 739-60

[121] Zeegers MP, Dorant E, Goldbohm RA, Van den Brandt PA. Are coffee tea and total fluid consumption associated with bladder cancer risk? Results from the Netherlands cohort study. Cancer Causes Control 2001; 12: 231-38

[122] Bianchi GD, Cerhan JR, Parker AS, Putnam SD, See WA, Lynch CF. *et al.* Tea consumption and risk of bladder and kidney cancers in a population-based case-control study. American journal of Epidemiology 2000; 151: 377-83

[123] Hakim IA, Harris RB, Weisburger UM. Tea intake and squamous cell carcinoma of the skin: influence of type of tea beverages. Cancer Epidemiology, Biomarkers & Prevention 2000; 9: 727-31

[124] Zhong L, Goldberg MS, Gao YT, Hanley JA, Parent ME, Jin F. A population-based case-control study of lung cancer and green tea consumption among women living in Shanghai, China. Epidemiology 2001; 12: 695-700

[125] Inoue M, Tajima K, Mizutani M, Iwata H, Iwase T, Miura S. *et al.* Regular consumption of green tea and the risk of breast cancer recurrence: follow-up study from the Hospital-based epidemiologic research programme at Aichi Cancer Centre (HERPACC), Japan. Cancer Letters 2001; 167: 175-82

[126] Michels KB, Holmberg L, Bergkvist L, Wolk A. Coffee, tea and caffeine consumption and breast cancer incidence in a cohort of Swedish women. Annals of Epidemiology 2002; 12: 21-26

[127] Nagano J, Kono S, Preston DL, Mabuchi KA. Prospective study of green tea consumption and cancer incidence, Hiroshima and Nagasaki (Japan). Cancer Causes Control 2001; 12: 501-08

[128] Duckworth SC, Duckworth R. The ingestion of fluoride in tea. British Dental Journal 1978; 145: 368-70

[129] Gulati P, Singh V, Gupta MK, Vaidya V, Dass S, Prakash S. Studies on the leaching of fluoride in tea infusions. The Science of the Total Environment 1993; 138: 213-22

Effects of Green and Black Tea on Glucose Tolerance, Serum Insulin and Antioxidant Enzyme Levels in Streptozotocin-Induced Diabetes Rats

Chua Yong Ruan Ray, Hsu A and Tan K H Benny*
Traditional Medicines and Natural Products Research Laboratory, Department of Pharmacology, Yong Loo Lin School of Medicine, National University of Singapore, 18 Medical Drive, Singapore 117597
**Corresponding email: phctankh@nus.edu.sg*

INTRODUCTION

Tea is one of the most popular beverages in the world. Since tea is widely consumed by hundreds of millions of people regularly, the possible effects of tea on human health are of particular importance in the field of medical, agricultural and food research.

As a plant product, tea has a highly complicated composition. The composition of tea may be different due to differences in cultivation and manufacturing processes. Different manufacturing processes lead to a great diversity of the composition among green tea, black tea and semi-fermented oolong tea. Green tea is absolutely non-fermented tea, while black tea is fully fermented tea. The fermentation process greatly determines the differences in composition between green tea and black tea. For instance, oxidation of catechins during the fermentation process produces two groups of polyphenol compounds, theaflavins and thearubigins, which are thought to be unique to black tea. Due to its highly diversified and complicated composition, tea has high bioactivity and therapeutic potential. Modern medical research has found that tea and tea products display a wide spectrum of bioactivity and show therapeutic effectiveness in a number of experimental disease models [1a,1b].

Numerous studies on antioxidative activity of tea have been carried out as the role of free radicals and reactive oxygen species in the pathogenesis of certain human diseases, including ageing, cardiovascular disease, diabetes and cancer becomes increasingly recognised.
Antioxidants are important in inhibiting lipid peroxidation and protecting

the damage of free radicals on membrane structures and organs [1b]. Studies by Salah et al. (1995) [2] and Sabu (2002) [3] have demonstrated *in vitro* antioxidative activity of tea components. Sabu (2002) [3] also showe dimproved glucose tolerance and antioxidative effects of green tea polyphenols (GTP) in normal and alloxan-diabetic rats. Gomes et al. (1994) [4] and Dashti and Morshedi (2000) [5] also reported that black tea has anti-hyperglycemic effect in rats.

Apart from these animal studies, the therapeutic potential of tea has also been investigated in humans, in whom diabetes mellitus is the third leading cause of death in the United States after heart disease and cancer [6]. Maxwell et al. (1997) [7] found that the diabetic patients they studied had significant defects in antioxidant protection; this may increase their vulnerability to oxidative damage. Davi et al. (2003) [8] reported that both high glucose levels and protein glycation enhanced the oxidation of low density lipoproteins (LDL) by metal ions.

This preliminary study was designed to evaluate the antioxidative and antidiabetic effects of green and black tea (if any) in an animal model of type 1 diabetes. Subsequent studies could then be done to isolate and identify the active constituents responsible for such effects.

MATERIALS AND METHODS

Animals

The experiments were carried out on male Sprague-Dawley (SD) rats weighing 150-200g. They were fed laboratory rat chow and were maintained under 12 h light/dark cycles at room temperature.

Source of tea

Both green tea and black tea bags were purchased from the local supermarket. Green tea was manufactured by Red Japan Ltd.; black tea was "blended and packed" by R. Twining & Company Limited, London, England.

Induction of type 1 diabetes in rats

Streptozotocin (STZ), purchased from Sigma Aldrich Chemicals, USA, was dissolved in distilled water. Rats were fasted overnight (15h) before administration of STZ. Each rat was administered 65mg/kg b.wt of STZ intraperitoneally. After seven days, serum glucose level was checked for each rat. Rats with serum glucose levels more than 200mg% were considered diabetic.

Preparation of hot water tea infusions

Ten tea bags (black or greentea, each weighing 2g) were soaked in 100mL of boiling water. After 5 min, the tea bags were removed and the infusion was fed intragastrically to rats. 1ml/100g b.wt of tea was fed twice a day to the diabetic rats with an equivalent volume of water fed to control rats. The dose used was in accord with that by Gòmes et al. (1994) [4].

Experimental protocols

I. Therapeutic study

Twelve STZ-diabetic rats were divided into three groups (each n=4), as follows:-

Control group: rats received 1mL/100g b.wt water twice a day for 16 days.

Green tea group: rats received 1mL/100g b.wt green tea twice a day for 16 days.

Black tea group: rats received 1mL/100g b.wt black tea twice a day for 16 days.

Serum glucose levels of all the STZ-treated rats were measured at the start of the study, one week later (day 7), and two weeks later (day 14) in order to monitor the hyperglycemic condition of the rats. On the 15th day, the oral glucose tolerance test (OGTT) [see below] was carried out on all the rats. The rats were sacrificed on day 17; serum was obtained for insulin assay and liver was sampled for tests of antioxidant enzyme and glucose-6-phosphatase activities.

II. Preventive study

Twelve normal rats were divided into three groups (4 rats per group) as follows:-

Control group: rats received 1mL/100g b.wt water twice a day for 21 days.

Green tea group: rats received 1mL/100g b.wt green tea twice a day for 21 days.

Black tea group: rats received 1mL/100g b.wt black tea twice a day for 21 days.

After 21 days (three weeks) of the respective treatments, all the rats were injected with STZ to induce diabetes. One week after induction of diabetes, OGTT was carried out, following which all the rats were sacrificed, serum was obtained for insulin assay and liver was obtained for assays of antioxidant enzyme and glucose-6-phosphatase activities.

Evaluation of antidiabetic activity

Oral glucose tolerance test (OGTT)

Rats were fasted overnight before OGTT. The rats were fed a dose of glucose (2g/kg b.wt) and blood samples were collected from them by bleeding from the tail vein. These blood samples were used for measurement of serum blood glucose level at 0h. Subsequently, blood samples were collected 1h and 2h after glucose feeding for measurement of serum glucose levels at these time points.

Serum insulin and hepatic glucose-6-phosphatase assays

The quantitative determination of insulin in serum was carried out using the Ultra Sensitive rat insulin ELISA kit from Crystal Chem INC., Illinois, USA. Hepatic glucose-6-phosphatase activity was assayed according to the method reported by Baginski et al. (1974) [9].

In vivo antioxidant activity

Experimental rats were sacrificed and liver was removed. For each rat, 0.5g of liver was homogenised in potassium phosphate buffer and the liver homogenate used for measuring superoxide dismutase (SOD) activity, catalase activity, glutathione (GSH) content, GSH peroxidase (GPx) activity and thiobarbituric acid-reactive substances (TBARs, particularly malonaldehyde) content. Methods for SOD, catalase, GSH and GPx assays were obtained from Beutler (1984) [10], while the TBARs assay was done as described by Ohkawa et al. (1979) [11].

Statistical analysis

Data were expressed as mean ± standard error of mean (each group, n=4). Statistical comparisons of body weight and data from the assays were performed with Student's t-test. Statistical comparisons of food and water intake were also performed with the Student's t-test. A probability level of $P<0.05$ was chosen as the limit for statistical significance.

RESULTS

In the therapeutic study, SOD, catalase and GPx activities as well as TBARs content were not significantly different (see Table 1) between the control group and the green tea- and black tea-treated groups. The mean GSH level in the black tea-treated group was relatively lower than that in the green tea-treated and control groups, but the difference was not statistically significant (P=0.057).

In the preventive study, the same parameters did not show significant differences between the control group and the green tea- and black tea-

Table 1. Effects of green and black tea on hepatic antioxidant enzyme activity and TBARs content in male STZ-diabetic rats

Therapeutic study	**SOD**[a]	**Catalase**[b]	**GSH**[c]	**GPx**[d]	**TBARs**[e]
Control	21.4±1.5	0.195±0.002	49.3±4.43	0.790±0.03	0.788±0.072
Green Tea	17.7±1.1	0.208±0.015	61.1±6.97	0.835±0.03	0.744±0.065
Black Tea	21.9±1.4	0.188±0.003	39.0±3.24#	0.916±0.083	0.850±0.084
Preventive study					
Control	26.3±0.59	0.188±0.024	31.4±3.12	0.820±0.059	0.649±0.046
Green Tea	26.2±2.32	0.194±0.008	32.0±1.63	0.738±0.102	0.529±0.042
Black Tea	22.8±0.20	0.172±0.012	20.5±2.36*	0.752±0.068	0.557±0.037

[a] Superoxide dismutase (SOD) activity (U/mg protein)
[b] Catalase activity (U/mg protein)
[c] Glutathione [GSH] (nmol/mg protein)
[d] Glutathione peroxidase [GPx] activity (U/mg protein)
[e] Thiobarbituric Acid-Reactive Substances (TBARS): Malonaldhyde content (nmol/mg protein)
* P<0.05 (compared to control and green tea-treated group)
# The value is low compared to control, but p=0.057

treated groups. However, GSH level in the black tea-treated group was significantly lower than in control and green tea-treated groups.

In the therapeutic study, the changes in mean serum glucose levels in the three groups over the two weeks showed that the increases at the end of weeks 1 and 2 were greatest in the green tea-treated rats (see Table 2). The black tea-treated group showed a marked decrease in blood sugar at the end of the 1st week compared to the control group; the decrease was less at the end of the second week.

Table 2. Changes in mean fasting serum glucose levels in green- and black tea-treated rats at the end of week 1 and week 2 (Therapeutic study)

Group	*Serum glucose level at week 0 (mg %)	*Serum glucose level at week 1 (mg%)	% change in glucose level at week 1 a#	*Serum glucose level at week 2 (mg %)	% change in glucose level at week 2 a#
Control	369±49	392±81	6.2	350±89	5.1
Green Tea	307±29	390±74	27.0	408±84	32.9
Black Tea	369±69	275±15	-25.5	329±60	-10.8

[a] % change in serum glucose level with reference to serum glucose level at week 0
* mean ± serum
\# compared to week 0

In the therapeutic study, serum glucose levels in the green tea-treated group showed the greatest increase at the 1st and 2nd hour compared to the control and black tea-treated groups (see Table 3). The decrease in serum glucose level from 1st hr to 2nd hr was significantly greater in the control group than in the green tea- (p value=0.012) and black tea-treated groups (p value=0.018), respectively.

In the preventive study, green and black tea-treated both the groups had lower increases in mean glucose level compared with the control group. However, because of high variability in both groups, the observed suppressed rises in glucose levels in these groups were not significantly different compared to the control group (p=0.138). At the 2nd hour, the serum glucose in the green tea-treated group had decreased compared to the increases in the control and black tea-treated groups.

In both therapeutic and preventive studies, there were no significant differences observed among the groups in glucose-6-phosphatase activity,

Table 3. Changes in mean serum glucose levels after oral glucose load in green tea- and black tea-treated diabetic rats (Therapeutic and Preventive studies)

	% change in mean serum glucose levels	
Therapeutic study	**1st hour [a]**	**2nd hour [b]**
Control	45.1	7.42
Green Tea	73.5	54.2
Black Tea	59.0	39.4
Preventive study		
Control	58.7	16.9
Green Tea	20.0	-4.17
Black Tea	20.9	12.5

[a] % change in serum glucose level at 1st hour after glucose intake with reference to 0hr glucose level.
[b] % change in serum glucose level at 2nd hour after glucose intake with reference to 0hr glucose level.

serum insulin levels as well as body weight (see Table 4). In the therapeutic study, the black tea-treated group showed a decrease in body weight of 3.7% compared to increases of 4.2% and 9.4% found in control and green tea-treated groups. The green tea-treated group had significantly higher mean food and water intakes (both $p<0.05$) compared to the control group.

In the preventive study, before the injection of STZ, both green tea- and black tea-treated groups had significantly higher mean intakes of water ($p=0.022$ and $p=4.418*10^{-6}$), respectively. Mean food intakes were not significantly different. During the 7 days after STZ injection, a significantly higher mean water intake ($p=0.041$) occurred in the green tea-treated group compared to the control group. The black tea-treated group had similar mean food and water intakes as the control group.

DISCUSSION

Glutathione (GSH) is an important antioxidant in the body. Its major function is to reductively eliminate hydrogen peroxides (precursor of hydroxyl free radicals) and organic peroxides, which are reactive oxygen metabolites that can irreversibly damage hemoglobin and cell membranes

Table 4. Effects of green and black tea on hepatic glucose-6-phosphatase (G-6P) activity, serum insulin level, body weight, mean food and water intakes in diabetic rats (Therapeutic and Preventive studies)

Therapeutic Study (each n=4)	**Glucose-6-Phosphatase**[a]	**Serum insulin (ng/ml)**	**% change in mean body weight**[b]		**Food intake (g/100g b.wt) (mean ± sem)**		**Water intake (g/100g b.wt) (mean ± sem)**	
			Day 7	**Day 14**				
Control	0.544±0.175	0.506±0.13	4.3	4.2	15.2±0.3		53.8±1.6	
Green Tea	0.489±0.086	0.358±0.09	7.9	9.4	17.6±0.4*		75.7±2.5*	
Black Tea	0.433±0.100	0.324±0.02	2.5	-3.7	14.6±0.2		53.3±1.2	
			Body weight (g) (mean ± sem)		**Food intake (g/100g b.wt) (mean ± sem)**		**Water intake (g/100g b.wt) (mean ± sem)**	
Preventive Study (each n=4)	**Glucose-6-Phosphatase**[a]	**Serum insulin (ng/ml)**	**Day 21 - before STZ injection**	**7 days after STZ injection**	**Before STZ injection (Days 1 - 21)**	**After STZ injection (Days 22 - 28)**	**Before STZ injection (Days 1 - 21)**	**After STZ injection (Days 22 - 28)**
Control	0.426±0.085	0.482±0.17	22.4±1.8	16.6±8.3	6.8±0.2	10.2±0.5	10.3±0.2	36.8±3.3
Green Tea	0.442±0.028	0.442±0.10	22.4±2.2	16.6±3.8	7.1±0.4	12.7±1.2	11.9±0.5*	56.0±6.5*
Black Tea	0.536±0.058	0.370±0.05	17.3±5.0	16.9±3.1	7.1±0.2	10.3±1.2	13.8±0.3*	48.5±5.8

[a] Glucose-6-Phosphatase activity, measured by rate and amount of Pi liberated (µmol/min/mg protein)
[b] % change in body weight, expressed with reference to initial (Day 1) body weight
* $P<0.05$ (compared to control group)

[6]. A significantly lower GSH level in the black tea-treated group in the preventive study and a lower GSH level in the black tea-treated group in the therapeutic study may suggest that constituents in black tea infusion may have a negative impact on the production of GSH *in vivo*. The production of GSH may result from synthesis from cysteine and glycine (catalysed by γ-glutamylcysteine synthetase and GSH synthetase), or regeneration from oxidized GSH (catalysed by glutathione reductase). Active constituents in black tea may thus inhibit either or both of these pathways. Further studies can be carried out with black tea to determine whether such an effect occurs.

In the OGTT of the therapeutic study, apart from the absence of significant differences in serum glucose levels at the 1st h and 2nd h among the groups, it was observed that there was a significantly greater decrease in serum glucose level from the 1st h to 2nd h in the control group compared to the green tea- and black tea-treated groups. These observations may suggest that green and black tea may not improve glucose tolerance in the diabetic state. In the preventive study, a suppressed rise in serum glucose level from 0h to the 1st h in the green tea- and black tea-treated groups compared to the control group was observed. However this finding was not significantly different. Further experiments using a larger number of animals are needed to confirm the prophylactic effect of green and black tea on glucose tolerance. An additional step, such as coupling the assay of serum glucose level with insulin at 0h, 1st h and 2nd h may also aid in the interpretation of such an effect on glucose tolerance.

Excessive drinking (polydipsia), a classic sign of diabetes, is related to the diuresis that occurs in diabetics. In addition, green and black tea may also induce diuresis. Hence, a significantly higher intake of water by the green tea-treated group in the therapeutic and preventive studies (after injection of STZ), may not necessarily indicate a relative worsening of the diabetic condition in the green tea-treated group. This requires further investigation.

To conclude, this preliminary study indicates that black tea infusion may reduce the hepatic level of the antioxidant, GSH, in STZ-diabetic rats, while green tea may cause greater thirst to occur in these rats. Overall, the findings of this study do not provide strong evidence for the antioxidative and antidiabetic effects of green and black tea that have been reported by other workers. Further studies are thus needed to clarify the effects of tea in the diabetic condition.

REFERENCES

[1] (a) Chen ML. Tea and Health – An Overview. In: Tea Bioactivity and Therapeutic Potential. (Yong-su Zhen, ZM Chen, SJ Cheng, ML Chen, eds). Taylor & Francis Group, Cornwall, UK, 2002. Pp.1-6
(b) Zong-mao Chen. The effects of tea on the Cardiovascular System. In: Tea Bioactivity and Therapeutic Potential. (Yong-su Zhen, ZM Chen, SJ Cheng, ML Chen, eds). Taylor & Francis Group, Cornwall, UK, 2002. Pp. 151-5

[2] Salah N, Miller NJ, Paganga G, Tijburg L, Bolwell GP, Rice-Evans C. Polyphenolic flavanols as scavengers of aqueous phase radicals and as chain-breaking antioxidants. Arch Biochem Biophys 1995; 322 (2): 339-46

[3] Sabu MC, Smitha K, R Kuttan. Anti-diabetic activity of green tea polyphenols and their role in reducing oxidative stress in experimental diabetes. J Ethnopharmacol 2002; 83: 109-16

[4] Gomes A, Vedasiromoni JR, Das M et al. Anti-hyperglycemic effect of black tea (*Camellia sinensis*) in rat. J Ethnopharmacol 1995; 45: 223-6

[5] Dashti MH, Morshedi A. A comparison between the effect of black tea and kombucha tea on blood glucose level in diabetic rat. Med J Islamic Acad Sci 2000; 13 (2): 83-7

[6] Voet D, Voet JG, Pratt CW. In Fundamentals of Biochemistry. John Wiley & Sons, Inc., New York, USA, 1999. Pp. 423, 526

[7] Maxwell SRJ, Thomason H, Sandler D et al. Antioxidant status in patients with uncomplicated insulin-dependent and non-insulin-dependent diabetes mellitus. Eur J Clin Invest 1997; 27(6): 484-90

[8] Davi G, Chiarelli F, Santilli F et al. Enhanced Lipid Peroxidation and Platelet Activation in the Early Phase of Type 1 Diabetes Mellitus. Circulation 2003; 107 (25): 3199-203

[9] Baginski ES, Foa PP, Zak B. Methods of Enzymatic Analysis. Academic Press Inc., New York, USA, 2004. Pp. 876-80

[10] Ernest Beutler. Red Cell Metabolism: A Manual of Biochemical Methods. Grune & Stratton, Inc., New York, 1984. Pp. 83-5

[11] Ohkawa H, Ohishi N, Yagi K. Assay for lipid peroxides in animal tissues by thiobarbituric acid reaction. Anal Biochem 1979; 95: 351-8

St John's Wort: A Precious Gift from the Saints?

Ying-Hui Li[a], Hongyan Du[a], Boon-Huat Bay[a] and Malini Olivo[b]*

[a]Department of Anatomy; National University of Singapore, 4 Medical Drive, Blk MD 10, Singapore 11757

[b]Department of Medical Science, National Cancer Centre, 11 Hospital Drive, Singapore 169610

**Corresponding email: dmsmcd@nccs.com.sg*

INTRODUCTION

St. John's wort (*Hypericum perforatum* L.) is a herbaceous perennial plant indigenous to Europe, West Asia, and North Africa which has been naturalized in the United States and Australia [1,2]. The plant derived its name from John the Baptist as its flowers bloom on 24th June, the day St John was beheaded. For centuries, this botanical plant has been traditionally used to combat bacterial and viral infections and as a herbal remedy for a variety of pathologies such as psychiatric disorders, inflammation, respiratory diseases, skin wounds and peptic ulcers [3,4]. In recent years, it has gained increasing interest for its potential as an alternative medicine for the treatment of mild to moderate depression.

At least ten classes of bioactive compounds have been identified in this medicinal herb. Its chemical constituents comprise of naphtodianthrone derivatives, flavonoids, phloroglucinol derivatives, procyanidines, tannins, essential oils, amino acids, phenylpropanes, xantones and other water-soluble compounds (organic acids, peptides and polysaccharides) [5]. Of these ten types of biochemical compounds, the first three groups are the most abundant and main contributors of the therapeutic applications of this plant.

HYPERICIN IN PHOTODYNAMIC THERAPY (PDT)

The naphtodianthrones, hypericin and its congener pseudohypericin, are polycyclic quinones found at the highest concentration in the flowering portions of St. John's wort and occur mainly in their proto-forms, protohypericin and protopseudohypericin [1,6]. Upon light stimulation, these biosynthetic precursors convert efficiently to their naphtodianthrone analogues, which are sometimes termed collectively as 'total hypericins' [6,7].

Hypericin is one of the most powerful photosensitizers found naturally to date and it has attracted immense scientific interest in the past decade for its potential application in antitumoral photodynamic therapy (PDT). Much of the clinical interest in hypercin stems from its ability to generate a high quantum yield of singlet oxygen (type II process) and other reactive oxygen species (ROS) including superoxide anions, hydrogen peroxide and hydroxyl radicals (type I process) efficiently upon light stimulation, with minimal genotoxicity and systemic toxicity [8-12]. The ROS released locally at the tumor site during photosensitization causes oxidative cellular damage, thereby leading to direct tumor cell killing and destruction of the tumor tissue.

This potent anti-neoplastic activity of hypericin-mediated PDT has been documented in a range of *in vitro* and *in vivo* model systems, thus stimulating great clinical interest in the manipulation of hypericin as a powerful anticancer agent [13-17]. To date, the use of hypericin-PDT has been found to be promising in the treatment of basal and squamous cell carcinoma as well as clinical trials of recurrent mesothelioma [18]. In addition, recent work by Olivo and colleagues in hypericin-PDT demonstrated the induction of tumor necrosis and lipid peroxidation in both *in vitro* and *in vivo* models of nasopharyngeal carcinoma (NPC) [19, 20]. An example of a HK1 NPC murine model with shrinkage of the tumor after hypericin-PDT is shown in Figure 1.

Taken together, these studies reaffirm the potential of this natural photosensitizer in antitumoral PDT, which has also been intensely investigated in the recent years as an alternative treatment modality in a variety of cancers, including malignant glioma, pituitary adenoma and cutaneous T cell lymphoma [21].

ST. JOHN'S WORT AS AN ANTIVIRAL AGENT

Besides the antineoplastic effects of photoactivated hypericin, this aromatic polycyclic dione has also been actively investigated as an antiviral agent, particularly for its potential application in the therapeutic management of retroviral-induced diseases such as acquired immunodeficiency syndrome (AIDS). In an early study by Meruelo and co-workers, hypericin and pseudohypericin were discovered to exhibit potent and highly efficacious antiretroviral activity against Friend leukemia virus (FV) and radiation leukemia virus (RadLV) in both *in vitro* and *in vivo* model systems [12]. Moreover, the antiviral effect was achieved at doses

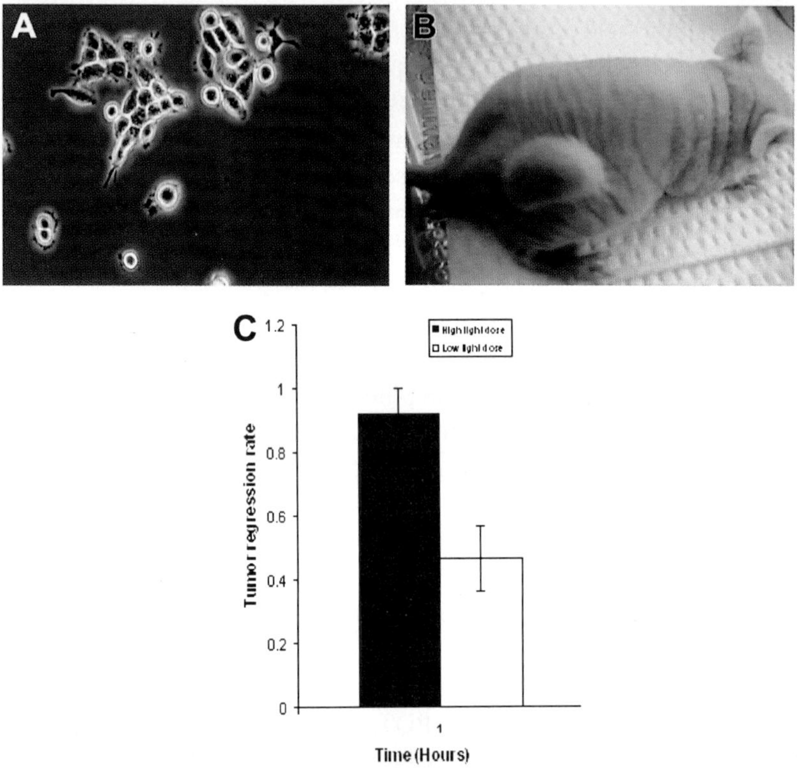

Figure 1. (A) Morphology of HK1 NPC cells in culture. (B) Balb/c mouse subcutaneously transplanted with HK1 cells. Approximately 1.5×10^6 HK-1 tumor cells suspended in 0.1 ml PBS was injected subcutaneously into the lower right flank of the mouse. (C) Tumor regression of NPC/HK1 tumor with different light dose. The animals were anesthetized 1 hour after hypericin administration at a dose of 2mg/kg through the tail veins. Tumors were then illuminated with a light with a dose of 14 J/cm^2 delivered at a fluence rate of 27 mW/cm^2, or 60 J/cm^2 delivered at 50 mW/cm^2. The tumor size was examined on the 10th day after light treatment. The regression of tumor was evaluated by $(V_{before}-V_{after})/V_{before}$, with V_{after} and V_{before} representing the tumor volume after and before PDT treatment. Values are represented as mean ± SE of 3 or 5 animals.

insufficient to cause any severe or toxic side effects *in vivo* or dramatic cytotoxicity *in vitro*, thus prompting enormous interest in the use of this herbal extract as a clinical antiviral drug.

The potential of hypericin as an antiretroviral agent was further substantiated in another investigation by Kraus and colleagues, wherein synthetic hypericin was found to markedly inhibit the replication of equine infectious anemia virus (EIAV)-infected cells [22]. In addition, several *in vitro* studies have reported the remarkable virucidal activity of hypericin against a broad range of encapsulated viruses and retroviruses, including herpes simplex types 1 and 2, influenza virus A, murine cytomegalovirus (MCMV), Sindbis virus (SV), Moloney murine leukemia virus (Mo-MuLV) and the human immunodeficiency virus type 1 (HIV-1) [23-26].

In contrast, non-enveloped viruses such as adenovirus and poliovirus were found to be resistant to hypericin, suggesting the dependence of the presence of a viral lipid envelope in the viral inactivation mechanism of this bioactive component of St. John's wort [23]. The virucidal properties of hypericin also appear to be photodependent, whereby the viral inhibitory/inactivation effect was substantially augmented upon illumination by visible light [26-28]. In view of the potent and broad spectrum antiviral activity of hypericin, this polycyclic quinone has been studied recently for its potential application as a photodependent blood sterilizer [29].

While the virucidal activity of hypericin has been well documented, a novel protein, $p27^{SJ}$, isolated from St. John's wort has also been implicated as an antiviral agent. This protein was identified from a callus culture of *H. perforatum* and found to inhibit transcription of the HIV-1 genome in a variety of human cell types including primary cultures of microglia and astrocytes [30]. Thus far, there is a huge potential in the use of St. John's wort as an antiviral tool, especially in the search for an efficacious drug for the treatment of AIDS.

ST. JOHN'S WORT FOR DEPRESSION

Of all the potential therapeutic applications of St. John's wort, the medicinal herb is most widely researched for its sedative properties and use in the treatment of depression. Its use as a remedy for psychiatric disorders was first established by the Swiss physician Paracelsus in 1952 and it was consequently used traditionally in the therapy of a variety of neurological conditions including neuralgia, neurosis, anxiety and depression in Europe [31]. To date, *Hypericum* is the most widely prescribed antidepressant in Germany and is fast gaining popularity over the conventional pharmaceutical drugs as an alternative remedy for mild to moderate depression in the United States and Europe [2,3].

In a Cochrane Systematic Review examining the efficacy of St. John's wort as an antidepressant, there was conclusive evidence that extracts of hypericum were either more effective than placebo or similar in efficacy as standard antidepressants [32]. Moreover, emerging evidence from more recent clinical studies also support the overall deductions of the Cochrane review, that St. John's wort was significantly superior to placebo and as efficacious as the older pharmaceutical antidepressants (e.g. fluoxetine [Prozac], imipramine [Tofranil], sertraline [Zoloft]) for treating mild to moderate depression [33-36]. This botanical was however found to be ineffective in the treatment of major depression [37,38].

Thus far, no severe or toxic adverse events associated with the use of *Hypericum* have been reported in clinical trials comparing its efficacy and safety with other antidepressants [31]. The adverse effects documented were generally mild, with gastrointestinal irritation, allergic reactions, tiredness, dizziness, and restlessness being the most common side effects reported [7,39]. Taken together, St. John's wort appears to be better tolerated and relatively safer with a more favorable side effect profile as compared to the first generation pharmaceutical antidepressants, which are usually associated with a range of serious drug reactions. Thus, *H. perforatum* serves as an attractive alternative remedy for the treatment of mild to moderate depression. However, more detailed studies involving well defined hypericum preparations are needed to establish any potential herb-drug interactions and contraindications associated with the clinical use of this botanical.

The primary mechanism of action of the antidepressant activity of *H. perforatum* has initially been thought to be the inhibition of monoamine oxidase by hypericin. However, it was only until recently that the prenylated phloroglucinol derivative hyperforin (instead of hypericin) was found to be the main constituent responsible for the antidepressant effect of St. John's wort [3,40,41]. Biochemical studies have suggested that this bioactive chemical exerts its antidepressant effect through blocking the synaptosomal re-uptake of a variety of neurotransmitters, namely serotonin, dopamine and norepinephrine, thereby elevating their synaptic concentrations [42,43].

HYPERFORIN AS A POTENTIAL INHIBITOR OF ANGIOGENESIS

Besides exhibiting an antidepressant effect, hyperforin also displays antiangiogenic activity. In a recent investigation by Martínez-Poveda and co-authors, this lipophilic compound was demonstrated to exhibit the characteristics of a multi-target antiangiogenic drug, interfering with key events in angiogenesis *in vitro*, as well as inhibiting *in vivo* angiogenesis [44]. Angiogenesis is a complex process involving the sprouting of new blood vessels [45]. Being identified as one of the hallmarks of cancer, angiogenesis is integral in the key processes of carcinogenesis - tumor growth, invasion and metastasis [46]. Thus, the discovery of hyperforin's inhibitory effects on angiogenesis has valuable implications in the current search for an efficacious and less toxic drug in the treatment of cancer as well as angiogenesis-related pathologies.

Hence, hyperforin has been proposed as an antineoplastic agent recently based on a plethora of studies documenting its anticarcinogenic, antiproliferant, and proapoptotic effects, as well as its ability to inhibit cancer invasion and metastatic growth [47-53]. Moreover, the recent finding of its antiangiogenic activity lends further support to its potential application in cancer therapy.

ANTIMICROBIAL AND WOUND HEALING EFFECTS OF ST. JOHN'S WORT

The antimicrobial properties of St. John's wort have been known since historic times, where it was traditionally used for the treatment of wounds and infections [1,45]. This antimicrobial activity can be attributed to the essential oil, phloroglucinols and flavonoids found in the botanical [1]. Of particular interest, hyperforin has been demonstrated to exert effective antibacterial activity against multidrug-resistant *Staphylococcus aureus* and other Gram-positive bacteria such as *Corynebacterium diphtheriae* and *Streptococcus pyogenes* [54]. It was however found to have no growth inhibitory effect on Gram-negative bacteria or *Candida albicans* [54].

Meanwhile, the flavonoid compounds including quercetin and hyperoside found in *H. perforatum* also contribute to the antiinflammatory properties of this herbaceous plant [55]. Thus, St. John's wort exhibits antimicrobial and antiinflammatory effects which are useful in the treatment of infections, wounds and inflammation. In addition, there is a possibility that this medicinal herb may be developed into a novel antibiotic which could alleviate the growing problem of antibiotic resistance.

CONCLUDING REMARKS

The traditional uses of St. John's wort are diverse, ranging from the treatment of bacterial and viral infections, inflammation and superficial wounds, to the therapeutic management of psychiatric disorders. Moreover, recent investigations of the major bioactive constituents of the medicinal plant have yielded promising evidence pointing to the immense potential of this botanical in the development of novel and more effective antidepressant, anticancer, antiviral and antimicrobial drugs. Hypericin-PDT could well become an alternative treatment modality for cancer in the new millennia [56, 57].

REFERENCES

[1] American Herbal Pharmacopeia. St. John's Wort. *Hypercium perforatum*. Quality control, analytical and therapeutic monograph, 3rd Edition. Texas: American Botanical Council; 1997.

[2] Lawvere S, Mahoney M C. St. John's Wort. Am Fam Physician 2005; 72: 2249-54

[3] Di Carlo G, Borrelli F, Ernst E, Izzo AA. St John's wort: Prozac from the plant kingdom. Trends Pharmacol Sci 2001; 22: 292-7

[4] Nathan PJ. The experimental and clinical pharmacology of St John's Wort (*Hypericum perforatum* L.). Mol Psychiatry 1999; 4: 333-8

[5] Greeson JM, Sanford B, Monti DA. St. John's wort (*Hypericum perforatum*): a review of the current pharmacological, toxicological and clinical literature. Psychopharmacol 2001; 153: 402-14

[6] Agostinis P, Vantieghem A, Merlevede W, de Witte PAM. Hypericin in cancer treatment: more light on the way. Int J Biochem Cell Biol 2002; 34: 221-41

[7] Barnes J, Anderson LA, Phillipson JD. St John's wort (*Hypericum perforatum* L.): a review of its chemistry, pharmacology and clinical properties. J Pharm Pharmacol 2001; 53: 583-600

[8] Ehrenberg B, Anderson JL, Foote CS. Kinetics and yield of singlet oxygen photosensitized by hypericin in organic and biological media. Photochem Photobiol 1998; 68: 135-40

[9] Hadjur C, Wagnieres G, Ihringer F, Monnier P, van den Bergh H. Production of the free radicals $^{\cdot}O_2^-$ and $^{\cdot}OH$ by irradiation of the photosensitizer zinc (II) phthalocyanine. J Photochem Photobiol B 1997; 38: 196-202

[10] Diwu Z, Lown JW. Photosensitization with anticancer agents: EPR studies of photodynamic action of hypericin: formation of semiquinone radical and activated oxygen species on illumination. Free Rad Biol Med 1993; 14: 209-15

[11] Ochsner M. Photophysical and photobiological processes in the photodynamic therapy of tumours. J Photochem Photobiol B 1997; 39: 1-18

[12] Meruelo D, Lavie G and Lavie D. Therapeutic agents with dramatic antiretroviral activity and little toxicity at effective doses – aromatic polycyclic diones hypericin and pseudohypericin. Proc Natl Acad Sci USA 1988; 85: 5230-4

[13] Colasanti A, Kisslinger A, Liuzzi R et al. Hypericin photosensitization of tumor and metastatic cell lines of human prostate. J Photochem Photobiol B 2000; 54: 103-7

[14] Kamuhabwa AR, Agostinis P, D'Hallewin MA, Kasran A, de Witte PAM. Photodynamic activity of hypericin in human urinary bladder carcinoma cells. Anticancer Res 2000; 20: 2579-84

[15] Chen B, de Witte PAM. PDT efficacy and tissue distribution of hypericin in a mouse P388 lymphoma tumor model. Cancer Lett 2000; 150: 111-7

[16] Chen B, Xu Y, Roskams T, Delaey E et al. Efficacy of antitumoral photodynamic therapy with hypericin: relationship between biodistribution and photodynamic effects in RIF-1 mouse tumor model. Int J Cancer 2001; 93: 275–82

[17] Chen B, Zupkó I, de Witte PAM. Photodynamic therapy with hypericin in a mouse P388 tumor model: vascular effects determine the efficacy. Int J Oncol 2001; 18: 737–42

[18] Alecu M, Ursaciuc C, Halalau F et al. Photodynamic treatment of basal cell carcinoma and squamous cell carcinoma with hypericin. Anticancer Res. 1998; 18: 4651–4

[19] Du H, Olivo M, Tan BKH and Bay BH. Hypericin-mediated photodynamic therapy induces lipid peroxidation and necrosis in nasopharyngeal cancer. Int J Oncol 2003; 23:1401-5

[20] Du H, Bay BH and Olivo M. Biodistribution and photodynamic therapy with hypericin in a human murine tumor model. Int J Oncol 2003; 22: 1019-24

[21] Lavie G, Meruelo D, Aroyo K, Mandel M. Inhibition of the CD8+ T cell-mediated cytotoxicity reaction by hypericin: potential for treatment of T cell-mediated diseases. Int Immunol 2000; 12: 479–86

[22] Kraus GA, Pratt D, Tossberg J, Carpenter S. Antiretroviral activity of synthetic hypericin and related analogs. Biochem Biophys Res Comm 1990; 172: 149-53

[23] Tang J, Colacino JM, Larsen SH, Spitzer W. Virucidal activity of hypericin against enveloped and non-enveloped DNA and RNA viruses. Antiviral Res 1990; 13: 313-26

[24] Wood S, Huffman J, Weber N et al. Antiviral activity of naturally occurring anthraquinone derivatives. Planta Med 1990; 56: 651-2

[25] Lavie G, Valentine F, Levin B et al. Studies of the mechanisms of action of the antiretroviral agents hypericin and pseudohypericin. Proc Natl Acad Sci USA 1989; 86: 5963-7

[26] Hudson JB, Lopez-Bazzocchi I, Towers GHN. Antiviral activities of hypericin. Antiviral Res 1991; 15: 101-12

[27] Lenard J, Rabson A, Vanderoef R. Photodynamic inactivation of infectivity of human immunodeficiency virus and other enveloped viruses using hypericin and rose bengal: Inhibition of fusion and syncytia formation. Proc Natl Acad Sci USA 1993; 90: 158-62

[28] Hudson JB, Harris L, Towers GHN. The importance of light in the anti-HIV effect of hypericin. Antiviral Res 1993; 20: 173-8

[29] Lavie G, Mazur Y, Lavie D et al. Hypericin as an inactivator of infectious viruses in blood components. Transfusion 1995; 35: 392-400

[30] Darbinian-Sarkissian N, Darbinyan A, Otte J et al. $p27^{SJ}$, a novel protein in St John's Wort, that suppresses expression in HIV-1 genome. Gene Ther 2006; 13: 288-95

[31] Bilia AR, Gallori S, Vincieri FF. St. John's wort and depression. Efficacy, safety and tolerability-an update. Life Sci 2002; 70: 3077-96

[32] Linde K, Ramirez G, Mulrow CD, Pauls A, Weidenhammer D, Melchart D. St. John's wort for depression-an overview and meta-analysis of randomised clinical trial. BMJ 1996; 313: 565-76

[33] Philipp M, Kohnen R, Hiller KO. Hypericum extract versus imipramine or placebo in patients with moderate depression: randomised multicentre study of treatment for eight weeks. BMJ 1999; 319: 1534-8

[34] Brenner R, Azbel V, Madhusoodanan S, Pawlowska M. Comparison of an extract of hypericum (LI 160) and sertraline in the treatment of depression: a double-blind, randomized pilot study. Clin Ther 2000; 22: 411-9

[35] Woelk H. Comparison of St John's wort and imipramine for treating depression: randomised controlled trial. BMJ 2000; 321: 536-9

[36] Schrader E. Equivalence of St John's wort extract (Ze 117) and fluoxetine: a randomized, controlled study in mild-moderate depression. Int Clin Psychopharmacol 2000; 15: 61-8

[37] Hypericum Depression Trial Study Group. Effect of Hypericum perforatum (St John's wort) in major depressive disorder: a randomized controlled trial. JAMA 2002; 287; 1807-14

[38] Shelton RC, Keller MB, Gelenberg A et al. Effectiveness of St John's wort in major depression: a randomized controlled trial. JAMA 2001; 285: 1978-86

[39] Miller AL. St. John's Wort (*Hypericum perforatum*): Clinical Effects on Depression and Other Conditions. Alt Med Rev 1998; 3: 18-26

[40] Briskin DP. Medicinal plants and pytomedicines. Linking plant biochemistry and physiology to human health. Plant Physiol 2000; 124: 507-14

[41] Muller WE. Current St John's wort research from mode of action to clinical efficacy. Pharmacol Res 2003; 47: 101-9

[42] Chatterjee SS, Bhattacharya SK, Wonneman M, Singer A, Muller WE. Hyperforin as a possible antidepressant component of Hypericum extracts. Life Sci 1998; 63: 499-510

[43] Jensen AG, Hansen SH, Nielsen EO. Adhyperforin as a contributor to the effect of Hypericum perforatum L. in biochemical models of antidepressant activity. Life Sci 2001; 14: 1593-605

[44] Martínez-Poveda B, Quesada AR, Medina MA. Hyperforin, a bio-active compound of St. John's Wort, is a new inhibitor of angiogenesis targeting several key steps of the process. Int J Cancer 2005; 117: 775-80

[45] Medina MA, Martínez-Poveda B, Amores-Sánchez MI, Quesada AR. Hyperforin: More than an antidepressant bioactive compound? Life Sci 2006 (Article online is ahead of print)

[46] Hanahan D, Weinberg RA. The hallmarks of cancer. Cell 2000; 100: 57-70

[47] Moore LB, Goodwin B, Jones SA et al. St. John's wort induces hepatic drug metabolism through activation of the pregnane X receptor. Proc Natl Acad Sci USA 2000; 97: 7500-2

[48] Schempp CM, Kirkin V, Simon-Haarhaus B et al. Inhibition of tumour cell growth by hyperforin, a novel anticancer drug from St. John's wort that acts by induction of apoptosis. Oncogene 2002; 21: 1242-50

[49] Gartner M, Muller T, Simon JC, Giannis A, Sleeman JP. Aristoforin, a novel stable derivative of hyperforin, is a potent anticancer agent. Chembiochem 2005; 6: 171-7

[50] Hostanska K, Reichling J, Bommer S, Weber M, Saller R. Hyperforin a constituent of St John's wort (*Hypericum perforatum* L.) extract induces apoptosis by triggering activation of caspases and with hypericin synergistically exerts cytotoxicity towards human malignant cell lines. Euro J Pharmaceutics Biopharmaceutics 2003; 56: 121-32

[51] Shay NF, Banz WJ. Regulation of gene transcription by botanicals: novel regulatory mechanisms. Ann Rev Nutr 2005; 25: 297-315

[52] Doná M, Dell'Aica I, Pezzato E et al. Hyperforin inhibits cancer invasion and metastasis. Cancer Res 2004; 64: 6225-32

[53] Schwarz D, Kisselev P, Roots I. St John's wort extracts and some of their constituents potently inhibit ultimate carcinogen formation from benzo[*a*]pyrene-7,8-dihydrodiol by human CYP1A1. Cancer Res 2003; 63: 8062-8

[54] Schempp CM, Pelz K, Wittner A, Schopf E, Simon JC. Antibacterial activity of hyperforin from St John's wort, against multiresistant *Staphylococcus aureus* and Gram-positive bacteria. Lancet 1999; 353: 2129

[55] Tedeschi E, Menegazzi M, Margotto D, Suzuki H, Forstermann U, Kleinert H. Anti-inflammatory actions of St. John's wort: inhibition of human inducible nitric-oxide synthase expression by down-regulating signal transducer and activator of transcription-1alpha (STAT-1alpha) activation. J Pharmacol Exp Therap 2003; 307: 254-61

[56] Olivo M, Chin WL. Perylenequinones in Photodynamic Therapy: Cellular versus vascular response. J Environ Toxicol Oncol 2006; 25: 223-37.

[57] Olivo M, Du HY, Bay BH. Hypericin lights up the way for the potential treatment of nasophharyngela cancer by PDT. Curr Clin Pharmacol. *In Press*.

From Medicine Man to Market: A Look at Natural Products and the Pharmaceutical Industries

Ronald E Young[a,*] **and Anthony Clayton**[b]
[a]*Department of Human & Comparative Physiology, Dean of the Faculty of Pure and Applied Sciences, The University of the West Indies, Mona, Jamaica*
[b]*Sir Arthur Lewis Institute for Social and Economic Studies, The University of the West Indies, Mona, Jamaica*
**Corresponding email: ronald.young@uwimona.edu.jm*

INTRODUCTION

In the constant quest to eat or be eaten, living organisms have, since the beginning of evolution, waged a battle of escalating measure *vs* counter-measure. Sessile and other organisms that have limited options for physical evasion or attack have been particularly adept at developing chemical defences, typically targeting vulnerable aspects of the metabolic pathways of their potential predators. These natural defence mechanisms, honed by millions of years of selective pressure, have provided traditional 'medicine men' with a gradually accumulated armamentarium of potions and poultices to deploy against the micro-organisms and other micro- or macro-parasites that can infect or afflict humans. The attempt to identify and exploit useful, naturally evolved chemical defence mechanisms is probably the longest-running research programme in human history. Chinese herbal prescriptions have been recorded since 1100 B.C., and Indian Ayurvedic writings since 1000 B.C.

Methodologies and technologies have advanced dramatically in recent decades, providing contemporary scientists with many new tools for understanding the functional systems in living organisms and thereby for intervening in their operation. Nevertheless, the firms that dominate the pharmaceutical industry today were built upon the identification, isolation, purification, characterization, modification and synthesis of useful active principles from plants and, to a lesser degree, from animals. Much of the early development of the modern industry was based, originally, on traditional knowledge, and some of the drugs derived this way (including morphine, strychnine, aspirin and atropine) were exceptionally profitable for the companies that developed them. Even today, 50% of the drugs most used in the United States are either natural products or derived from them.

The pharmaceutical industry today has become relatively closed to new entrants. This is partly because of the advanced technical capacity and high levels of capital investment now required to develop new products and meet increasingly stringent regulatory requirements, and partly because of the large-scale marketing and distribution networks generally required to bring these products to global markets. It has been estimated that the cost of marketing a new drug is twice the cost of the background research. This combination of factors tends to eliminate all but the largest and best financed enterprises. Europe currently accounts for 37% of global pharmaceutical research and development, the United States 36% and Japan 19%, which means that 92% of the global research and development capacity is now held inside these three major markets [1].

The high barriers to entry present a particular problem to developing nations, who are less likely to have either the technical capacity or the distribution networks needed to develop and market new products. This in turn presents a global dilemma, as some of the same developing nations (*e.g.* Central and South America, East Africa and Oceania) currently possess the greatest remaining untapped biological resources, in terms of endemic species of largely unknown potential, as a source of the next generation of medicines. It is therefore vital to develop models of prospecting, assaying, development and marketing that will allow these resources to be developed in a way that is acceptable and profitable for all stakeholders.

BARRIERS TO DEVELOPMENT

Some barriers to entry – such as trade access and market receptiveness – are extrinsic to the pharmaceutical industry. Others – such as technical capacity – are intrinsic; they are inherent in the structure of an industry.[1] Porter et al. (2003) [2] point out, for example, that the pharmaceutical industry is more science-intensive than any comparable industrial sector, noting that in 1997 the majority of patents in this industry already cited at least one peer-reviewed scientific article. Despite the myriad molecules invented by biological organisms for their defence, these molecules were not necessarily designed to operate in human beings or to target human diseases. The pharmaceutical industry often has to modify the natural products to optimize their effectiveness (or at least to claim novelty). This is achieved by employing diverse research strategies, including combinatorial chemistry, to generate vast numbers of targeted molecular variations through automated techniques; genetic recombination to improve proteins or create new ones, and basic research aimed at developing a more detailed understanding of particular biological mechanisms in order to more effectively manipulate them. 'Dry lab' computational analysis and modeling techniques now closely parallel and complement traditional 'wet lab' research.

Far-sighted companies must invest in researching new cross-boundary synergies such as the emerging field of pharmacogenomics, which may eventually allow prescriptions and dosages to be tailored to individual genotypes

The industry operates in an exceptionally tightly regulated and controlled environment, and very high levels of investment in product development, clinical trials and evaluation are required in order to obtain regulatory approval. The environment is also increasingly litigious (particularly in the US, the largest single market[2]), and there can be very costly penalties for unanticipated side-effects, or failures in product efficacy or quality. The standards of quality assurance and therapeutic probity now required have pushed up the total cost and time involved in the production of a marketable pharmaceutical product to the point where only the most highly organized, stable and very well-funded organizations can contemplate taking on such an enterprise [3].[3]

A pharmaceutical company will typically start assessing 10,000 molecules for every one that finally gets brought to market as a product, and success is not guaranteed even at that stage. Parlange (1999) [4] has pointed out that a new drug takes, on average, 12 years and costs between \$300 to \$400 million to develop. This assessment is supported by the IMS Health 2002 Annual Review of New Active Substances (NAS), which found that development times (measured from priority product patent application to first world launch) ranged from a little under 6 years to over

23 years, depending largely on the type of product [5]. Some estimates of cost run to $1.2 billion from initial testing to first sales [6].

Whilst most drugs are cheap to manufacture once fully specified (a point highlighted by HIV/AIDS campaigners), there are enormous 'up-front' costs for discovery, R&D, regulatory approval and marketing a new product.

The current cost structure of the pharmaceutical industry, heavily weighted by expensive, long-term research programmes and the need for competitive marketing, has several unfortunate consequences. One is to downplay emphasis on the development of drugs needed primarily in less lucrative markets. Another is to favour investments in 'me-too' copies and line extensions, rather than new drug development. The IMS Health 2002 Annual Review notes, for example, that the industry launched just 36 new active substances (NASs) in 2002; the lowest output in 20 years. This is a matter of concern, both for the industry and society as a whole, because ageing populations and increased global mobility are resulting in increases in the prevalence of both chronic and infectious diseases (*e.g.* SARS outbreaks), which in turn highlight the importance of developing a more innovative and proactive pharmaceutical research capacity.

THE STATUS OF DEVELOPING COUNTRIES

Some strongly developing countries that are rapidly becoming major industrial powers (such as China) are also developing strong biotechnology industries, partly in order to address domestic problems, and partly to expand into new export markets. The under-developing countries (such as Jamaica) tend to have a more defensive position, based on concerns about the inability of domestic firms to compete in external markets, partly because of the rate of technological progress and change, and partly because of the changing terms of trade. Diminishing prospects for current agricultural and manufactured products creates an urgent need to find new export markets.

So the problem, in essence, is that the pharmaceutical companies with the research capacity and distribution networks to develop and market new products need to accelerate innovation. Neither the scientific community nor the pharmaceutical industry is in any doubt about the importance of natural products in the production of innovative drugs.

Pharmaceutical companies thus need to gain access to as diverse an array as possible, of as yet unexploited biological resources. Paradoxically, it is the developing nations that are most likely to provide those reservoirs of high levels of untapped endemicity and of traditional knowledge and expertise in herbal medicine which have developed over generations outside of the Western, science-based knowledge system. Much of this genetic resource and traditional expertise, moreover, is concentrated in areas which are being threatened simultaneously by the synergistic pressures of over-population, deforestation and climatic change, so the need to document and preserve these natural evolutionary inventions assumes great urgency. What the developing countries lack is the capacity to undertake the diverse, complex and managerially challenging activities that are required for entry into the major markets. The task, therefore, is to find a robust solution to these problems.

Some developing nations *e.g.* India and Brazil [7] have in part focused on manufacturing variants of drugs that have been developed by others, but this, obviously, does not help to develop genuinely new products.[4] One possible solution for developing countries might be to specialize in product lines based on traditional formulations that avoid claims for any specific, targeted therapeutic effectiveness. The standards required are less stringent, and product development is therefore quicker and less costly. The market is still sizable. Over 80% of the world's population still relies at least in part upon traditional, folklore-based medication for primary health care, partly due to custom and partly to cost (so that *per capita* expenditure is obviously also much lower).

The problem, of course, is that the concentration, efficacy and safety of the active agents in traditional herbal remedies cannot be stated to a degree which would satisfy the criteria required by science-based regulations, and this is becoming increasingly unacceptable. There was a serious controversy in the UK, for example, when it was discovered that certain traditional Chinese herbal remedies were being used in concentrations that had hepatotoxic effects.

THE REGULATORY ENVIRONMENT

A more advanced option, therefore, which lies midway between the unregulated use of poorly characterized herbal remedies and the costly development of fully tested pharmaceuticals, would be to produce nutraceuticals, functional foods, or related formulations, with regulated levels of toxicity testing and appropriate levels of safety assurance [8].[5] These are significant markets, both lucrative and high-growth [9, 10].[6]

The nutraceutical or functional foods market, however, is likely to provide only a temporary solution. The dangers of uncontrolled levels of active ingredients in some formulations and the possibility of antagonistic or synergistic interactions between standard treatment regimes and herbal formulations are being increasingly recognized. The result is that the need for setting higher standards for approval and regulation of such formulations is becoming increasingly evident and more widely accepted. The authors believe that this is a positive development. 'Western' medicine and agricultural practices have attained their current status through the application of scientific principles and high standards of quality assurance in the interest of consumer protection. It is therefore likely that traditional systems of medicine and related activities will also have to be prepared to stand up to similar criteria.[7] An obvious consequence, however, may be that by the time some developing countries achieve the capacity to exploit their indigenous resources, the investment required to meet rising regulatory standards may have escalated to the point where the barriers become similar to those for conventional pharmaceuticals.

THE CHALLENGE

The more fundamental long-term challenge, therefore, is to develop a business model that allows all parties to both contribute to innovation and profit from successful outcomes; *i.e.* a stable and mutually beneficent arrangement that can sustain a high rate of new drug development, a good return to the country that makes the primary material available, and a healthy dividend for the scientists, firms and investors that are involved in the various stages of developing the product.

There are examples of such good practice. For example, the UK drug company Phytopharm is currently developing an anti-obesity drug, P57, from actives found in *Hoodia gordonii,* a cactus from the Kalahari Desert.[8] Hoodia's properties were known by the Kung or Xhomani Sans Bushmen of the Kalahari, who are believed to have used it for centuries to suppress the appetite during long hunting trips. Phytopharm wanted to avoid the stigma of being seen to exploit a traditional medicine without rewarding the original discoverers, and therefore signed a deal in 1997 with the South African government for a percentage of the royalties. One reason why this is a shrewd investment by the firm is that it is thought that the Kung know of some 300 other plants with potentially valuable properties, which highlights the value of a good, long-term relationship.

Unfortunately, this model is not currently the norm. If anything, the bioprospecting that has been carried out to date has been a source of friction. For example, several wealthy US-based universities now have a strong presence in areas which boast high endemicity and under-explored traditional medicinal and related practices, such as Brazil, Haiti, Central America, Madagascar and others. These institutions do not always seek collaborations with local universities, or with indigenous peoples, who also have a vested interest in seeing to the documentation and preservation of the knowledge and expertise embedded in the local folklore and practices.[9]

Whilst both the local and external agencies therefore have interests in the monitoring, documentation and conservation of the indigenous knowledge and biodiversity, the relationship is inherently asymmetrical, and the different parties involved may have divergent objectives and aspirations. The indigenous traditional practitioners may wish to conserve their time-honored practices and knowledge which give them status, power and independence in their communities. Monetary recompense and a promise of a share in intellectual property may not be adequate compensation when the outcome may mean the replacement of affordable remedies by expensive drugs and the introduction of a system of health care which may demolish the power base of the traditional practitioner. On the other hand, in the push to realize reasonable returns on investment by the shortest and least costly means, the external agencies could easily be swayed to use their political, financial and legal power to patent plants and products and to push for the development of drugs and other preparations which could, directly or indirectly, wittingly or unwittingly, lead to the disenfranchisement of the indigenous practitioners whose knowledge base fueled the discovery in the first instance.

The issues involved are undoubtedly complex and far-reaching, and while the solutions need not be monolithic, they must take into account the local history and prevailing conditions. In large countries such as China,

India and South Africa which have a large, established and vigorous mainstream marketplace for traditional medicaments, and a philosophical framework guiding their use, it might be possible to gradually increase quality control standards as increasing research documenting the efficacy, mode of action or the dangers of particular products and/or practices becomes available. This would allow development of scientifically sound regulations for protection of consumers and a locally based development of the industry. Possible additional outcomes could be the evolution of traditional practices toward Western norms or of Western norms to accommodate alternative systems of medicine.

In smaller countries with less well developed systems of either traditional medicine or scientific research, however, the need for collaboration becomes greater, but the danger of detrimental and asymmetrical exploitation also increases. There are increasingly strong calls from the developed world, through the Convention on Biodiversity, for developing countries to recognize their biological endowment as part of the general heritage of mankind, and to document and make this accessible as a pre-requisite for staking any IP claims. This is governed by a well designed regulatory framework, but it is not clear whether the aspect of enforcement against abuse is as well developed, and whether a poor country would be able to sustain an expensive litigation process to establish a claim against a large multinational corporation.

Perhaps the most thorough attempt to date at a balanced solution to the issues of equity, intellectual property rights and biodiversity conservation is represented by the carefully formulated guidelines for the International Cooperative Biodiversity Group. The group started in 1992 by collaboration between six NIH agencies, the NSF and USAID, and is dedicated to fostering the development of natural products, drug discovery, biodiversity conservation and sustainable economic growth, in such a way that local communities and organizations can benefit from the host nation's biodiversity and traditional knowledge. The awards granted support collaborative projects between public and private institutions (US and local) including Universities, environmental organizations and pharmaceutical companies. The underlying philosophy is to ensure that the prior rights of indigenous peoples over their environment and their right to self determination are protected, that equitable compensation and profit sharing agreements are reached, and that indigenous research capacity is supported and developed. However, this model too has had some implementation problems; the fact that the guidelines emphasize the need to respect the rights, autonomy, confidentiality and prior ownership by the indigenous people did not prevent the breakdown of one of the early Mexican projects, involving accusations of 'bio-piracy' from some local organizations. This highlights the complexity and delicacy of the interrelationships involved in such collaborations.

CONCLUSION

The authors believe that an important part of the solution to the fundamental problem of developing a robust and productive business model for the development of new natural products lies in achieving greater clarity about the various roles involved at different stages of development. The precise arrangement will, of course, vary according to the local capacity to take on the more advanced tasks further along the value chain. In general terms, however, a fairly common initial division of roles might involve researchers in developing countries in surveying (including assessing indigenous knowledge), assaying and first-round screening, while the pharmaceutical majors would probably lead in the later stages of product development, including the clinical trials, second and third-round screening, and submission for regulatory approval.[10]

The development of new products from natural resources however, will involve more than simply being able to screen widely and rapidly, and being able to call upon reliable and able collaborators. It will also involve a clear focus upon what products to develop and in what way, which requires the detailed understanding of the marketplace held by the pharmaceutical majors. The ideal relationship between the international firms and the local participants would therefore be both dynamic and interactive, with ideas flowing in both directions.

Such a relationship would confer clear benefits. Informed, first-order screening, for example, brings a very significant cost-saving. The success rate for identifying active ingredients by screening plants identified based upon traditional knowledge is about 25%, compared to about 6% based upon random sampling, probably because the traditional herbalists have already had some trial-and-error based success on selecting plants with useful active ingredients. The particular significance of this contribution is that there is an exponential increase in the investment required at each stage of the development of a new drug, so it is very important to eliminate any harmful, useless or unpromising compounds as early as possible. This means that a four-fold increase in accuracy at an early stage represents a greater hidden cost saving than could be achieved at later stages.

In conclusion, a glaring paradox is obvious when the relatively dry pipelines bringing innovative pharmaceuticals to market, are viewed against the high rate of discovery of new active molecules reported by Universities in both developed and developing countries and the rapid improvement in techniques for High Throughput Screening being used by the pharmaceutical companies. A solution to this dysjunction must be found. There is evidently much to be gained by devising a robust model that will support the effective and sustainable exploitation of indigenous biodiversity. This will involve concerted collaboration between a diverse array of players, both local and international, irrespective of whether the product to be developed is a pharmaceutical, a nutraceutical, an agrochemical or a flavour. The choice will be determined by complex considerations involving both short and long-term evaluation of all factors, from land-use and cultivation of crops, to the identification and cultivation of a sustainable market, and the equitable sharing of profits through all levels of the production chain, so that at each level of the chain, all parties can both contribute to development and profit from successful outcomes. This can provide a basis for a stable and mutually beneficial arrangement that can sustain a high rate of new drug development, a good return to the country that makes the primary material available, and a healthy dividend for the scientists, firms and investors, both local and foreign, that are involved in the various stages of developing the product.

Endnotes

[1] Intrinsic barriers may appear more accessible, in that they are in principle under local control, but they can actually represent far greater obstacles. The barriers to pharmaceutical markets, for example, can only be resolved by building the technical capacity and quality control mechanisms needed to achieve internationally-recognized standards, which may take many years.

[2] The US and Europe combined currently account for about two-thirds of total world pharmaceutical sales.

[3] This is partly why there were so many mergers between pharmaceutical firms in the 1990s, as they tried to achieve greater economies of scale, but most of these recent mergers (perhaps more driven by competitive pressure than informed by a strategic vision) have as yet failed to produce a significant increase in the rate of development of new drugs in the 1990s. Some of the large companies then tried to take over new biotech start-up companies in order to absorb their innovative capacity and embryonic products, but any gains were outweighed by the even faster increase in the difficulty and cost of finding and developing new drugs.

[4] In August 2003 a deal was brokered by the WTO that allowed poor countries that had broken the patents on life-saving drugs to export their cheap generics to other poor countries, which cut into corporate profits and increased fears about the viability of global intellectual property (The Economist 1st September 2003). This may persuade some firms to quietly reduce their exposure to areas where they now expect to get their patents broken, thus undermining the private element of the research effort.

[5] Nutraceuticals or functional foods are products for which claims of having specific health effects have been made. The Rutgers' Nutraceutical Institute adds "health promoting" (with dubious acceptability), "disease preventing or medicinal properties." Clayton (2001) identifies nine key groups of protective nutrients that, in combination, are claimed to afford some health protection. These include:

- Anti-oxidants (such as vitamins E and C, the flavonoid group, selenium and co-enzyme Q10), the carotenoids (such as beta carotene, lutein, lycopene, cryptoxanthin and astaxanthin), flavonoids (such as pycnogenol) and isoflavones (such as genistein).
- Omega 3 fatty acids.
- The methyl donors (such as betaine).
- The pre-biotics (such as fructo-oligosaccharides)
- Co-enzyme Q10 and glucosamine.

The "actives" concerned can be supplied as minimally processed preparations or, more recently, as refined extracts which contain primarily the desired actives. Traditional herbalism uses materials with little or no processing and which therefore contain unknown quantities of desirable (and undesirable) actives.

[6] Using a strict definition of functional foods (limited to food and drinks that make specific health claims of some kind on the packaging or in advertising), the functional food and drinks market in the five major European markets, the US, Japan and Australia had a combined value of US$9.925bn in 2003, with the largest single market
– valued at US$4.5bn – in Japan. If a broader definition is used (this includes a wide range of products that do not necessarily make specific health claims, but that are still (or perceived to be) "functional," the market rises to nearly US$24.2bn,

with Japan accounting for over half of the total (Leatherhead Food Research Institute 2004).

[7] There is, of course, more than one possible outcome. One is that there will be an evolution and modification of current 'Western' approaches (as has been seen in the increasingly widespread use of acupuncture). Another is that it will be necessary to accept the modification or demise of those traditional methods and beliefs that fail to make the grade.

[8] P57 works by stimulating the glucose receptors in the hypothalamus, about 10,000 times more effectively than glucose itself, which creates a feeling of satiety for about 24 hours and thereby induces a voluntary reduction in calorie intake - typically by about 30-40% per day.

[9] Small countries also have to face the problem of sustainable exploitation of the limited resource-base of endemic material. An elegant solution to this problem was recently seen when researchers at the University of the West Indies devised a method for sustainably reaping and allowing regeneration of colonies of the mangrove swamp ascidian *Ecteinascidia turbinata* from which a potent anti-tumor agent ecteinascidin has been isolated by American scientists and developed by a Spanish pharmaceutical company. As might be expected, congeners of the active compounds have since been synthesized by chemists at Harvard University, so that the ascidian source population is no longer under threat. Another local organism, the sea whip *Plexaura homomalla*

[10] There are examples of such relationships. The University of the West Indies is in fact at the moment involved in an ICBG planning grant with the University of Mississippi National Centre for Natural Products Research to investigate the biodiversity and potential therapeutic value of coral reef organisms in Jamaica. We hope through this to build strong international linkages and to expand our capacity for the development of a wide array of rapid screening methods for evaluating activity in natural materials.
which was found to be a good source of certain prostaglandins, did not fare so well, and the population was rapidly decimated by indiscriminate collectors. In neither case was there any intellectual property agreement with the UWI, nor was there expectation of any. The involvement of the UWI scientists was circumscribed and appropriate recompense was agreed upon.

REFERENCES

[1] PhRMA facts and figures. Washington, D.C.: Pharmaceutical Research and Manufacturers of America, August 1997

[2] Porter A, Ashton B, Clar G, Coates J, Cuhls K, Cunningham S, Ducatel K, vd Duin P, Georghiou L, Gordon T, Linstone H, Marchau V, Massari G, Miles I, Mogee M, Salo A, Scapolo F, Smits R, Thissen W. Technology Futures Analysis: *Toward Integration of the Field and New Methods.* Technology Futures Analysis Methods Working Group Discussion Paper, November 5th 2003

[3] Angell M. The Pharmaceutical Industry – to whom is it accountable? The New England Journal of Medicine June 22, 2000; Vol. 342, No. 25: 1902-1904, citing:

- Tanouye E. Drug dependency: U.S. has developed an expensive habit: now, how to pay for it? Wall Street Journal. November 16, 1998:1
- Bernstein S. Drug makers face evolving marketplace. Los Angeles Times. January 31, 1999: A1
- The pharmaceutical industry. The Economist. February 21, 1998.
- Anderson C. Drug firms said to pay less in taxes. Boston Globe. December 26, 1999

[4] Parlange M. *Eco-nomics.* New Scientist no 2172, 06/02/1999

[5] Data from the *IMS Health 2002 Annual Review of New Active Substances* cited from IMS LifeCycle. Presented at the 9th Economist Annual Pharmaceutical Conference, London February 12-13, 2003

[6] Murray-West R. *Are the drug giants in danger of bleeding themselves dry?* Money Telegraph, 9 October 2004

[7] The Economist. The right fix? India and Brazil are good at making cheap copies of life-saving drugs. Now they are allowed to export them too. 1 September 2003

[8] Clayton P. Health Defence, 2001. Accelerated Learning Systems, Aylesbury, UK.

[9] Leatherhead Food Research Institute 2004

[10] Clayton A and Staple-Ebanks C. Nutraceuticals and functional foods: a development opportunity for Jamaica, June 2002. Technical report for the National Commission on Science and Technology: market-scoping study. Environmental Foundation of Jamaica, Kingston, Jamaica

Index of Contributors

M

O

P

R

S

T

W

X

Y

Z